Joseph Morrison

An Elementary Treatise on Plane Trigonometry

with numerous examples and applications

Joseph Morrison

An Elementary Treatise on Plane Trigonometry
with numerous examples and applications

ISBN/EAN: 9783337276416

Printed in Europe, USA, Canada, Australia, Japan

Cover: Foto ©berggeist007 / pixelio.de

More available books at **www.hansebooks.com**

AN

ELEMENTARY TREATISE

ON

PLANE TRIGONOMETRY,

WITH NUMEROUS EXAMPLES AND APPLICATIONS.

Designed for the use of High Schools and Colleges.

BY

J. MORRISON, M.D., M.A.,

PRINCIPAL OF THE WALKERTON HIGH SCHOOL; EX-PRINCIPAL OF THE NEWMARKET
HIGH SCHOOL; LATE MEMBER OF THE MEDICAL COUNCIL AND EXAMINER
ON THEORETICAL AND PRACTICAL CHEMISTRY IN THE COLLEGE
OF PHYSICIANS AND SURGEONS OF ONTARIO.

TORONTO:
CANADA PUBLISHING CO. (LIMITED).
1880.

Entered according to Act of Parliament of Canada, in the year one thousand eight hundred and eighty, by THE CANADA PUBLISHING COMPANY (LIMITED), in the office of the Minister of Agriculture.

C. B. ROBINSON,
PRINTER,
JORDAN STREET, TORONTO.

PREFACE.

THE following course of Trigonometry has been prepared with the view of meeting the wants not only of High School but also of Collegiate classes, and in pursuance of this end, the more elementary portions—such as are required for the matriculation examination in the University of Toronto—are printed in a larger type, and form a connected treatise independently of the portions printed in the smaller type, which are intended for candidates reading for University honors.

In discussing the trigonometrical functions I have adopted the method of ratios or the Cambridge method; but on the recommendation of several eminent teachers in our colleges, I have given a full account of the older method or that of the "line definitions," and have employed the latter method in solving right-angled triangles and in tracing the variations in sign and magnitude of the several functions in the various quadrants. Of course, teachers who prefer to use the "ratio definitions" only, can omit the portions treating of the "line definitions" without breaking the continuity of the course.

The solution of right-angled triangles by the *natural* functions, and the application of Trigonometry to the determination of heights and distances, have been introduced at a very early stage, with the view of leading the student to take a deeper interest in the study of the science at the outset.

Numerous examples and solutions, chiefly of a practical character, are given, at frequent intervals throughout the work, for the purpose of fixing in the mind of the student the principles of the text, and of illustrating the practical bearing and utility of Trigonometry in Surveying, Navigation and Astronomy.

The more advanced portions of the work have been discussed from an elementary point of view, and will be found to include all that is required on this subject from University students generally.

As my aim has been to produce a work of utility, I have not hesitated to consult, in the preparation of this volume, the various English and foreign works on the subject; but those to which I am most indebted for suggestions, are the works of Hind, Todhunter, Snowball, Wiegand and Chauvenet.

The examples have been obtained from various sources; some are either entirely original or original in the form in which they are presented; some have been selected from other trigonometries, and some from examination papers set during the last twenty years in the Universities of Cambridge, London, and Toronto, and in McGill College, Montreal, and Harvard College, U.S. The answers have been given wherever necessary, and great care has been taken to secure their accuracy; but it is quite possible that some errors have crept in, and I shall be very thankful for any corrections anyone may send to myself or the publishers. I shall also feel very grateful to my professional confrères for any remarks regarding difficulties or omissions in the text or examples. Great care has been taken to render the text clear and explicit, and I am sure my readers will unite with me in an expression of thanks to the publishers for the almost faultless style in which they have executed the typographical portion of the work.

A second part, on Spherical Trigonometry, with numerous examples and applications to Nautical and Spherical Astronomy and Geodesy, is in course of preparation, and will appear if the present work meet with a favorable reception.

J. MORRISON.

HIGH SCHOOL, WALKERTON,
June 1st, 1880.

CONTENTS

CHAPTER I.

	PAGE
Definitions	1
Sexagesimal Division of Angles	3
Centesimal Division of Angles	4
Circular Measure	4

CHAPTER II.

Definitions of the Trigonometrical Functions	8
General Formulæ	10
Line Definitions of the Trigonometrical Functions	12
Functions of 30°, 60°, 45°, 18°, 54°, &c.	18
Funtions of half an Angle in terms of the whole Angle	21
Examples	24

CHAPTER III.

Solution of Right-angled Triangles	28
Heights and Distances	32

CHAPTER IV.

Extension of the Definitions of the Trigonometrical Functions, use of the signs + and −	37
Variations in Sign and Magnitude of the Trigonometrical Functions	38
Line Definitions	43
Functions of $180° - A$	45
Functions of $90° + A$	46
Functions of Negative Angles	47
Groups of Angles having the same sine, &c.	48
Examples	50

CHAPTER V.

	PAGE
Functions of the sum and difference of two Angles	54
Functions of Compound Angles	60
General Formulæ	61
Functions of the Multiple and Submultiple of an Angle	64
Functions of mA	66
Functions of $(A+B+C)$, $(-A+B+C)$, &c.	67
General Formulæ	69
Formulæ of Verification	71
Examples	72

CHAPTER VI.

Circular Measure of an Angle $>$ sine, but $<$ tangent	80
Limiting values of $\dfrac{\sin \theta}{\theta}$ and $\dfrac{\tan \theta}{\theta}$	81
$\sin \theta > \theta - \dfrac{\theta^3}{4}$	82
Calculation of the Numerical Values of the Trigonometrical Functions	82
Trigonometrical Tables	86
Increments of the Trigonometrical Functions corresponding to a given Increment of the Angle	89
Examples	92
Dip of the Horizon	94
Examples	96

CHAPTER VII.

Properties and Uses of Logarithms	99
Modulus of a System of Logarithms	103
Relation between the Bases of two Systems	104
Common Logarithms	105
Determination of the Characteristic	106
Logarithmic Tables	108
Arithmetical operations	112
Arithmetical Complement of a Logarithm	114
Logarithms of the Trigonometrical Functions	115
Logarithmic Functions of Angles near the limits of the Quadrant	120
Examples	123

CONTENTS.

CHAPTER VIII.

	PAGE
Formulæ for solving Right-angled Triangles	127
Formulæ for solving Oblique-angled Triangles	129
Area of a Triangle	139
Perpendicular of a Triangle	140
Examples	140

CHAPTER IX.

Solution of Right-angled Triangles	151
Solution of Oblique-angled Triangles	154
Examples	169

CHAPTER X.

Application to Surveying, Navigation and Astronomy	176
Mariner's Compass	184
Length of a Degree of Longitude	185
Moon's Distance from the Earth	186
Parallax	187
Examples	190

CHAPTER XI.

Radius of the Inscribed Circle of a Triangle	201
Radii of the Escribed Circles	203
Radius of the Circumscribed Circle	207
Distance between the Centres of the Inscribed and Circumscribed Circles	209
Distance between the Centres of the Escribed and Circumscribed Circles	210
Perimeter and Area of Inscribed and Circumscribed Polygons	212
Circumference and Area of a Circle	213
Area and Angles of the Inscribed Quadrilateral	214
Examples	215

CHAPTER XII.

Inverse Trigonometrical Functions	225
Examples	227

CHAPTER XIII.

	PAGE
Division of Angles	231
Solution of Equations	235
Auxiliary Angles	238
Examples	241
Quadratic Equations	242
Cubic Equations	244
Elimination of Trigonometrical Functions	246
Examples	248

CHAPTER XIV.

The Exponential Theorem	250
The Logarithmic Series	252
The Napierian Base	255
Calculation of Napierian Logarithms	257
Calculation of Common Logarithms	258
Theory of Proportional Parts	258
Examples	262

CHAPTER XV.

De Moivre's Theorem	263
Trigonometrical Expansions	266
The sine and cosine of an Angle expressed in terms of the Circular Measure	268
Expansion of the Integral Powers of the cosine	269
Expansion of the Integral Powers of the sine	270
Expansion of $\cos n\theta$ in terms of Descending Powers of $\cos \theta$	272
Expansion of $\cos n\theta$ in terms of Ascending Powers of $\cos \theta$	274
Expansion of $\sin x$ and $\cos x$ in terms of the Circular Measure, independently of De Moivre's Theorem	275
Sines and tangents of Small Angles	277
Examples	278

CHAPTER XVI.

Exponential Values of the sin, cos and tan	280
Circular Measure expressed in terms of the tangent—Gregory's Series	281

CONTENTS. xi.

	PAGE
Euler's Series for computing π	282
Machin's Series for computing π	283
Solution of Equations by Series	284
Summation of Trigonometrical Series	290
Examples	295

CHAPTER XVII.

Resolution of $x^{2n} - 2\cos\phi \cdot x^n + 1 = 0$ into Simple and Quadratic Factors	299
Resolution of $x^n - 1 = 0$, n being odd	302
Resolution of $x^n - 1 = 0$, n being even	303
Resolution of $x^n + 1 = 0$, n being odd	304
Resolution of $x^n + 1 = 0$, n being even	304
Resolution of $\sin x$ and $\cos x$ into Quadratic Factors	307
Computation of Logarithmic sine and cosine	308
Miscellaneous Examples	310
Appendix	328
Numbers often used in Calculations	332

THE GREEK ALPHABET.

The Greek Alphabet is here inserted to aid those unacquainted with Greek in reading the parts of the text in which its letters are used.

Α α Alpha.	Ι ι Iota.	Ρ ρ Rho.
Β β Beta.	Κ κ Kappa.	Σ σ Sigma.
Γ γ Gamma.	Λ λ Lambda.	Τ τ Tau.
Δ δ Delta.	Μ μ Mu.	Υ υ Upsilon.
Ε ε Epsilon.	Ν ν Nu.	Φ φ Phi.
Ζ ζ Zeta.	Ξ ξ Xi.	Χ χ Chi.
Η η Eta.	Ο ο Omicron.	Ψ ψ Psi.
Θ θ Theta.	Π π Pi.	Ω ω Omega.

PLANE TRIGONOMETRY.

CHAPTER I.

ON THE MEASUREMENT OF ANGLES.

Article 1.—Trigonometry (from τρίγωνον, a triangle, and μετρέω, I measure) is that branch of Mathematics which treats of the methods of determining the unknown parts of a triangle when a sufficient number of them is given; but in its more extended signification, it embraces the investigation of the various relations of angles in general, such investigations being carried on by means of certain quantities called trigonometrical ratios or functions. It is divided into two branches,—*Plane Trigonometry* and *Spherical Trigonometry*,—the former treating of angles and triangles drawn on a *plane*, and the latter of those on the surface of a *sphere*.

2.—Definition of an Angle.

In Plane Geometry an angle is the inclination of two straight lines to each other, and must therefore be always less than two right angles; but in Trigonometry an angle may be of any magnitude whatever, and is conceived to be described by a straight line revolving about a given point from one position to another. Thus, the lines AC, BC, would, in Geometry, bound only one angle ACB; but in accordance with the more extended definition of an angle, 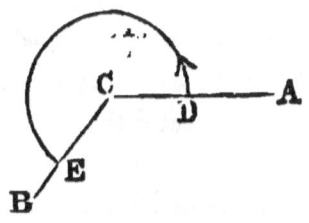 they may also be regarded as containing an angle subtended by the arc DE, and described by a line revolving about C, in the direction of the arrow-head from the position AC to the position CB. Again, in order to effect this, the revolving line may be supposed to make any number of revolutions. Thus, the minute hand of a watch, at half-past one o'clock, will, since twelve o'clock, have described an angle whose magnitude is six right angles.

3. Let AC and BD be drawn at right angles to each other, then the whole angular space about O is divided into four right angles, AOB, BOC, COD, DOA, which are respectively called the *first, second, third* and *fourth quadrants*. Let OP_1, OP_2, OP_3 and OP_4 be different positions of the revolving line OP, by the revolution of which about O from its primitive position AO, in the direction from A to B, the various angles at O are

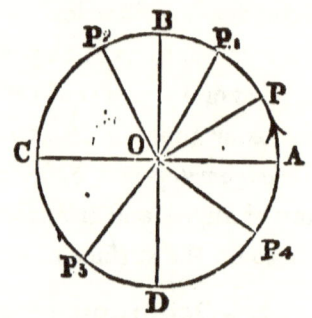

described. The extremity of the revolving line will evidently describe the circumference of a circle. When the revolving line coincides with AO, it is said to make with it an angle *zero;* when it reaches the position OP_1, it will have described an angle AOP_1, which is called an angle in the *first quadrant*; when it coincides with BO, it will have described the right angle AOB; when it takes the position OP_2, it will have described the angle AOP_2, greater than one right angle, but less than two, which is called an angle in the *second quadrant;* when it coincides with CO, it will have described the *trigonometrical* angle AOC, or two right angles. In like manner, when it takes the position OP_3, it will have described the trigonometrical angle AOP_3, subtended by the arc $ABCP_3$, which is an angle in the *third quadrant;* when it coincides with OP_4, it will have described the trigonometrical angle AOP_4, subtended by the arc $ABCDP_4$

which is an angle in the *fourth quadrant*, and when it coincides with AO, it will have described four right angles.

If the revolution be still continued until the position OP_1 is again reached, an angle greater than four and less than five right angles will be described, and so on. The angles described by the progressive revolution of OP in the direction from A to B, are regarded as +, or *positive;* while those described by OP revolving in the *opposite* direction from A to D, are regarded as −, or *negative*, the signs + and − here indicating merely contrariety of direction. Positive and negative angles are always understood in this sense, unless the contrary be stated.

4.—Sexagesimal Division of the Right Angle.

A *degree* is the ninetieth part of a right angle; a *minute* is the sixtieth part of a degree; and a *second* the sixtieth part of a minute, and so on according to a sexagesimal subdivision, but in practice all subdivisions beyond seconds are expressed as decimal parts of a second. The characters used to denote degrees, minutes and seconds are °, ′, ″; thus 17° 5′ 24″.4, represent seventeen degrees, five minutes and twenty-four and four-tenths seconds.

The *complement* of an angle containing A degrees is the remainder obtained by subtracting it from 90°, and is written $90° - A°$. Thus the complement of 40° 10′ is 49° 50′. If the angle is greater than 90°, its complement will be negative; thus, the complement of 130° is $90° - 130° = -40°$.

In every right-angled triangle, each of the acute angles is the complement of the other.

The *supplement* of an angle containing A degrees is the remainder obtained by subtracting it from 180°, and is written $180° - A°$; thus the supplement of 70° is $180° - 70 = 110°$; and of 204°, − 24°.

Since the sum of the angles of a triangle is equal to two

right angles, each of the angles is the supplement of the sum of the other two.

5.—Centesimal Division of the Right Angle.

By some of the French mathematicians the right angle was divided into 100 equal angles, called *grades;* each grade into 100 minutes, each minute into 100 seconds, and so on according to a centesimal subdivision.

If D denote the number of degrees in an angle, and G the number of grades in the same, then, since 90 degrees = 100 grades, we have

$$D : G :: 90 : 100$$
$$\text{or, } G = \tfrac{10}{9} D, \text{ and } D = \tfrac{9}{10} G. \qquad (1)$$

The centesimal division is now abandoned even in France, as its general adoption would involve a change in our tables, and in the graduation of astronomical and other instruments.

6. From Article 4 it will be seen that the unit of angular measure there adopted, viz: 1°, cannot be directly compared with any lineal unit, such as an inch, a foot, &c., since they are magnitudes of *different* kinds. In the following articles it will be shown how a lineal unit may be applied to determine the magnitude of an angle.

7.—Circular Measure.

The circumference of a circle varies directly as its radius.

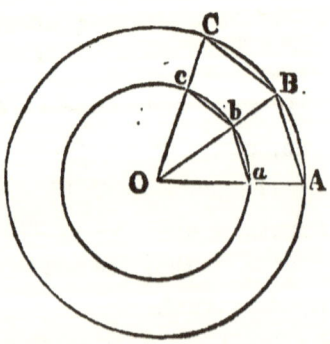

In the two concentric circles inscribe regular polygons $ABC... abc...$ having the same number of sides; then from the similarity of the triangles ABO, abO, we have

$$\frac{AB}{ab} = \frac{OA}{Oa}$$

and since the polygons are equilateral,
$$\frac{\text{perimeter of the polygon } ABC\ldots}{\text{perimeter of the polygon } abc\ldots} = \frac{OA}{Oa}.$$

If the number of the sides of the polygons be increased, their perimeters will approach more nearly to those of the circles in which they are inscribed. Now, let the number of the sides be increased indefinitely, then the perimeters of the polygons will ultimately coincide with the perimeters of the circles, and, since the above proportion will still hold true, we shall have

$$\frac{\text{circumference of the circle } ABC\ldots}{\text{circumference of the circle } abc\ldots} = \frac{\text{radius } OA}{\text{radius } Oa}.$$

Therefore the circumference of a circle varies directly as its radius, and therefore, also, as the diameter; hence it follows that the ratio, $\dfrac{\text{circumference}}{\text{diameter}}$, has a certain invariable value.

It will be shewn hereafter that this ratio is, to five places of decimals, the number 3·14159 which is usually denoted by the Greek letter π.

If, then, r denote the radius of a circle, its circumference will be $2\pi r$.

8. In the circle BCD, take any angle $BAC = A°$: let CB the arc which subtends it $= a$, and the radius $AB = r$. The semi-circumference BCD will be πr which subtends the angle $180°$; and since "the angles at the centre of a circle are proportional to the arcs which subtend them" (*Euc.* VI. 33), we have

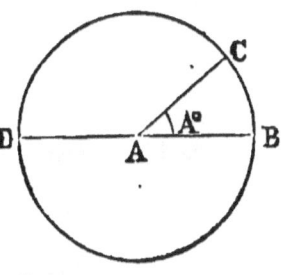

$$180° : A° :: \text{arc } BCD : \text{arc } BC$$
$$:: \pi r : a$$

Whence
$$A° = \frac{180°}{\pi} \cdot \frac{a}{r}$$
$$= \frac{180°}{3 \cdot 14159 \text{ &c.}} \times \frac{a}{r}$$
$$= 57° \cdot 2957795\ldots \times \frac{a}{r}. \qquad (2)$$

Therefore if a and r be given in *lineal* measure, the number of degrees in the corresponding angle A can be obtained.

The invariable angle $57° \cdot 2957795\ldots$ is usually denoted by $\omega°$, therefore (2) may be written

$$A° = \omega° \times \frac{a}{r}. \qquad (3)$$

9. If $a = r$, then $A° = \omega°$, that is, the angle which in any circle is subtended by an arc equal to the radius, contains $\omega°$ or $57° \cdot 2957795\ldots$, which is called the *unit* of circular measure, and the fraction $\dfrac{\text{arc}}{\text{radius}}$ is called the *circular measure* of the angle.

From (3) we have

$$r = \frac{\omega°}{A°} a \qquad (4)$$

and
$$a = \frac{A°}{\omega°} r. \qquad (5)$$

10. If the circular measure $\dfrac{\text{arc}}{\text{radius}}$ be denoted by θ, we have

$$\left. \begin{array}{l} A° = 57° \cdot 2957795\ldots \times \theta \\ A' = 3437' \cdot 74677 \ldots \times \theta \\ A'' = 206264'' \cdot 806\ldots \times \theta \end{array} \right\} \qquad (6)$$

If r denote the radius of a circle, its circumference is $2\pi r$, and therefore the circular measure of four right angles is $\dfrac{2\pi r}{r}$ or 2π.

Hence it follows that the circular measure of two right angles is π, and of one right angle $\dfrac{\pi}{2}$.

Examples.

1. Find the number of degrees in an angle which is subtended by an arc of 5 feet, the radius being 2 yards.

Ans. 47° 44′ 47″·34

2. Find the circular measure of 42°.

Ans. ·73303

3. How many degrees in an angle whose circular measure is $\tfrac{11}{21}$?

Ans. 30° 0′ 43″·45.

4. Find the radius when an arc of 1° measures 10 feet.

Ans. 572·957795 feet.

5. The arc of a railway curve is 250 yards, and subtends an angle of $6\tfrac{1}{4}$° at the centre. Find the radius.

Ans. 1·302 miles.

6. How many degrees are in an angle subtended by an arc of one inch to a radius of one foot?

Ans. 4° 46′ 29″.

7. What angle will an arc of one inch subtend to a radius of one mile?

Ans. 3″·254.

CHAPTER II.

TRIGONOMETRICAL RATIOS—FUNDAMENTAL FORMULÆ.

11. Let QAP be any angle, and in AP take any points B, B_1, B_2, &c., and draw BC, B_1C_1, &c., perpendicular to AQ, then the right-angled triangles ABC, AB_1C_1, AB_2C_2, are similar, and we have by Geometry,

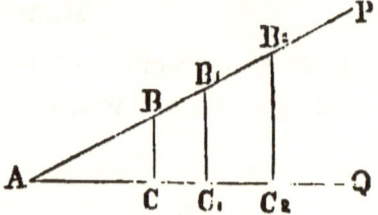

$$BC : AB :: B_1C_1 : AB_1 :: B_2C_2 : AB_2$$

or

$$\frac{BC}{AB} = \frac{B_1C_1}{AB_1} = \frac{B_2C_2}{AB_2}$$

and also

$$\frac{BC}{AC} = \frac{B_1C_1}{AC_1} = \frac{B_2C_2}{AC_2},$$

$$\frac{AB}{AC} = \frac{AB_1}{AC_1} = \frac{AB_2}{AC_2}.$$

These ratios, then, depend entirely on the magnitude of the angle, and not at all on the absolute lengths of the sides containing it. Both they and their reciprocals are, therefore, indices or functions of *the angle*, and have received special names as follows:

Let QAP be any angle in the first quadrant; in AP take any point B and draw BC perpendicular to AQ. Represent the angles of the right-angled triangle

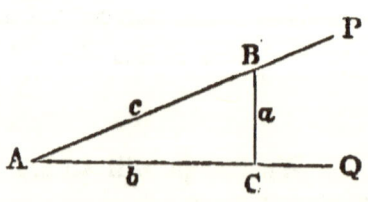

ABC by A, B and C, and the sides opposite them by the small

THE TRIGONOMETRICAL RATIOS.

letters a, b and c respectively: then whatever may be the absolute lengths of the sides:

(1) The sine of the angle A is the quotient of the opposite side by the hypothenuse.

Thus, $\sin A = \dfrac{a}{c}$, and similarly $\sin B = \dfrac{b}{c}$.

(2) The tangent of the angle A is the quotient of the opposite side by the adjacent side.

Thus, $\tan A = \dfrac{a}{b}$, and similarly $\tan B = \dfrac{b}{a}$.

(3) The secant of the angle A is the quotient of the hypothenuse by the adjacent side.

Thus, $\sec A = \dfrac{c}{b}$, and similarly $\sec B = \dfrac{c}{a}$.

(4) The cosine of the angle A is the quotient of the adjacent side by the hypothenuse.

Thus, $\cos A = \dfrac{b}{c}$, and similarly $\cos B = \dfrac{a}{c}$.

(5) The cotangent of the angle A is the quotient of the adjacent side by the opposite side.

Thus, $\cot A = \dfrac{b}{a}$, and similarly $\cot B = \dfrac{a}{b}$.

(6) The cosecant of the angle A is the quotient of the hypothenuse by the opposite side.

Thus, $\operatorname{cosec} A = \dfrac{c}{a}$, and similarly $\operatorname{cosec} B = \dfrac{c}{b}$.

12. Since the angle B is the complement of the angle A, Art. 4, we observe from the above six definitions of the Trigonometrical ratios, that the cosine, cotangent and cosecant of an angle are respectively the sine, tangent and secant of its complement. In fact it is from this circumstance that the terms cosine, cotangent and cosecant derive their names, being respectively abbreviations for "sine of the complement," "tangent of the complement," and "secant of the complement."

Hence we have from the last article:

$$\left.\begin{array}{ll} \sin A = \cos B = \dfrac{a}{c}. & \cos A = \sin B = \dfrac{b}{c}. \\\\ \tan A = \cot B = \dfrac{a}{b}. & \cot A = \tan B = \dfrac{b}{a}. \\\\ \sec A = \operatorname{cosec} B = \dfrac{c}{b}. & \operatorname{cosec} A = \sec B = \dfrac{c}{a}. \end{array}\right\} \quad (7)$$

From these equations it is seen that the sine and cosecant of the same angle are reciprocals; so also are the cosine and secant, the tangent and cotangent, reciprocals. That is,

$$\left.\begin{array}{ll} \sin A = \dfrac{1}{\operatorname{cosec} A}. & \operatorname{cosec} A = \dfrac{1}{\sin A}. \\\\ \cos A = \dfrac{1}{\sec A}. & \sec A = \dfrac{1}{\cos A}. \\\\ \tan A = \dfrac{1}{\cot A}. & \cot A = \dfrac{1}{\tan A}. \end{array}\right\} \quad (8)$$

or thus,

$$\sin A \operatorname{cosec} A = \cos A \sec A = \tan A \cot A = 1. \quad (9)$$

13. From the figure of Art. 11, we have (*Euc.* I. 47.)

$$a^2 + b^2 = c^2,$$

or

$$\dfrac{a^2}{c^2} + \dfrac{b^2}{c^2} = 1,$$

or, by the definition of the sine and cosine (7)

$$\sin^2 A + \cos^2 A = 1. \quad (10)$$

where $\sin^2 A$ signifies "the square of the sine of A," &c.

FUNDAMENTAL FORMULÆ.

Hence we find
$$\sin A = \sqrt{1 - \cos^2 A}$$
$$= \sqrt{(1 + \cos A)(1 - \cos A)}. \quad (11)$$

and also
$$\cos A = \sqrt{1 - \sin^2 A}$$
$$= \sqrt{(1 + \sin A)(1 - \sin A)}. \quad (12)$$

14. From (7), we have
$$\tan A = \frac{a}{b}$$
$$= \frac{\dfrac{a}{c}}{\dfrac{b}{c}}$$
$$= \frac{\sin A}{\cos A}. \quad (13)$$

And since the cotangent is the reciprocal of the tangent,
$$\cot A = \frac{\cos A}{\sin A}. \quad (14)$$

15. The figure of Art. 11 gives
$$c^2 = b^2 + a^2,$$
or
$$\frac{c^2}{b^2} = 1 + \frac{a^2}{b^2},$$

and therefore by the definition of the secant and tangent (7)
$$\sec^2 A = 1 + \tan^2 A. \quad (15)$$

16. Again, dividing by a^2 gives
$$\frac{c^2}{a^2} = \frac{b^2}{a^2} + 1,$$
or
$$\csc^2 A = \cot^2 A + 1. \quad (16)$$

Line Definitions.

17. The Trigonometrical ratios defined in Art. 11 may be represented geometrically by means of a circle, as follows:

Let DHK be a circle whose radius $= 1$, and let DAB be any angle in the first quadrant DAH. From B, the extremity of the radius AB, draw BC perpendicular to AD, and from D draw DE perpendicular to AD, to meet AB produced in E. Also draw BG and HF perpendicular to AH, the latter meeting AB produced in F. 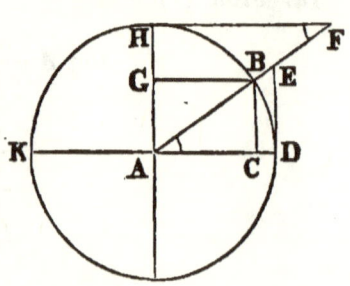 Representing the angle DAB by A, we have

$$\sin A = \frac{BC}{AB} = \frac{BC}{1} = BC.$$

(1) That is, to a radius of *unity*, the sine of an angle is the perpendicular drawn from one extremity of the arc subtending it, to the diameter passing through the other extremity.

$$\cos A = \frac{AC}{AB} = \frac{AC}{1} = AC = GB,$$

(2) or, the cosine of an angle is the distance from the centre to the foot of the sine, or the sine of the complement.

$$\tan A = \frac{ED}{AD} = \frac{ED}{1} = ED,$$

(3) or, the tangent of an angle is the line touching one extremity of the arc subtending it, and terminated by the diameter produced through the other extremity.

$$\cot A = \tan BAH, \quad \text{(Art. 12)}$$
$$= \frac{HF}{AH} = \frac{HF}{1} = HF,$$

(4) or, the cotangent of an angle is the tangent of the complement.

$$\sec A = \frac{AE}{AD} = \frac{AE}{1} = AE,$$

(5) or, the secant of an angle is the produced radius drawn through one extremity of the arc subtending it, and terminated by the tangent drawn from the other extremity.

$$\operatorname{cosec} A = \sec BAH, \quad \text{(Art. 12)}$$
$$= \frac{AF}{AH} = \frac{AF}{1} = AF,$$

(6) or, the cosecant of an angle is the secant of the complement.

18. From the similar triangles ABG, AFH, we have

$$AG : AH :: AB :: AF,$$

or, since $\quad AG = BC = \sin A$, and $AF = \operatorname{cosec} A$,

$$\sin A : 1 :: 1 : \operatorname{cosec} A.$$

Hence $\quad \sin A = \dfrac{1}{\operatorname{cosec} A}.$

From the similar triangles ABC, AED, we have

$$AC : AD :: AB : AE,$$

or, $\quad \cos A : 1 :: 1 : \sec A$

Hence $\quad \cos A = \dfrac{1}{\sec A}$

The angle $AFH =$ the angle BAD, therefore the triangles ADE, AHF, are similar, and we have

$$DE : AH :: AD : HF,$$

or,
$$\tan A : 1 :: 1 : \cot A.$$

Hence
$$\tan A = \frac{1}{\cot A}.$$

The results of this article agree with (8).

19. From the similar triangles ABC, AED, we have

$$AC : AD :: BC : DE,$$

or,
$$\cos A : 1 :: \sin A : \tan A.$$

Hence
$$\tan A = \frac{\sin A}{\cos A}, \text{ which agrees with (13).}$$

Again, from the similar triangles AGB, AHF,

$$\cot A = \frac{\cos A}{\sin A}.$$

From the right-angled triangles ABC, AED, AHF, we obtain, by *Euc.* I. 47,

$$\sin^2 A + \cos^2 A = 1$$
$$\sec^2 A = 1 + \tan^2 A$$
$$\operatorname{cosec}^2 A = 1 + \cot^2 A.$$

20. Besides the ratios already defined, CD, the portion of the diameter intercepted between the foot of the sine and the extremity of the arc, is called the versed-sine, abbreviated "versin;" and HG the versed-sine of the complement is the coversed-sine (coversin).

The versin and coversin can have no existence according to "ratio definitions" given in Art. 11, since the ratios there defined are independent of the radius; nevertheless the terms versin and coversin are used as convenient abbreviations for 1 - cosine and 1 - sine, respectively.

21. By means of the formulæ (8) (16) any ratio of an angle may be expressed in terms of any other; or if any ratio is given, all the others may be found.

Ex.—Express all the trigonometrical ratios in terms of the tangent.

$$\sec A = \pm \sqrt{(1 + \tan^2 A)} \qquad \cos A = \frac{1}{\sec A} = \frac{1}{\pm \sqrt{(1 + \tan^2 A)}}.$$

$$\sin A = \pm \sqrt{(1 - \cos^2 A)} = \pm \frac{\tan A}{\sqrt{(1 + \tan^2 A)}}.$$

$$\operatorname{cosec} A = \frac{1}{\sin A} = \pm \frac{\sqrt{(1 + \tan^2 A)}}{\tan A}.$$

$$\cot A = \frac{1}{\tan A}.$$

The significance of the double sign will be explained hereafter.

Examples.

1. Given $\tan A = \dfrac{5}{12}$, find the sine, cosine and secant.

$$\textit{Ans. } \sin A = \frac{5}{13}, \cos A = \frac{12}{13}, \sec A = \frac{13}{12}.$$

2. If $\tan A = \dfrac{2}{5}$, find the versin and cosec.

$$\textit{Ans. } \left(1 \pm \frac{5}{\sqrt{29}}\right) \text{ and } \tfrac{1}{2}\sqrt{29}.$$

3. Given $\sec \theta = 4$, find $\sin \theta$ and $\cot \theta$.

$$\textit{Ans. } \sin \theta = \frac{\pm \sqrt{15}}{4}, \cot \theta = \pm \frac{1}{\sqrt{15}}.$$

4. Given $\cot \theta = 2$, find $\sec \theta$ and $\operatorname{cosec} \theta$.

$$\textit{Ans. } \sec \theta = \pm \frac{\sqrt{5}}{2}, \operatorname{cosec} \theta = \pm \sqrt{5}.$$

5. Given versin $\theta = \dfrac{1}{3}$, find $\tan \theta$.

$$\text{Ans. } \tan \theta = \pm \dfrac{\sqrt{5}}{2}.$$

6. If $\cos \theta = \tan \theta$, find $\sin \theta$.

$$\text{Ans. } \sin \theta = -\dfrac{\pm\sqrt{5}-1}{2}.$$

7. If $\tan \theta + 3 \cot \theta = 4$, find $\cos \theta$.

$$\text{Ans. } \cos \theta = \pm \dfrac{\sqrt{2}}{2} \text{ or } \pm \dfrac{1}{\sqrt{10}}.$$

8. Prove $\cot^2 \theta - \cos^2 \theta = \cot^2 \theta \cos^2 \theta$.

9. If $\dfrac{\cot \theta - \sec \theta}{\cot \theta} = \dfrac{1}{16}$, find $\sin \theta$.

$$\text{Ans. } \sin \theta = \pm \dfrac{3}{5}.$$

10. Prove $\dfrac{\sec \theta \cot \theta - \operatorname{cosec} \theta \tan \theta}{\cos \theta - \sin \theta} = \sec \theta \operatorname{cosec} \theta$.

11. If $\tan \theta + \cot \theta = 2$, find the value of $\sin \theta + \cos \theta$.

$$\text{Ans. } \pm \sqrt{2}.$$

12. Determine $\sin a$ from the equation

$$9 \sin^2 a - 4 \tan^2 a = \dfrac{1}{2}.$$

$$\text{Ans. } \sin a = \pm \dfrac{1}{3} \text{ or } \pm \dfrac{\sqrt{2}}{2}.$$

13. If $\sin x = m \sin y$, and $\tan x = n \tan y$, shew that

$$\cos^2 x = \dfrac{m^2 - 1}{n^2 - 1}.$$

14. Prove that $\dfrac{(\operatorname{cosec} x + \sec x)^2}{\sec^2 x + \operatorname{cosec}^2 x} = (\sin x + \cos x)^2$.

EXAMPLES.

15. If $2 \tan a = \cos a$, find $\sin a$.

    Ans. $\sin a = \sqrt{2} - 1$.

16. If $\sin \theta - \cos \theta = \dfrac{\sqrt{3}-1}{2}$, find $\tan \theta$.

    Ans. $\tan \theta = \sqrt{3}$.

17. Given $\sec \theta - \tan \theta = \dfrac{1}{2}$, find $\sin \theta$.

    Ans. $\sin \theta = \dfrac{3}{5}$.

22. To radius *unity*, it will be seen that the results of Articles 18 and 19, deduced from the "line definitions," agree exactly with those previously obtained from the "ratio definitions." If, however, the radius is any other number than unity, the sine, tangent, &c., of any angle have not a fixed value as they have in the "ratio definitions" or in the "line definitions" when the radius is unity, but vary with the radius employed.

Thus, if AB in the figure of Art. 17 be represented by r, then BC, ED, AE, &c., become the sine, tangent, secant, &c., respectively of the angle BAD to the radius r, and therefore the formula

$$\sin^2 A + \cos^2 A = 1, \text{ becomes, to radius } r,$$
$$\sin^2 A + \cos^2 A = r^2.$$

Also, $\quad \tan A = \dfrac{\sin A}{\cos A}$, becomes, to radius r,

$$\tan A = \dfrac{r \sin A}{\cos A}.$$

and $\quad \sin A = \dfrac{1}{\operatorname{cosec} A}$, becomes, to radius r,

$$\sin A = \dfrac{r^2}{\operatorname{cosec} A},$$

and so on for all the other formulæ of Articles 18 and 19.

This inconvenience, arising from the introduction of the radius, has led to the almost general adoption of the "ratio definitions," by which the trigonometrical ratios (or functions, as they are also called), become abstract numerical quantities, dependent only on the magnitude of the angle, and altogether independent of the radius.

The "line definitions" to radius *unity*, however, give rise to no inconvenience, and may be employed with advantage in the solution of problems, and especially in illustrating geometrically the changes in magnitude of the different ratios through the four quadrants.

23. *To find the sine, cosine, &c., of 30° and 60°.*

Let ABC be an equilateral triangle, each of whose angles is therefore 60°. Draw AD perpendicular to BC, then the angle $BAD = 30°$, and $AB = 2\,BD$.

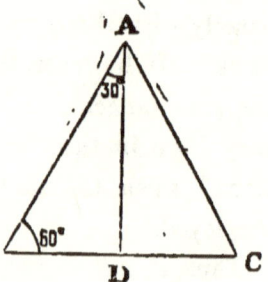

$$\sin 30° = \frac{BD}{AB} = \frac{BD}{2\,BD} = \frac{1}{2} = \cos 60°. \quad \text{Art 12.}$$

$$\cos 30° = \sqrt{1 - \sin^2 30°} = \frac{\sqrt{3}}{2} = \sin 60°.$$

$$\tan 30° = \frac{\sin 30°}{\cos 30°} = \frac{\frac{1}{2}}{\frac{\sqrt{3}}{2}} = \frac{1}{\sqrt{3}} = \cot 60°.$$

$$\cot 30° = \frac{1}{\tan 30°} = \sqrt{3} = \tan 60°.$$

$$\sec 30° = \frac{1}{\cos 30°} = \frac{2}{\sqrt{3}} = \csc 60°.$$

$$\csc 30° = \frac{1}{\sin 30°} = 2 = \sec 60°.$$

TRIGONOMETRICAL FUNCTIONS OF 18°, 36°, ETC.

24. *To find the sine, cosine, &c., of* 45°.

Since 45° is the complement of 45°, we have

$$\sin 45° = \cos 45°;$$

and since by (10) $\sin^2 A + \cos^2 A = 1$, we have

$$\sin^2 45° + \cos^2 45° = 2 \sin^2 45° = 2 \cos^2 45° = 1$$

$$\sin^2 45° = \cos^2 45° = \tfrac{1}{2},$$

and

$$\sin 45° = \cos 45° = \sqrt{\tfrac{1}{2}} = \frac{\sqrt{2}}{2}.$$

$$\tan 45° = \frac{\sin 45°}{\cos 45°} = 1.$$

$$\cot 45° = \frac{1}{\tan 45°} = 1.$$

$$\sec 45° = \frac{1}{\cos 45°} = \sqrt{2}.$$

$$\csc 45° = \frac{1}{\sin 45°} = \sqrt{2}.$$

25. *To find the sine, cosine, &c., of* 18°, 36°, 54° *and* 72°.

Let ABD be an isosceles triangle having each of the angles, ABD, ADB, double of the angle BAD; hence, it follows that the angle BAD is the fifth of two right angles, and therefore contains 36°; therefore each of the angles ABD, ADB, contains 72°. Draw AE perpendicular to BD, then the angle BAE contains 18°.

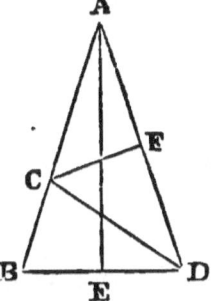

Now, if AB be divided in C, such that $AB \cdot BC = AC^2$, then by Euc. IV. 10, $AC = BD = DC$, and therefore if CF be drawn perpendicular to AD, the angle $ACF = 54°$.

Since $AB \cdot BC = AC^2$, we have

$$AB(AB - AC) = AC^2,$$

or,
$$\frac{AC^2}{AB^2} + \frac{AC}{AB} = 1.$$

By solving this quadratic, we obtain
$$\frac{AC}{AB} = \frac{-1 \pm \sqrt{5}}{2}.$$

Since $\frac{AC}{AB}$ is a positive quantity, the upper sign must be taken;

therefore,
$$\frac{AC}{AB} = \frac{\sqrt{5}-1}{2}.$$

Now, $\sin 18° = \dfrac{BE}{AB} = \dfrac{2\,BE}{2\,AB} = \dfrac{AC}{2\,AB}$

$$= \frac{\sqrt{5}-1}{4} = \cos 72°.$$

$\cos 18° = \sqrt{(1 - \sin^2 18°)}$

$$= \frac{\sqrt{(10 + 2\sqrt{5})}}{4} = \sin 72°.$$

Again, $\cos 36° = \dfrac{AF}{AC} = \dfrac{2\,AF}{2\,AC} = \dfrac{AB}{2\,AC}$

$$= \frac{1}{\sqrt{5}-1} = \frac{\sqrt{5}+1}{(\sqrt{5}-1)(\sqrt{5}+1)} = \frac{\sqrt{5}+1}{4}.$$

$= \sin 54°.$

$\sin 36° = \sqrt{(1 - \cos^2 36°)}$

$$= \tfrac{1}{4}\sqrt{(10 - 2\sqrt{5})} = \cos 54°.$$

The other trigonometrical functions of these angles can now be found. The student should verify the following:

1. $\cot 18° = \tfrac{1}{4}\sqrt{5}\sqrt{2+\sqrt{5}}$ $= 3\cdot 07768.$ $= \tan 72°.$

2. $\sec 72° = \sqrt{5}+1.$ $= 3\cdot 23607.$ $= \operatorname{cosec} 18°.$

THE FUNCTIONS OF HALF AN ANGLE.

3. $\tan 54° = \sqrt{(\frac{1}{2} + \frac{2}{5}\sqrt{5})}.\quad = 1\cdot37638.\quad = \cot 36°.$
4. $\operatorname{cosec} 36° = \sqrt{(2 + \frac{2}{5}\sqrt{5})}.\quad = 1\cdot70130.\quad = \sec 54°.$
5. $\tan 36° = \sqrt{(5 - 2\sqrt{5})}.\quad = \cdot72654.\quad = \cot 54°.$
6. $\tan 18° = \sqrt{(1 - \frac{2}{5}\sqrt{5})}.\quad = \cdot32492.\quad = \cot 72°.$

26. *To express the functions of half an angle in terms of those of the whole angle, and vice versa.*

Let BAC be any angle in the first quadrant, represent it by A, and in AC take any point P, and with A as a centre and AP as radius, describe a circle; produce BA to meet the circumference in D, join DP and draw PE perpendicular to AB.

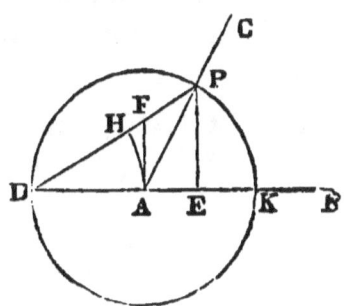

By *Euc.* III. 20, the angle BDP is half of the angle A;

then
$$\tan \frac{A}{2} = \tan BDP$$
$$= \frac{PE}{DE}$$
$$= \frac{PE}{AD + AE} = \frac{PE}{AP + AE}$$
$$= \frac{\frac{PE}{AP}}{1 + \frac{AE}{AP}}$$
$$= \frac{\sin A}{1 + \cos A} \qquad (17)$$

Or thus, by the "line definitions." Take AP as unity and with D as a centre and AD as radius, describe the arc AH and draw AF perpendicular to AB, then $AF = \tan\frac{A}{2}$, $PE = \sin A$, and $AE = \cos A$. From the similar triangles DAF, DEP, we have

$$DE : AD :: EP : AF,$$

or

$$1 + \cos A : 1 :: \sin A : \tan\frac{A}{2},$$

whence

$$\tan\frac{A}{2} = \frac{\sin A}{1 + \cos A}, \text{ as before.}$$

27. The formula just proved enables us to find the trigonometrical functions of 15° and 75°, from having given those of 30°.

Thus, let $A = 30°$, then we have

$$\tan 15° = \frac{\sin 30°}{1 + \cos 30°} = \frac{\frac{1}{2}}{1 + \frac{\sqrt{3}}{2}}$$

$$= \frac{1}{2 + \sqrt{3}} = 2 - \sqrt{3} = \cot 75°.$$

Hence, by Art. 21, we easily find

$$\cot 15° = 2 + \sqrt{3} \quad\quad = \tan 75°.$$
$$\sec 15° = \sqrt{2}(\sqrt{3} - 1) = \operatorname{cosec} 75°.$$
$$\operatorname{cosec} 15° = \sqrt{2}(\sqrt{3} + 1) = \sec 75°.$$
$$\sin 15° = \frac{\sqrt{3} - 1}{2\sqrt{2}} \quad\quad = \cos 75°.$$
$$\cos 15° = \frac{\sqrt{3} + 1}{2\sqrt{2}} \quad\quad = \sin 75°.$$

28. Squaring (17) we have

THE FUNCTIONS OF HALF AN ANGLE.

$$\tan^2\frac{A}{2} = \frac{\sin^2 A}{(1+\cos A)^2}$$

$$= \frac{1-\cos^2 A}{(1+\cos A)^2}$$

$$= \frac{1-\cos A}{1+\cos A}. \tag{18}$$

$$\sec^2\frac{A}{2} = 1 + \tan^2\frac{A}{2}$$

$$= 1 + \frac{1-\cos A}{1+\cos A}$$

$$= \frac{2}{1+\cos A}$$

therefore $\quad\cos^2\dfrac{A}{2} = \dfrac{1+\cos A}{2}. \tag{19}$

$$\sin^2\frac{A}{2} = 1 - \cos^2\frac{A}{2}$$

$$= 1 - \frac{1+\cos A}{2}$$

$$= \frac{1-\cos A}{2}. \tag{20}$$

From (18) we easily find

$$\cos A = \frac{1-\tan^2\dfrac{A}{2}}{1+\tan^2\dfrac{A}{2}}. \tag{21}$$

Multiplying (19) by (20), and extracting the square root, we find

$$\sin A = 2\sin\frac{A}{2}\cos\frac{A}{2}. \tag{22}$$

Subtracting (20) from (19) we have

$$\cos A = \cos^2\frac{A}{2} - \sin^2\frac{A}{2}. \tag{23}$$

If we write $2A$ for A, which we are evidently at liberty to do, then (22) and (23) become

$$\sin 2A = 2 \sin A \cos A. \qquad (24)$$

and
$$\begin{aligned}\cos 2A &= \cos^2 A - \sin^2 A \\ &= 1 - 2 \sin^2 A \\ &= 2 \cos^2 A - 1.\end{aligned} \qquad (25)$$

Also,
$$\begin{aligned}\tan 2A &= \frac{\sin 2A}{\cos 2A} \\ &= \frac{2 \sin A \cos A}{\cos^2 A - \sin^2 A} \\ &= \frac{\dfrac{2 \sin A}{\cos A}}{1 - \dfrac{\sin^2 A}{\cos^2 A}} \\ &= \frac{2 \tan A}{1 - \tan^2 A}.\end{aligned} \qquad (26)$$

29. The formulæ of the last three Articles will be found extremely useful hereafter, and the student is requested to pay particular attention to them. They will be again deduced by a more general process in a subsequent chapter.

Examples.

1. Find A from the equation $\tan^2 A - 4 \tan A + 1 = 0$.

Adding 3 to both members we have

$$\tan^2 A - 4 \tan A + 4 = 3$$

hence
$$\tan A - 2 = \pm \sqrt{3}$$
$$\tan A = 2 + \sqrt{3} \text{ or } 2 - \sqrt{3}$$
$$= \tan 75° \text{ or } \tan 15°$$

therefore
$$A = 75° \text{ or } 15°.$$

EXAMPLES. 25

2. Given $\operatorname{cosec} \theta - \sin \theta = \tan\dfrac{\theta}{2}$, find $\cos \theta$.

By (8) and (17), the given equation becomes

$$\dfrac{1}{\sin \theta} - \sin \theta = \dfrac{\sin \theta}{1 + \cos \theta}$$

$$1 - \sin^2 \theta = \dfrac{\sin^2 \theta}{1 + \cos \theta}$$

or $\qquad \cos^2 \theta = \dfrac{1 - \cos^2 \theta}{1 + \cos \theta} = 1 - \cos \theta$

$$\cos^2 \theta + \cos \theta = 1$$

$$\cos^2 \theta + \cos \theta + \tfrac{1}{4} = \dfrac{5}{4}$$

$$\cos \theta + \tfrac{1}{2} = \pm \dfrac{\sqrt{5}}{2}$$

hence $\qquad \cos \theta = \dfrac{\pm \sqrt{5} - 1}{2}.$

3. Given $\tan \theta + \cot \theta = 2 \sec \theta$, find θ.

By (8), (13) and (14), the given equation becomes

$$\dfrac{\sin \theta}{\cos \theta} + \dfrac{\cos \theta}{\sin \theta} = \dfrac{2}{\cos \theta}$$

$$\sin^2 \theta + \cos^2 \theta = 2 \sin \theta$$

or $\qquad 1 = 2 \sin \theta,$

hence $\qquad \sin \theta = \tfrac{1}{2} = \sin 30°,$

therefore $\qquad \theta = 30°.$

4. Prove that $\operatorname{cosec} \theta = \tfrac{1}{2} \left(\cot \dfrac{\theta}{2} + \tan \dfrac{\theta}{2} \right).$

$$\operatorname{cosec} \theta = \dfrac{1}{\sin \theta} = \dfrac{\cos^2 \dfrac{\theta}{2} + \sin^2 \dfrac{\theta}{2}}{2 \sin \dfrac{\theta}{2} \cos \dfrac{\theta}{2}}, \text{ by (10) and (22)}$$

$$= \tfrac{1}{2} \left(\cot \dfrac{\theta}{2} + \tan \dfrac{\theta}{2} \right).$$

5. Prove that $\tan \dfrac{\theta}{2} = \dfrac{1-\cos\theta}{\sin\theta} = \dfrac{\sec\theta - 1}{\tan\theta}$.

6. Prove that $\cot\theta = \tfrac{1}{2}\left(\cot\dfrac{\theta}{2} - \tan\dfrac{\theta}{2}\right)$.

7. Given $\sec x \tan x = 2\sqrt{3}$, find x.

   Ans. $x = 60°$.

8. Given $6\cot^2 x - 4\cos^2 x = 1$, find x.

   Ans. $x = 60°$.

9. If $\sin A \sec B = \sqrt{\dfrac{3}{2}}$, and $\cos A \csc B = \dfrac{\sqrt{2}}{2}$, find A and B.

   Ans. $A = 60°$, $B = 45°$.

10. If $\cos(2A + B) = \dfrac{\sqrt{3}}{2}$, and $\sin(3A - B) = \dfrac{1}{2}$, find A and B.

    Ans. $A = 12°$, $B = 6°$.

11. If $\tan(3A + 2B) = 2 + \sqrt{3}$, and $\tan(5A - B) = 2 - \sqrt{3}$, find A and B.

    Ans. $A = 8\tfrac{1}{13}°$, $B = 25\tfrac{5}{13}°$.

12. Find the tangent of $7\tfrac{1}{2}°$.

    Ans. $\sqrt{6} - \sqrt{3} + \sqrt{2} - 2$.

13. Find the tangent of $37\tfrac{1}{2}°$.

    Ans. $\sqrt{6} + \sqrt{3} - \sqrt{2} - 2$.

14. Given $2\csc\theta - \cot\theta = \sqrt{3}$, find θ.

    Ans. $\theta = 60°$.

15. Prove that $\sin A = \dfrac{1}{\cot\dfrac{A}{2} - \cot A}$.

16. If $\tan\theta = \sqrt{2} - 1$, find $\cos 2\theta$.

    Ans. $\cos 2\theta = \dfrac{1}{\sqrt{2}}$.

17. If $\cos\theta = \dfrac{a\cos\phi - b}{a - b\cos\phi}$, shew that $\tan\dfrac{\theta}{2} = \sqrt{\dfrac{a+b}{a-b}}\tan\dfrac{\phi}{2}$.

18. Prove that $\tan\theta - \tan\dfrac{\theta}{2} = \tan\dfrac{\theta}{2}\sec\theta$.

19. Prove that $1 + \tan A \tan\dfrac{A}{2} = \sec A$.

20. If $\sec\theta + \operatorname{cosec}\theta = m$, and $\sec\theta - \operatorname{cosec}\theta = n$, find $\tan\theta$.

Ans. $\tan\theta = \dfrac{m+n}{m-n}$.

21. Given $\sin\theta\cos\theta = \dfrac{1}{2\sqrt{2}}$, find θ.

Ans. $\theta = 22\tfrac{1}{2}°$.

22. In the fig. of Art. 26, join PK, and prove geometrically that

$$\tan\dfrac{A}{2} = \dfrac{1 - \cos A}{\sin A},$$

$$\sin A = 2\sin\dfrac{A}{2}\cos\dfrac{A}{2},$$

$$1 + \cos A = 2\cos^2\dfrac{A}{2},$$

$$1 - \cos A = 2\sin^2\dfrac{A}{2}.$$

23. Prove that

$\sin A(1 + \tan A) + \cos A(1 + \cot A) = \operatorname{cosec} A + \sec A$.

24. Prove that

$$\dfrac{\tan\theta + \sec\theta - 1}{\tan\theta - \sec\theta + 1} = \tan\theta + \sec\theta,$$

and

$$\dfrac{\tan\theta - \sin\theta}{\sin^3\theta} = \dfrac{\sec\theta}{1 + \cos\theta}.$$

CHAPTER III.

APPLICATION OF THE PRECEDING FORMULÆ.

30. The formulæ of Article 12 are directly available for the solution of right-angled triangles. Thus, if ABC be a right-angled triangle, C being the right angle, and a, b, c the sides opposite the angles A, B, C respectively, we have according to the definitions of the sine, &c.,

$$\sin A = \frac{a}{c}; \quad \cos A = \frac{b}{c}; \quad \tan A = \frac{a}{b}.$$

Each of these equations expresses a relation between three parts—an angle and two sides—hence, when any two of these parts are given, the other can be found.

In order, however, to solve a triangle trigonometrically, it will be necessary to know the values of the sine, cosine, &c., of any given angle. The trigonometrical tables supply these values for every minute, and sometimes for every ten seconds, from 0° to 90°. At present, however, we will use only those trigonometrical ratios which we have already computed in the preceding chapter. The construction and use of the tables will be explained in a subsequent part of this work.

31. The solution of right-angled triangles presents the following cases:

Case I.—*Given the hypothenuse and one angle, as c and A.*

To find a, we have

$$\sin A = \frac{a}{c}, \text{ whence } a = c \sin A. \qquad (27)$$

SOLUTION OF RIGHT-ANGLED TRIANGLES.

To find b, we have

$$\cos A = \frac{b}{c}, \text{ whence } b = c \cos A. \qquad (28)$$

To find B, we have $B = 90° - A$.

Example.—Given $c = 24$, and the angle $A = 60°$, to find the other parts.

By (27) we have

$$a = 24 \times \sin 60°$$
$$= 24 \times \frac{\sqrt{3}}{2} = 12 \sqrt{3}.$$

By (28) we have

$$b = 24 \cos 60° = 24 \times \tfrac{1}{2} = 12.$$
and $\qquad B = 90° - 60° = 30°.$

If the angle B had been given instead of A, we might first find A, thus, $A = 90° - B$, and then proceed as above.

Case II.—*Given the hypothenuse and one side; c and a.*

To find A, we have

$$\sin A = \frac{a}{c}. \qquad (29)$$

To find B, we have $B = 90° - A$.

To find b, we have

$$\cos A = \frac{b}{c}, \text{ whence } b = c \cos A. \qquad (30)$$

or $\qquad b = \sqrt{c^2 - a^2} = \sqrt{(c+a)(c-a)}. \qquad (31)$

Example.—Given $c = 10$ and $a = 5\sqrt{2}$, find other parts.

By (29) $\qquad \sin A = \frac{5\sqrt{2}}{10} = \frac{1}{\sqrt{2}} = \sin 45°,$

hence $\qquad A = 45°$ and $B = 45°.$

By (30) $\qquad b = 10 \cos 45° = 10 \times \frac{1}{\sqrt{2}} = 5\sqrt{2}.$

Case III.—*Given an angle and its adjacent side; A and b.*

To find a, we have

$$\tan A = \frac{a}{b}, \text{ whence } a = b \tan A. \tag{32}$$

To find c, we have

$$\cos A = \frac{b}{c}, \text{ whence } c = \frac{b}{\cos A} = b \sec A. \tag{33}$$

Example.—Given $A = 75°$ and $b = 15$, find a and c.

By (32) $a = 15 \tan 75°$
$$= 15 (2 + \sqrt{3}) = 55\cdot98.$$

By (33) $c = 15 \sec 75°$
$$= 15 \cdot \frac{2\sqrt{2}}{\sqrt{3}-1} = 15(\sqrt{6} + \sqrt{2}) = 57\cdot648.$$

Case IV.—*Given an angle and its opposite side; A and a.*

We may first find B, which reduces this to Case III., or we may proceed thus:

To find b, we have

$$\tan A = \frac{a}{b}, \text{ whence } b = \frac{a}{\tan A} = a \cot A. \tag{34}$$

To find c, we have

$$\sin A = \frac{a}{c}, \text{ whence } c = \frac{a}{\sin A} = a \csc A. \tag{35}$$

Or, using b just found, we have

$$\cos A = \frac{b}{c}, \text{ whence } c = b \sec A. \tag{36}$$

Case V.—*Given the two sides; a and b.*

To find A and B, we have

$$\tan A = \cot B = \frac{a}{b}. \tag{37}$$

SOLUTION OF RIGHT-ANGLED TRIANGLES.

To find c, we have

$$\sin A = \frac{a}{c}, \text{ whence } c = \frac{a}{\sin A} = a \text{ cosec } A. \quad (38)$$

32. The formulæ of the last Article may be illustrated geometrically by the "line definitions," as follows:

With A as a centre and a radius unity, describe an arc DG, draw DE and GF perpendicular to AC; then, Art. 17, $DE = \sin A$, $AE = \cos A$, $GF = \tan A$, and $AF = \sec A$. From the similar right-angled triangles, AED, ACB, we have

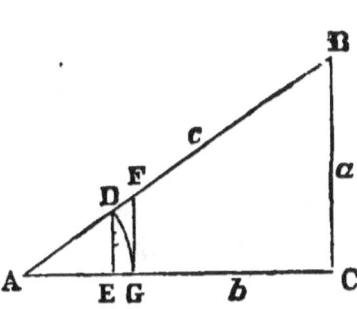

$$AD : AB :: DE : BC$$

that is, $1 : c :: \sin A : a,$

whence $a = c \sin A,$ which is (27),

and $\sin A = \dfrac{a}{c},$ which is (29).

Again $AD : AB :: AE : AC$

that is, $1 : c :: \cos A : b,$

whence $b = c \cos A,$ which is (28),

and $\cos A = \dfrac{b}{c},$ which is (36).

Again, from the similar right-angled triangles, AGF, ACB, we have

$$AG : AC :: GF : CB$$

that is, $1 : b :: \tan A : a,$

whence $a = b \tan A,$ which is (32),

and $\tan A = \dfrac{a}{b},$ which is (37).

Examples.

1. The radius of a circle is 10 feet; find the side of the inscribed regular pentagon.

Ans. $5(\sqrt{5}-1)$ ft.

2. The base of an isosceles triangle is 20 yards, and the vertical angle is 108°; find the sides.

Ans. $10(\sqrt{5}-1)$ yds.

3. The side of a regular octagon is 14 feet; find the radius of the inscribed circle.

Ans. $7(\sqrt{2}+1)$ ft.

4. If the earth be a sphere whose diameter is 7912 miles, find the radius of the 45th parallel of latitude.

Ans. $1978\sqrt{2}$ miles.

5. The angular elevation of a tower at a place A due south of it is 45°, and at a place B due west of A, and at a distance a yards from it, the elevation is 30°; find the height of the tower and the distance from B to the foot of the tower.

Ans. Height $= \dfrac{a}{2}\sqrt{2}$ yds.; distance from $B = \dfrac{a}{2}\sqrt{6}$ yds.

6. A person observes the angle of elevation of a mountain to be 27°, and approaching one mile nearer, the elevation is 54°; find its height and distance from the first station.

Ans. Height $= \tfrac{1}{4}(\sqrt{5}+1)$ miles.
Distance from first station $= \tfrac{1}{4}(4+\sqrt{10-2\sqrt{5}})$ miles.

7. A tower 60 feet high casts a shadow $20\sqrt{3}$ feet in length; find the altitude of the sun above the horizon.

Ans. 60°.

EXAMPLES.

8. In a triangle ABC, $\sin A = \dfrac{\sqrt{5}}{3}$, $\sin B = \dfrac{\sqrt{3}}{2}$ and the sides opposite to these angles are $\sqrt{15}$ and $4\frac{1}{2}$; find the other side.

Ans. $3 + \frac{1}{2}\sqrt{15}$.

9. A church spire subtends an angle of 60° at a certain distance from its base, and 80 feet farther off it subtends an angle of 45°; find its height, allowing 5 feet for the observer's height.

Ans. $125 + 40\sqrt{3}$ feet.

10. The angular elevation of a steeple standing on a horizontal plane is observed, and at a point 10 yards nearer to it the angular elevation is found to be the complement of the former. On advancing 3 yards nearer, the elevation is double of the first. Find the height of the steeple, the distance of the first point from its foot, and the tangent of the angle of elevation at the third point.

Ans. Height = 12 yds.; distance = 18 yds.; tan of the angle of elevation at third point = 2·4.

11. Wishing to know the distance from a given point A, to an inaccessible point B, on the opposite bank of a river, a base line AC, 84 rods, is measured along the bank of the river at right-angles to AB, and from its extremity C, equal distances CD, CE, 6 rods, are measured on the lines CB, CA. Finally, 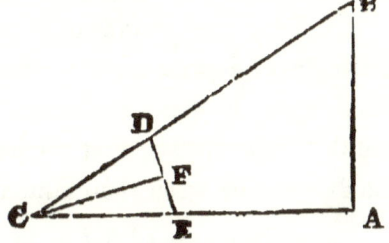 DE is measured and found to be 4 rods. Find the distance between A and B, and the tangent of the angle ACB.

Bisect DE in F, join CF; then, in the right-angled triangle CDF,

$$\sin DCF = \frac{DF}{CD} = \frac{2}{6} = \frac{1}{3},$$

hence $\quad \cos DCF = \sqrt{1-\tfrac{1}{9}} = \frac{2\sqrt{2}}{3},$

$$\tan DCF = \frac{\sin DCF}{\cos DCF} = \frac{1}{3} \cdot \frac{3}{2\sqrt{2}} = \frac{\sqrt{2}}{4}.$$

The angle DCE is double of the angle DCF, therefore, by (26),

$$\tan ACB = \frac{2 \tan DCF}{1 - \tan^2 DCF} = \frac{\frac{\sqrt{2}}{2}}{1 - \frac{2}{16}} = \frac{4\sqrt{2}}{7}.$$

Hence, by (32), $\quad AB = AC \tan ACB$

$$= 84 \times \frac{4\sqrt{2}}{7} = 48\sqrt{2} \text{ rods.}$$

12. If, in the figure of the last problem, $CB = 25$ yards, and $CD = CE = 5$ yards, and $DE = 4$ yards, find AB and the tangent of B.

Ans. $AB = 4\sqrt{21}$ yards; $\tan B = \frac{17}{84}\sqrt{21}$.

13. From the summit of a tower whose height is 108 feet, the angles of depression of the top and bottom of another tower standing on the same horizontal plane, are observed to be 30° and 60° respectively; find the height of the second tower, and the distance between their summits.

Ans. Height = 72 ft.; distance between summits = 72 ft.

14. One side of a right-angled triangle is 60 rods, and the

cosine of the adjacent angle equals the cotangent of the opposite angle; find the hypotenuse and the other side.

Ans. Hypotenuse $= 30\sqrt{2(\sqrt{5}+1)}$; side $= 30\sqrt{2(\sqrt{5}-1)}$.

15. The base of a right-angled triangle is 60 yards, and the tangent of the opposite angle equals three times the cosine of the adjacent angle; find the perpendicular.

Ans. $15\sqrt{2}$ yds.

16. In a right-angled triangle whose right angle is C, shew by (17) that $\tan\dfrac{A}{2} = \dfrac{a}{b+c} = \dfrac{c-b}{a}$;

and by (20) that $\sin\dfrac{A}{2} = \sqrt{\dfrac{c-b}{2c}}$.

17. In any right-angled triangle, given A and c, shew that the perpendicular from the right-angle on the hypothenuse is $\dfrac{c}{2}\sin 2A$.

18. The sides of a right-angled triangle are 3, 4 and 5, find the length of the perpendicular from the right angle on the hypothenuse.

Ans. 2.4.

19. A person in a line with two towers, and at distances of 100 and 150 yards from them, observes that their apparent altitudes are the same. He then walks towards them a distance of 60 yards, and finds that the angle of elevation of the nearer is just double of that of the former. Find the heights of the towers.

Ans. The nearer, $\dfrac{20}{3}\sqrt{7}$ yds.; the other, $10\sqrt{7}$ yds.

20. A tower of unknown height stands on a horizontal plane, and at a distance of 80 yards from the foot of the tower

a mark which is known to be 50 feet high, has an angular elevation just half of that of the tower. Find the height of tower, the observer's eye being on a level with the plane.

Ans. 104·537 feet.

21. The sides of a right-angled triangle are in arithmetical progression, and the area is 108; find the sides.

Ans. 9, 12 and 15.

22. Standing straight in front of a house, opposite one corner, I find that its length subtends an angle whose sine is $\frac{2}{5}\sqrt{5}$, while its height subtends an angle whose tangent is $\frac{3}{5}$; the height of the house is 51 feet; find its length.

Ans. 170 ft.

23. When the altitude of the sun is 54°, the shadow cast by a church spire is 150 feet. Find the height of the spire.

Ans. 68.819 yds.

24. The base of a right-angled triangle is one-third of the hypothenuse, and the perpendicular from the right angle on the hypothenuse is $6\sqrt{2}$; find the sides.

Ans. 9, $18\sqrt{2}$ and 27.

25. A May-pole being broken off by the wind, its top struck the ground at an angle of 36°, and at a distance of 30 feet from the foot of the pole; find its whole height.

Ans. 58·878 ft.

CHAPTER IV.

EXTENSION OF THE DEFINITIONS OF THE TRIGONOMETRICAL FUNCTIONS—CHANGES IN THEIR SIGN AND MAGNITUDE.

33. In chapter II., the sine, tangent, &c., have been defined for angles in the first quadrant only, or for those less than 90°. We will now extend these definitions so as to include all angles.

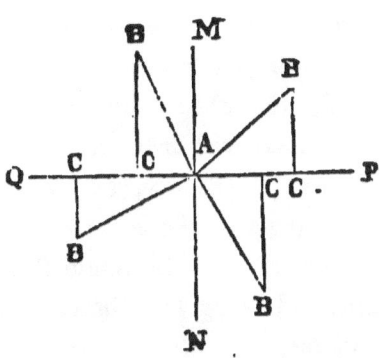

Draw the lines PQ, MN, intersecting at right angles in A; let AB be different positions of the revolving line in the first, second, third and fourth quadrants, and from any point of AB as B draw BC perpendicular to PQ. In the first and second quadrants BC lies *above*, and in the third and fourth quadrants, *below PQ;* while in the first and fourth quadrants, AC lies to the *right*, and in second and third quadrants, to the *left* of MN. These opposite directions are indicated by the signs + and - ; thus, lines which lie *above PQ* are regarded as *positive*, consequently those which lie *below* it must be regarded as *negative*. Again, lines which lie to the *right* of MN are regarded as positive, consequently those which lie to the *left* must be regarded as *negative*. Hence BC is *positive* in the first and second quadrants, and *negative* in the third and fourth, and AC is *positive* in the first and fourth quadrants, and *negative* in the second and third.

The revolving line AB, is always regarded as *positive*, being measured off in the direction from A towards B; if, however, AB be produced backwards, the produced part must be considered *negative*. The student will not fail to perceive that the signs $+$ and $-$, as thus employed, indicate, as in Art. 3, merely contrariety of direction. So it is in geography and astronomy— if north latitudes and east longitudes be considered positive, south latitudes and west longitudes must be considered negative. The student must be careful, however, to observe that no absolutely *fixed* direction is here meant. Any line whatever may be regarded as the initial one, the object of the convention being merely to indicate that lines measured in opposite directions are affected with *opposite signs*. Thus, in the obtuse-angled triangle ABC, the obtuse angle C is measured from BC as the initial line, and is an angle in the *second* 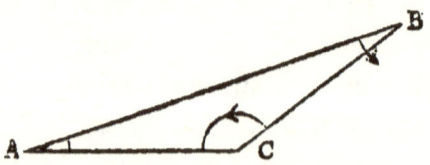 quadrant; while the angle B is measured from AB as the initial line, and being less than a right angle, is an angle in the *first* quadrant.

34. **To trace the Variations in Sign and Magnitude of the sine, cosine and tangent of an Angle, as it varies from 0° to 360°.**

First Quadrant.—Let a, b, c represent the sides of the right-angled triangle ABC, opposite to the angles A, B, C respectively.

VARIATIONS OF TRIGONOMETRICAL FUNCTIONS. 39

$\sin A = \dfrac{+a}{c}$, which is therefore *positive*.

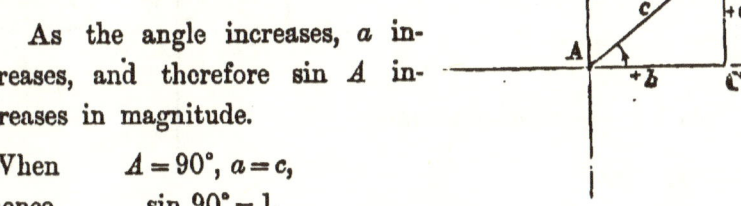

When $A = 0°$, $a = 0$,
hence $\sin 0° = 0$.

As the angle increases, a increases, and therefore $\sin A$ increases in magnitude.

When $A = 90°$, $a = c$,
hence $\sin 90° = 1$.

$\cos A = \dfrac{+b}{c}$, which is therefore *positive*.

When $A = 0°$, $b = c$,
hence $\cos 0° = 1$.

As A increases, b decreases, and therefore $\cos A$ decreases in magnitude.

When $A = 90°$, $b = 0$,
hence $\cos 90° = 0$.

$\tan A = \dfrac{+a}{+b}$, which is therefore *positive*.

When $A = 0°$, $a = 0$ and $b = c$,
hence $\tan 0° = 0$.

As A increases, a increases and b decreases, and therefore $\tan A$ increases in magnitude.

When $A = 90°$, $a = c$ and $b = 0$,
hence $\tan 90° = \infty$.

Second Quadrant.

$\sin A = \dfrac{+a}{c}$, which is therefore *positive*.

As A increases, a decreases, and therefore sin A decreases in magnitude.

When $A = 180°$, $a = 0$,
hence sin $180° = 0$.

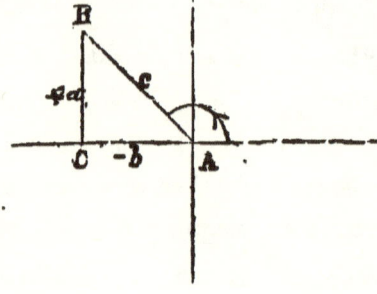

$\cos A = \dfrac{-b}{c}$, which is therefore *negative*.

As A increases, b increases, and therefore cos A increases in magnitude.

When $A = 180°$, $b = c$,
hence $\cos 180° = \dfrac{-c}{c} = -1$.

$\tan A = \dfrac{+a}{-b}$, which is therefore *negative*.

As A increases, a decreases and b increases, therefore tan A decreases in magnitude.

When $A = 180°$, $a = 0$ and $b = c$,
hence $\tan 180° = 0$.

Third Quadrant.

$\sin A = \dfrac{-a}{c}$, which is therefore *negative*.

As A increases, a increases, and therefore sin A increases in magnitude.

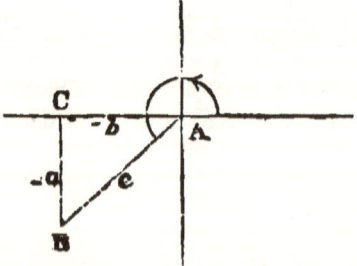

When $A = 270°$ (3 right angles), $a = c$,
hence $\sin 270° = \dfrac{-c}{c} = -1$.

$\cos A = \dfrac{-b}{c}$, which is therefore *negative*.

As A increases, b decreases, and therefore $\cos A$ decreases in magnitude.

When $\quad\quad\quad\quad A = 270°, b = 0,$
hence $\quad\quad\quad\quad \cos 270° = 0.$

$\tan A = \dfrac{-a}{-b} = +\dfrac{a}{b}$, which is therefore *positive*.

As A increases, a increases and b decreases, and therefore $\tan A$ increases in magnitude.

When $\quad\quad\quad\quad A = 270°, a = c$ and $b = 0,$
hence $\quad\quad\quad\quad \tan 270° = \infty.$

Fourth Quadrant.

$\sin A = \dfrac{-a}{c}$, which is therefore *negative*.

As A increases, a decreases, and therefore $\sin A$ decreases in magnitude.

When $\quad A = 360°, a = 0,$
hence $\quad \sin 360° = 0.$

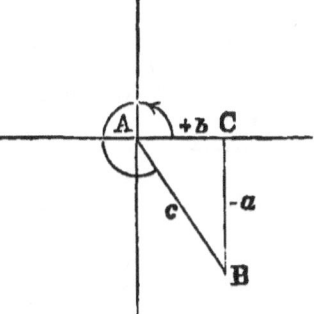

$\cos A = \dfrac{+b}{c}$, which is therefore *positive*.

As A increases, b increases, and therefore $\cos A$ increases in magnitude.

When $\quad\quad\quad\quad A = 360°, b = c,$
hence $\quad\quad\quad\quad \cos 360° = 1.$

$\tan A = \dfrac{-a}{b}$, which is therefore *negative*.

As A increases, a decreases and b increases, therefore $\tan A$ decreases in magnitude.

When $\qquad A = 360°$, $a = 0$ and $b = c$,
hence $\qquad\qquad \tan 360° = 0.$

35. As the cosecant, secant and cotangent are the reciprocals of the sine, cosine and tangent respectively, they require no special examination.

The variations of all the trigonometrical functions, both in sign and magnitude, are given in the following tabular form:

	1st Quadrant.	2nd Quadrant.	3rd Quadrant.	4th Quadrant.
	0° to 90°	90° to 180°	180° to 270°	270° to 360°
sine	positive. 0 to 1	positive. 1 to 0	negative. 0 to -1	negative. -1 to 0
cosine.....	positive. 1 to 0	negative. 0 to -1	negative. -1 to 0	positive. 0 to 1
tangent....	positive. 0 to ∞	negative. ∞ to 0	positive. 0 to ∞	negative. ∞ to 0
cosecant...	positive. ∞ to 1	positive. 1 to ∞	negative. ∞ to -1	negative. -1 to ∞
secant.....	positive. 1 to ∞	negative. ∞ to -1	negative. -1 to ∞	positive. ∞ to 1
cotangent..	positive. ∞ to 0	negative. 0 to ∞	positive. ∞ to 0	negative. 0 to ∞

36. From the above table it is seen that the values of the sine and cosine always lie between 0 and ± 1, and those of the secant and cosecant between $+1$ and $+\infty$, and between -1 and $-\infty$; while those of the tangent and cotangent lie between

0 and $\pm\infty$, and therefore may be of any magnitude whatever, positive or negative. It is also seen that each of the functions changes its sign when its value passes through zero or infinity; hence we may write $\cos 90° = +0$ or -0, and $\tan 90° = +\infty$ or $-\infty$, &c.

37. The preceding results of this chapter are easily obtained by means of the line definitions. In a circle of radius unity, draw two diameters, PQ and MN, at right angles to each other, and let PAB be an angle taken in each of the four quadrants. Draw the various lines according to the definitions of Art. 17, and let, as usual, the positions of the lines in the first quadrant be taken as the positive positions.

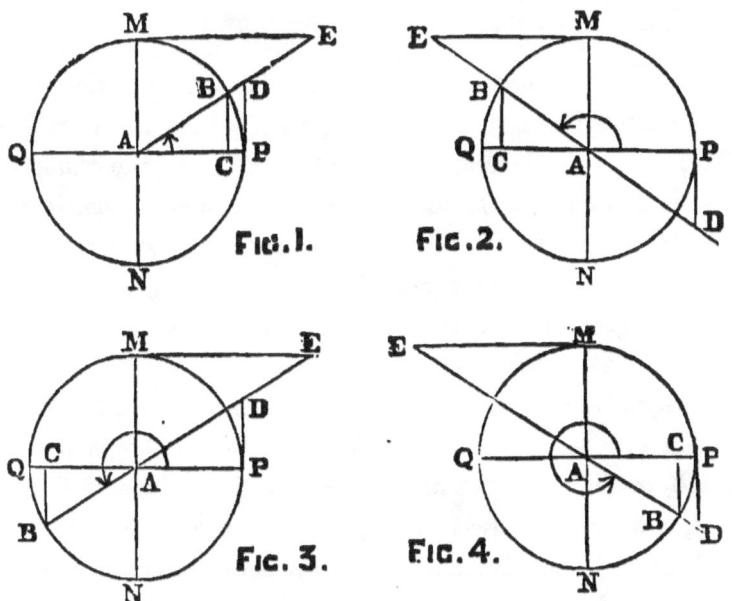

The sine BC is *positive* in the first and second quadrants, and *negative* in the third and fourth.

The cosine AC is *positive* in the first and fourth quadrants, and *negative* in the second and third.

The tangent PD is *positive* in the first and third quadrants, and *negative* in the second and fourth.

The cotangent ME is *positive* in the first and third quadrants, and *negative* in the second and fourth.

The secant AD, in the first and fourth quadrants, is obtained by producing the revolving radius AB *forwards;* while in the second and third, it is found by producing AB *backwards.* It is therefore *positive* in the first and fourth quadrants, and *negative* in the second and third.

The cosecant AE, in the first and second quadrants, is obtained by producing AB *forwards;* while in the third and fourth it is formed by producing AB *backwards.* It is therefore *positive* in the first and second quadrants, and *negative* in the third and fourth.

38. With regard to magnitude, it is seen from the figures of the last Article, that as the angle PAB increases from 0° to 90°, the sine BC increases from 0 to AM or 1; the cosine AC decreases from 1 to 0; the tangent PD increases continually from 0, and when the secant AD coincides nearly with AM both the tangent and secant become exceedingly great, and infinitely great when AD actually coincides with AM, or when the angle $PAB = 90°$, AD being then parallel to PD. Hence the tangent varies from 0 to ∞, and the secant from 1 to ∞.

Again, when the angle PAB begins at 0°, AE is parallel to ME, or the cotangent and cosecant begin by being infinitely great, and then decrease continually as the angle increases, until at 90° the cotangent ME vanishes, and the cosecant becomes AM or 1.

In the same way we may proceed to find the magnitude of the various functions in the other quadrants; but this we leave to the student. The results will be found to agree exactly with those obtained in Art. 35.

FUNCTIONS OF 180° − A.

From the figures of the last Article it is seen that the versed-sine CP, which equals $1 - AC$, or $1 -$ cosine, is always positive, and varies from 0 to PQ or 2, in the first two quadrants, and from 2 to 0 in the other two quadrants.

The coversed-sine, which equals $1 -$ sine, is also always positive, and varies from 1 at 0° to 0 at 90°, then from 0 at 90° to 1 at 180°. In the third and fourth quadrants it varies from 1 at 180° to 2 at 270°, and then from 2 to 1 at 360° or 0°.

39. To find the Trigonometrical Functions of 180° − A.

Let PAB be any angle A; produce PA to Q, and make the angle $B'AQ =$ to the angle PAB; then PAB' is the supplement of PAB or PAB' $= 180° - A$. Take $AB' = AB$, and draw $B'C''$ and BC perpendicular to PQ; then $B'C'' = BC$, and $AC'' = AC$.

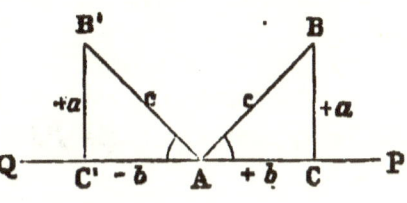

$$\sin PAB' = \frac{+a}{c} = \sin PAB.$$

$$\cos PAB' = \frac{-b}{c} = -\cos PAB.$$

$$\tan PAB' = \frac{+a}{-b} = -\tan PAB.$$

or, we may write these results thus:

$$\sin (180° - A) = \sin A.$$
$$\cos (180° - A) = -\cos A.$$
$$\tan (180° - A) = -\tan A.$$

The cosecant, secant and cotangent being the reciprocals of the sine, cosine and tangent respectively, we have

$$\operatorname{cosec}(180° - A) = \frac{1}{\sin(180° - A)} = \frac{1}{\sin A} = \operatorname{cosec} A.$$

$$\sec(180° - A) = \frac{1}{\cos(180° - A)} = \frac{1}{-\cos A} = -\sec A.$$

$$\cot(180° - A) = \frac{1}{\tan(180° - A)} = \frac{1}{-\tan A} = -\cot A.$$

The same results are readily obtained from Figs. 1 and 2 of Art. 37, where it is seen that BC is the sine of PAB, and also of BAQ, which are supplementary angles.

Also, $\cos PAB = AC = \cos BAQ$, but with the contrary sign.

Hence, $\sin 70° = \sin 110°$; $\cos 160° = -\cos 20°$; $\tan 170° = -\tan 10°$, &c.

40. To find the Trigonometrical Functions of $90° + A$.

In the figure, let the angle $MAB = A$, then the angle $PAB = 90° + A$.

Now, the radius being unity,

$$\sin(90° + A) = +BC$$
$$= +AD = \cos A.$$

$$\cos(90° + A) = -AC$$
$$= -BD = -\sin A.$$

$$\tan(90° + A) = \frac{\sin(90° + A)}{\cos(90° + A)} = \frac{\cos A}{-\sin A} = -\cot A.$$

$$\operatorname{cosec}(90° + A) = \frac{1}{\sin(90° + A)} = \frac{1}{\cos A} = \sec A.$$

$$\sec(90° + A) = \frac{1}{\cos(90° + A)} = \frac{1}{-\sin A} = -\operatorname{cosec} A.$$

$$\cot(90° + A) = \frac{1}{\tan(90° + A)} = \frac{1}{-\cot A} = -\tan A.$$

These results, as well as those of the preceding Article, will be deduced by a more general process in the next chapter.

41. To find the Trigonometrical Functions of a Negative Angle.

In the last figure, let the negative angle PAK be denoted by $-A$. Draw KH perpendicular to PQ; then, radius being unity, we have

$$\sin(-A) = HK, \text{ which is } \textit{negative} \quad (\text{Art. 33.})$$
$$= -\sin A.$$
$$\cos(-A) = AH, \text{ which is } \textit{positive}$$
$$= \cos A.$$
$$\tan(-A) = \frac{\sin(-A)}{\cos(-A)} = \frac{-\sin A}{\cos A} = -\tan A.$$
$$\csc(-A) = \frac{1}{\sin(-A)} = \frac{1}{-\sin A} = -\csc A.$$
$$\sec(-A) = \frac{1}{\cos(-A)} = \frac{1}{\cos A} = \sec A.$$
$$\cot(-A) = \frac{1}{\tan(-A)} = \frac{1}{-\tan A} = -\cot A.$$

In the same way, let the student find the various functions of negative angles in the other quadrants.

42. It is evident that any given angle has only *one* set of trigonometrical functions; but from the preceding Articles it is seen that, for any given function, there are more than one corresponding angle. We will now proceed to determine the groups of angles corresponding to any assigned value of each of the trigonometrical functions.

To find a general expression for all Angles which have the same sine.

In a circle of radius unity, let PAB be an angle whose sine BC is *given*. Make the angle $B'AQ =$ to the angle PAB, and draw

$B'C'$ perpendicular to PQ; then $B'C' = BC$, and therefore the angles PAB, PAB' have the same sine. Moreover, if to each of these angles we add 360°, 720°, &c., the positions of the revolving lines AB, AB' will be the same.

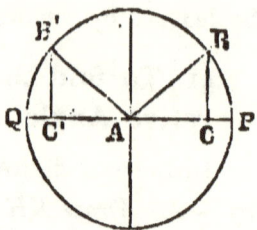

Hence, if we represent the circular measure of the angle PAB by a, and that of two right angles or 180°, by π, the positive angles which have the same sine are, a, $\pi - a$, and these increased by $2n\pi$ where n is any integer positive or negative, including *zero*.

For negative angles, the revolving line, in passing from the position AP to AB', describes an angle whose magnitude is $-(\pi + a)$, the negative sign being prefixed to shew the *direction* of revolution. When it attains the position AB, the angle described will be $-(2\pi - a)$. Hence the negative angles which have the same sine are, $-(\pi + a)$, $-(2\pi - a)$, and these increased by $2n\pi$.

Therefore the positive angles are

$2n\pi + a$ and $2n\pi + \pi - a$, or, $2n\pi + a$ and $(2n+1)\pi - a$.

The negative angles are

$-2n\pi - (\pi + a)$ and $-2n\pi - (2\pi - a)$, or $-(2n+1)\pi - a$ and $-2(n+1)\pi + a$.

Now, observe that when a is positive the multiple of π is even, and either positive or negative; and when a is negative, the multiple of π is odd, and either positive or negative. Hence all these angles are included in the general expression,

$$n\pi + (-1)^n a,$$

since $(-1)^n = +1$ or -1, according as n is even or odd.

Therefore $\quad \sin a = \sin (n\pi + (-1)^n a)$. \hfill (39)

43. To find a general expression for all Angles which have the same cosine.

Let PAB be the angle whose cosine AC is equal to the *given* one, and let a be its circular measure as before. Make the angle PAB'

equal to PAB; then, since the cosine is positive in the first and fourth quadrants, AC, the given cosine, is the cosine of all angles corresponding to the positions AB, AB' of the revolving line.

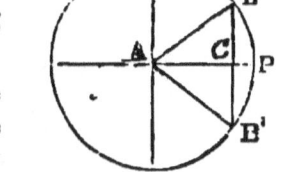

Therefore the positive angles which have the same cosine are, a, $2\pi - a$, and these increased by $2n\pi$.

The negative angles which have the same cosine are, $-a$, $-(2\pi - a)$, and these increased by $2n\pi$.

The positive angles are therefore

$$2n\pi + a \text{ and } 2n\pi + 2\pi - a, \text{ or } 2n\pi + a \text{ and } 2(n+1)\pi - a.$$

The negative angles are

$$-2n\pi - a \text{ and } -2n\pi - (2\pi - a), \text{ or } -2n\pi - a \text{ and } -2(n+1)\pi + a,$$

which are all included in the general expression,

$$2n\pi \pm a.$$

Therefore $\qquad \cos a = \cos (2n\pi \pm a).$ \hfill (40)

44. To find a general expression for all Angles which have the same tangent.

Let PAB be the angle whose tangent PD is equal to the *given* tangent. The positive angles which have the same tangent are evidently PAB and PAB', or a and $\pi + a$, and these increased by $2n\pi$.

The negative angles are manifestly $-(\pi - a)$ and $-(2\pi - a)$, and these increased by $2n\pi$.

Therefore we have for positive angles,

$$2n\pi + a \text{ and } 2n\pi + \pi + a, \text{ or } 2n\pi + a \text{ and } (2n+1)\pi + a,$$

And for negative angles,

$$-2n\pi - (\pi - a) \text{ and } -2n\pi - (2\pi - a), \text{ or } -(2n+1)\pi + a \text{ and } -2(n+1)\pi + a,$$

which are all included in the general expression,

$$n\pi + a.$$

Therefore $\quad \tan a = \tan(n\pi + a).\quad$ (41)

45. From the last three results we evidently have

$$\operatorname{cosec} a = \operatorname{cosec}(n\pi + (-1)^n a). \quad (42)$$
$$\sec a = \sec(2n\pi \pm a). \quad (43)$$
$$\cot a = \cot(n\pi + a.) \quad (44)$$

In deducing these expressions, the least positive angle which has the given value for the trigonometrical functions, has been employed; but they are equally true for *any* angle which has the given value for its sine, cosine, etc., as the case may be.

Examples.

1. Shew that $\sin A$, when determined from $\cos A$, has two values, equal in magnitude but opposite in sign.

From (10) we have

$$\sin A = \pm \sqrt{1 - \cos^2 A}.$$

This is also seen from the figure of Art. 43, where, if AC be the given cosine, $BC = +\sin A$, and $B'C = -\sin A$.

2. Given $\tan \theta = \sqrt{3}$, shew that $\sin \theta$ will have two values. From Art. 21 we have

$$\sin \theta = \frac{\tan \theta}{\pm \sqrt{(1 + \tan^2 \theta)}} = \pm \frac{\sqrt{3}}{2}$$
$$= \sin 60° \text{ or } \sin 240°,$$

since the tangent is positive in the *first* and *third* quadrants.

3. The cosine of an angle $= -\frac{3}{5}$, and the tangent $1\frac{1}{3}$, in what quadrant is the angle?

Ans. The third.

EXAMPLES.

4. Given $\sec \theta = -2$, find the general value of θ.

Here $\theta = 120° = \frac{2}{3}\pi$; hence we have from (43)

$$\theta = (2n\pi \pm \frac{2\pi}{3}) \text{ or } (6n \pm 2)\frac{\pi}{3}.$$

5. Given $\tan 5\theta = \sqrt{3}$, find the general value of θ.

Here we find $5\theta = 60°$ or $\frac{\pi}{3}$, and the general expression for all angles which have the same tangent is $n\pi + a$, where a is the circular measure.

Hence, the general value $= \frac{1}{5}(n\pi + \frac{\pi}{3}) = (3n+1)\frac{\pi}{15}$.

6. Find the sines and cosines of 300° and 162°.

Ans. $\sin 300° = -\frac{\sqrt{3}}{2}$, $\cos 300° = \frac{1}{2}$,

$\sin 162° = \frac{\sqrt{5}-1}{4}$, $\cos 162° = -\frac{1}{4}\sqrt{10 + 2\sqrt{5}}$.

7. Find the trigonometrical functions of 1098°.

Ans. The same as those of 18°.

8. The sine and tangent of an angle are both negative, and the tangent $= 2$ sine; find a general expression for all angles having this property. Ans. $(6n-1)\frac{\pi}{3}$.

9. Express in terms of θ all the angles whose sine is $-\sin \theta$.

Ans. $n\pi + (-1)^{n+1}\theta$.

10. Find the values of θ which satisfy each of the following equations:

(1) $\cos \theta = -1$. Ans. $(2n+1)\pi$.

(2) $\tan \theta = -1$. Ans. $n\pi - \frac{\pi}{4}$.

(3) $\sin^2 \theta = \frac{1}{4}$. Ans. $(6n \pm 1)\frac{\pi}{6}$.

(4) $\sec \theta = 2 \tan \theta$. Ans. $(6n + (-1)^n)\frac{\pi}{6}$.

(5) $2 \operatorname{cosec} \theta - \cot \theta = \sqrt{3}$. Ans. $(6n \pm 1)\frac{\pi}{3}$.

(6) $\tan \theta + \cot \theta = 2 \sec \theta$. Ans. $(6n + (-1)^n)\frac{\pi}{6}$.

11. Trace the variations in the sign and magnitude of the tangent and secant, as the angle varies from 90° to 270°, by means of the line definitions.

12. Prove that $\left(\sec \frac{\pi}{6} + \tan \frac{\pi}{6}\right)\left(\operatorname{cosec} \frac{\pi}{3} + \tan \frac{\pi}{3}\right) = 5$.

13. Find the general value of θ in $\sin \theta + \operatorname{cosec} \theta = 2$.

Ans. $\theta = (4n + 1)\frac{\pi}{2}$.

14. Shew that $\tan 52\frac{1}{2}° = (\sqrt{3} + \sqrt{2})(\sqrt{2} - 1)$.

15. Shew that $\sec 2\theta - \tan 2\theta = \frac{1 - \tan \theta}{1 + \tan \theta}$.

16. Shew that $\tan \theta = \dfrac{2}{\cot \dfrac{\theta}{2} - \tan \dfrac{\theta}{2}}$.

17. Shew that $\tan \dfrac{A}{2} = \dfrac{\sin 2A}{1 + \cos 2A} \cdot \dfrac{\cos A}{1 + \cos A}$.

18. Shew that $\sin \theta = \dfrac{2}{\tan \dfrac{\theta}{2} + \cot \dfrac{\theta}{2}} = \dfrac{1}{\tan \dfrac{\theta}{2} + \cot \theta}$.

19. Shew that $\cos \theta = \sin \theta \cot \dfrac{\theta}{2} - 1$.

20. Find all the values of θ which satisfy $1 - \cos \theta = 2 \sin^2 \theta$.

Ans. $\theta = \{(2n + 1) \pm 1\}\dfrac{\pi}{3}$.

EXAMPLES.

21. Trace the variations in sign and magnitude of the secant, cosecant and cotangent, in the third and fourth quadrants, by the line definitions.

22. Shew that $\cos 11\frac{1}{4}° = \frac{1}{2}\sqrt{2 + \sqrt{2 + \sqrt{2}}}$.

23. Shew that $\sin \frac{A}{2}$ and $\cos \frac{A}{2}$, when determined in terms of $\cos A$, will each have *two* values, but in terms of $\sin A$, *four* values.

24. Shew that $\cos 105° = \dfrac{1 - \sqrt{3}}{2\sqrt{2}}$.

25. Find $\tan 165°$ and $\sin 165°$.

$$\text{Ans. } \sqrt{3} - 2, \ -\frac{\sqrt{3} - 1}{2\sqrt{2}}.$$

26. If $\tan \theta = \sqrt{2} + 1$, find $\cos 2\theta$, and the general value of θ.

$$\text{Ans. } \cos 2\theta = -\frac{1}{\sqrt{2}}, \ \theta = n\pi \pm \frac{3\pi}{8}.$$

27. Find the general values of the limits between which θ lies when $\sin^2 \theta$ is greater than $\cos^2 \theta$.

$$\text{Ans. Between } 2n\pi + \frac{\pi}{4} \text{ and } 2n\pi + \frac{3}{4}\pi;$$

$$\text{and between } 2n\pi + \frac{5}{4}\pi \text{ and } 2n\pi + \frac{7}{4}\pi.$$

28. Prove that $\sec^4 \theta + \tan^4 \theta = 1 + 2\sec^2 \theta \tan^2 \theta$.

CHAPTER V.

TRIGONOMETRICAL FUNCTIONS OF THE SUM AND DIFFERENCE OF TWO ANGLES, AND OF MULTIPLES AND SUBMULTIPLES OF ANGLES.

45. To find sin (A + B) in terms of the sines and cosines of A and B.

Let the angle $BAC = A$ and the angle $CAD = B$, then the angle $BAD = A + B$.

In AD, one of the bounding lines of the compound angle $(A + B)$, take any point P, and draw PM, PQ perpendicular to AB and AC respectively, and from Q draw QK perpendicular to PM; then the angle $QPK =$ the angle $KQA = A$.

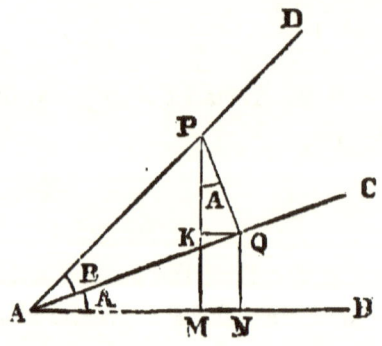

$$\sin (A + B) = \frac{PM}{AP}$$

$$= \frac{MK + PK}{AP}$$

$$= \frac{QN}{AP} + \frac{PK}{AP}$$

$$= \frac{QN}{AQ} \cdot \frac{AQ}{AP} + \frac{PK}{PQ} \cdot \frac{PQ}{AP}$$

$$= \sin A \cos B + \cos A \sin B. \qquad (45)$$

FUNCTIONS OF $(A+B)$.

46. To find cos $(A+B)$ in terms of the sines and cosines of A and B.

Employing the figure of the last Article, we have

$$\cos(A+B) = \frac{AM}{AP}$$

$$= \frac{AN - NM}{AP}$$

$$= \frac{AN}{AP} - \frac{QK}{AP}$$

$$= \frac{AN}{AQ} \cdot \frac{AQ}{AP} - \frac{QK}{QP} \cdot \frac{QP}{AP}$$

$$= \cos A \cos B - \sin A \sin B. \qquad (46)$$

47. To find tan $(A+B)$ in terms of the tangents of A and B.

$$\tan(A+B) = \frac{\sin(A+B)}{\cos(A+B)}$$

$$= \frac{\sin A \cos B + \cos A \sin B}{\cos A \cos B - \sin A \sin B}.$$

Dividing the numerator and denominator of this fraction by $\cos A \cos B$, we have

$$\tan(A+B) = \frac{\dfrac{\sin A}{\cos A} + \dfrac{\sin B}{\cos B}}{1 - \dfrac{\sin A \sin B}{\cos A \cos B}}$$

$$= \frac{\tan A + \tan B}{1 - \tan A \tan B}. \qquad (47)$$

48. To find sin (A – B) in terms of the sines and cosines of A and B.

Let the angle $BAC = A$ and $CAD = B$, then $BAD = A - B$.

In AD, one of the bounding lines of the compound angle $(A - B)$, take any point P, and draw PM perpendicular to AB, PQ perpendicular to AC, QN perpendicular to AB, and PK perpendicular to QN; then the angle PQK being the complement of AQN, is equal to BAC or A.

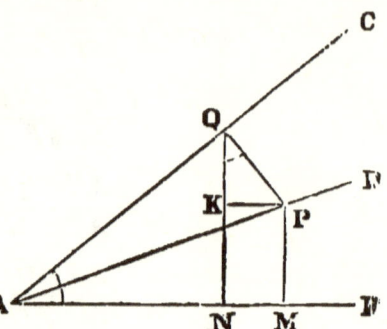

$$\sin (A - B) = \frac{PM}{AP}$$

$$= \frac{QN - QK}{AP}$$

$$= \frac{QN}{AQ} \cdot \frac{AQ}{AP} - \frac{QK}{PQ} \cdot \frac{PQ}{AP}$$

$$= \sin A \cos B - \cos A \sin B. \qquad (48)$$

49. To find cos (A – B) in terms of the sines and cosines of A and B.

From the last figure we have

$$\cos (A - B) = \frac{AM}{AP}$$

$$= \frac{AN + NM}{AP}$$

$$= \frac{AN}{AP} + \frac{PK}{AP}$$

$$= \frac{AN}{AQ} \cdot \frac{AQ}{AP} + \frac{PK}{PQ} \cdot \frac{PQ}{AP}$$

$$= \cos A \cos B + \sin A \sin B. \qquad (49)$$

FUNCTIONS OF $(A+B)$ AND $(A-B)$.

50. To find tan $(A-B)$ in terms of the tangents of A and B.

$$\tan(A-B) = \frac{\sin(A-B)}{\cos(A-B)}$$

$$= \frac{\sin A \cos B - \sin B \cos A}{\cos A \cos B + \sin A \sin B}.$$

Dividing numerator and denominator by $\cos A \cos B$, we have

$$\tan(A-B) = \frac{\tan A - \tan B}{1 + \tan A \tan B}. \qquad (50)$$

51. Formulæ (46) (48) and (49) can be easily deduced from (45), as follows:

In (45) write $-B$ for B and we have

$$\sin(A-B) = \sin A \cos(-B) + \cos A \sin(-B)$$
$$= \sin A \cos B - \cos A \sin B. \qquad \text{(Art. 41.)}$$

The complement of $(A+B)$ is $90° - (A+B)$, and since the cosine of an angle is the sine of the complement (Arts. 12 and 19), we have

$$\cos(A+B) = \sin\{90° - (A+B)\}$$
$$= \sin\{(90° - A) - B\}$$
$$= \sin(90° - A)\cos B - \cos(90° - A)\sin B,$$
$$= \cos A \cos B - \sin A \sin B.$$

If we write $-B$ for B in the last formula, we have

$$\cos(A-B) = \cos A \cos(-B) - \sin A \sin(-B)$$
$$= \cos A \cos B + \sin A \sin B.$$

In a similar manner, from any one of the four formulæ (45), (46), (48) and (49), the others may be deduced.

52. The preceding formulæ of this chapter have been established only when $A+B$ is less than $90°$. We will now proceed to shew that they are true for all angles, whether positive or negative; and from the results of the last Article it is evident that it will be sufficient to shew this in the case of (45) alone.

I. Let $A+B$ be $> 90°$, and let $A = 90° - x$ and $B = 90° - y$, and hence $x = 90° - A$ and $y = 90° - B$.
Then $A+B = 180° - (x+y)$, and therefore $(x+y) < 90°$.
Hence the formula (45) is true for $(x+y)$.

Now, $\sin(A+B) = \sin\{180° - (x+y)\}$
$= \sin(x+y)$ (Art. 39.)
$= \sin x \cos y + \cos x \sin y$
$= \sin(90° - A)\cos(90° - B) + \cos(90° - A)\sin(90° - B)$
$= \cos A \sin B + \sin A \cos B.$

Hence the formulæ are true when $(A+B) > 90°$.

II. Again, let $x = 90° + A$, and therefore $A = -(90° - x)$.
Then $x + B = 90° + A + B$
and $\sin(x+B) = \sin\{90° + (A+B)\}$
$= \cos(A+B)$ (Art. 40.)
$= \cos A \cos B - \sin A \sin B$
$= \cos\{-(90° - x)\}\cos B - \sin\{-(90° - x)\}\sin B$
$= \cos(90° - x)\cos B + \sin(90° - x)\sin B$
$\sin x \cos B + \cos x \sin B.$

Hence the formulæ are true for $90° + A$ and B.

In the same manner it may be shewn that they are true for A and $90° + B$; hence they are true if $90°$, or any multiple of $90°$, be added to either or to both, that is, they are true for all angles, since by adding multiples of $90°$ we may increase the angles A and B to any magnitude whatever.

In a similar manner we may show that the formulæ are true for all negative angles.*

53. The preceding formulæ furnish us at once with the trigonometrical functions of all compound angles.

---

* For a geometrical proof in any assigned case, see Appendix.

FUNCTIONS OF $(90° + B)$ AND $(180° - B)$.

In (45) and (46) let $A = 90°$, then

$$\sin (90° + B) = \sin 90° \cos B + \cos 90° \sin B$$
$$= 1 \times \cos B + 0 \times \sin B$$
$$= \cos B.$$

$$\cos (90° + B) = \cos 90° \cos B - \sin 90° \sin B$$
$$= 0 \times \cos B - 1 \times \sin B$$
$$= - \sin B.$$

$$\tan (90° + B) = \frac{\sin (90° + B)}{\cos (90° + B)} = \frac{\cos B}{- \sin B} = - \cot B.$$

$$\cot (90° + B) = \frac{1}{\tan (90° + B)} = \frac{1}{- \cot B} = - \tan B.$$

$$\sec (90° + B) = \frac{1}{\cos (90° + B)} = \frac{1}{- \sin B} = - \operatorname{cosec} B.$$

$$\operatorname{cosec} (90° + B) = \frac{1}{\sin (90° + B)} = \frac{1}{\cos B} = \sec B.$$

These results agree with those of Art. 40.

54. In (48) and (49) let $A = 180°$, then

$$\sin (180° - B) = \sin 180° \cos B - \cos 180° \sin B$$
$$= 0 \times \cos B - (-1) \times \sin B$$
$$= \sin B.$$

$$\cos (180° - B) = \cos 180° \cos B + \sin 180° \sin B$$
$$= -1 \times \cos B + 0 \times \sin B$$
$$= - \cos B.$$

Hence $\tan (180° - B) = - \tan B,$ $\quad \cot (180° - B) = - \cot B,$
$\sec (180° - B) = - \sec B,$ $\quad \operatorname{cosec} (180° - B) = \operatorname{cosec} B.$

These results agree with those of Art. 39.

Here we may observe that if B be less than 90°, its supplement $180° - B$ is greater than 90° or obtuse, and that the sine and cosecant of an obtuse angle are *positive*, while the cosine, tangent, cotangent and secant are *negative*.

55. In (45) and (46) let $A = 180°$, then

$\sin(180° + B) = -\sin B$, $\cos(180° + B) = -\cos B$,
$\tan(180° + B) = \tan B$, $\cot(180° + B) = \cot B$,
$\sec(180° + B) = -\sec B$, $\operatorname{cosec}(180° + B) = -\operatorname{cosec} B$.

56. In (48) and (49) let $A = 270°$, then

$\sin(270° - B) = -\cos B$, $\cos(270° - B) = -\sin B$,
$\tan(270° - B) = \cot B$, $\cot(270° - B) = \tan B$,
$\sec(270° - B) = -\operatorname{cosec} B$, $\operatorname{cosec}(270° - B) = -\sec B$.

57. In (45) and (46) let $A = 270°$, then

$\sin(270° + B) = -\cos B$, $\cos(270° + B) = \sin B$,
$\tan(270° + B) = -\cot B$, $\cot(270° + B) = -\tan B$,
$\sec(270° + B) = \operatorname{cosec} B$, $\operatorname{cosec}(270° + B) = -\sec B$.

58. In (48) and (49) let $A = 360°$, then

$\sin(360° - B) = -\sin B$, $\cos(360° - B) = \cos B$,
$\tan(360° - B) = -\tan B$, $\cot(360° - B) = -\cot B$,
$\sec(360° - B) = \sec B$, $\operatorname{cosec}(360° - B) = -\operatorname{cosec} B$,

or, the functions of $360° - B$ are the same as those of $-B$.

59. In (45) and (46) let $A = 360°$, then

$\sin(360° + B) = \sin B$, $\cos(360° + B) = \cos B$,

or, the functions of an angle greater than 360° are the same as those of the excess above 360°.

60. General Formulæ involving the Functions of two Angles.

From (45), (46), (48) and (49) we have

$$\sin(A+B) + \sin(A-B) = 2 \sin A \cos B. \quad (51)$$
$$\sin(A+B) - \sin(A-B) = 2 \cos A \sin B. \quad (52)$$
$$\cos(A+B) + \cos(A-B) = 2 \cos A \cos B. \quad (53)$$
$$\cos(A+B) - \cos(A-B) = -2 \sin A \sin B. \quad (54)$$

GENERAL FORMULÆ.

If we put $A + B = A'$ and $A - B = B'$,
we have $2A = A' + B'$ and $2B = A' - B'$,
whence $A = \tfrac{1}{2}(A' + B')$ and $B = \tfrac{1}{2}(A' - B')$.

Making these substitutions in the above group, and omitting the accents, since A' and B' may be any angles whatever, we have

$$\sin A + \sin B = 2 \sin \tfrac{1}{2}(A+B) \cos \tfrac{1}{2}(A-B). \quad (55)$$
$$\sin A - \sin B = 2 \cos \tfrac{1}{2}(A+B) \sin \tfrac{1}{2}(A-B). \quad (56)$$
$$\cos A + \cos B = 2 \cos \tfrac{1}{2}(A+B) \cos \tfrac{1}{2}(A-B). \quad (57)$$
$$\cos A - \cos B = -2 \sin \tfrac{1}{2}(A+B) \sin \tfrac{1}{2}(A-B). \quad (58)$$

These formulæ are of very great importance, and the student should make himself quite familiar with them under this form. They are especially valuable in trigonometrical computations in transforming a sum or difference into a product, and *vice versa*. Each of them may be expressed as a theorem; thus, "*The sum of the sines of any two angles is equal to twice the sine of half the sum of the angles multiplied by the cosine of half their difference,*" and similarly for the others.

61. Divide (55) by (56), then we have by (13) and (14)

$$\frac{\sin A + \sin B}{\sin A - \sin B} = \frac{2 \sin \tfrac{1}{2}(A+B) \cos \tfrac{1}{2}(A-B)}{2 \cos \tfrac{1}{2}(A+B) \sin \tfrac{1}{2}(A-B)}$$
$$= \tan \tfrac{1}{2}(A+B) \cot \tfrac{1}{2}(A-B)$$
$$= \frac{\tan \tfrac{1}{2}(A+B)}{\tan \tfrac{1}{2}(A-B)}. \quad (59)$$

In the same manner let the student verify the following:

$$\frac{\cos A - \cos B}{\cos A + \cos B} = -\tan \tfrac{1}{2}(A+B) \tan \tfrac{1}{2}(A-B). \quad (60)$$

$$\frac{\sin A + \sin B}{\cos A + \cos B} = \tan \tfrac{1}{2}(A+B). \quad (61)$$

$$\frac{\sin A - \sin B}{\cos A - \cos B} = -\cot \tfrac{1}{2}(A+B). \tag{62}$$

$$\frac{\sin A - \sin B}{\cos A + \cos B} = \tan \tfrac{1}{2}(A-B). \tag{63}$$

$$\frac{\sin A + \sin B}{\cos A - \cos B} = -\cot \tfrac{1}{2}(A+B). \tag{64}$$

62. Divide (45), (46), (48) and (49) by $\cos A \cos B$, then we have by (13)

$$\frac{\sin(A+B)}{\cos A \cos B} = \frac{\sin A \cos B + \cos A \sin B}{\cos A \cos B}$$
$$= \tan A + \tan B. \tag{65}$$

$$\frac{\cos(A+B)}{\cos A \cos B} = 1 - \tan A \tan B. \tag{66}$$

$$\frac{\sin(A-B)}{\cos A \cos B} = \tan A - \tan B. \tag{67}$$

$$\frac{\cos(A-B)}{\cos A \cos B} = 1 + \tan A \tan B. \tag{68}$$

In a similar manner we find

$$\frac{\sin(A+B)}{\sin A \sin B} = \cot B + \cot A. \tag{69}$$

$$\frac{\cos(A+B)}{\sin A \sin B} = \cot A \cot B - 1. \tag{70}$$

$$\frac{\sin(A-B)}{\sin A \sin B} = \cot B - \cot A. \tag{71}$$

$$\frac{\cos(A-B)}{\sin A \sin B} = \cot A \cot B + 1. \tag{72}$$

$$\frac{\sin(A+B)}{\sin A \cos B} = 1 + \cot A \tan B. \tag{73}$$

$$\frac{\cos(A+B)}{\sin A \cos B} = \cot A - \tan B. \tag{74}$$

GENERAL FORMULÆ.

$$\frac{\sin (A - B)}{\sin A \cos B} = 1 - \cot A \tan B. \qquad (75)$$

$$\frac{\cos (A - B)}{\sin A \cos B} = \cot A + \tan B. \qquad (76)$$

63. Divide (45) by (48) and (46) by (49), then we have

$$\frac{\sin (A + B)}{\sin (A - B)} = \frac{\sin A \cos B + \cos A \sin B}{\sin A \cos B - \cos A \sin B}$$

$$= \frac{\tan A + \tan B}{\tan A - \tan B}. \qquad (77)$$

$$= \frac{\cot B + \cot A}{\cot B - \cot A}. \qquad (78)$$

and $\quad \dfrac{\cos (A + B)}{\cos (A - B)} = \dfrac{1 - \tan A \tan B}{1 + \tan A \tan B}. \qquad (79)$

$$= \frac{\cot A \cot B - 1}{\cot A \cot B + 1}. \qquad (80)$$

64. In (47) and (50) let $A = 45°$, then

$$\tan (45° \pm B) = \frac{\tan 45° \pm \tan B}{1 \mp \tan 45° \tan B}$$

$$= \frac{1 \pm \tan B}{1 \mp \tan B}. \qquad (81)$$

$$= \frac{\cot B \pm 1}{\cot B \mp 1}. \qquad (82)$$

In (50) let $B = 45°$, then

$$\tan (A - 45°) = \frac{\tan A - 1}{\tan A + 1}. \qquad (83)$$

$$= \frac{1 - \cot A}{1 + \cot A}. \qquad (84)$$

65. From (47) and (50) we have by (8)

$$\cot (A \pm B) = \frac{\cot B \cot A \mp 1}{\cot B \pm \cot A}. \qquad (85)$$

The secant and cosecant of $(A+B)$ or $(A-B)$ are easily derived from those of the cosine and sine respectively, by means of (8); thus we have

$$\sec(A+B) = \frac{1}{\cos(A+B)}$$

$$= \frac{1}{\cos A \cos B - \sin A \sin B}$$

$$= \frac{\sec A \sec B}{1 - \tan A \tan B}, \tag{86}$$

by multiplying numerator and denominator by $\sec A \sec B$. In a similar manner the others may be derived.

66. The product of (45) and (48), and of (46) and (49), are

$$\sin(A+B)\sin(A-B) = \sin^2 A \cos^2 B - \cos^2 A \sin^2 B,$$
$$\cos(A+B)\cos(A-B) = \cos^2 A \cos^2 B - \sin^2 A \sin^2 B.$$

But $\cos^2 A = 1 - \sin^2 A$, and $\cos^2 B = 1 - \sin^2 B$,

therefore
$$\sin(A+B)\sin(A-B) = \sin^2 A - \sin^2 B. \tag{87}$$
$$= \cos^2 B - \cos^2 A. \tag{88}$$

and
$$\cos(A+B)\cos(A-B) = \cos^2 A - \sin^2 B. \tag{89}$$
$$= \cos^2 B - \sin^2 A. \tag{90}$$

67. General Formulæ involving the Functions of the Multiples and Submultiples of an Angle.

In (45), (46) and (47), let $B = A$, then

$$\sin 2A = \sin A \cos A + \cos A \sin A$$
$$= 2 \sin A \cos A. \tag{91}$$

$$\cos 2A = \cos A \cos A - \sin A \sin A$$
$$= \cos^2 A - \sin^2 A. \tag{92}$$

$$\tan 2A = \frac{\tan A + \tan A}{1 - \tan A \tan A}$$
$$= \frac{2 \tan A}{1 - \tan^2 A}. \tag{93}$$

GENERAL FORMULÆ.

If, in the last three equations, we write A for $2A$, and therefore $\tfrac{1}{2}A$ for A, which we are at liberty to do since A is any angle whatever, we have

$$\sin A = 2 \sin \frac{A}{2} \cos \frac{A}{2}. \tag{94}$$

$$\cos A = \cos^2 \frac{A}{2} - \sin^2 \frac{A}{2}. \tag{95}$$

$$\tan A = \frac{2 \tan \dfrac{A}{2}}{1 - \tan^2 \dfrac{A}{2}}. \tag{96}$$

From (10) and (92) we have

$$\cos^2 A + \sin^2 A = 1$$
$$\cos^2 A - \sin^2 A = \cos 2A,$$

the sum and difference of which are

$$2 \cos^2 A = 1 + \cos 2A \tag{97}$$
$$2 \sin^2 A = 1 - \cos 2A. \tag{98}$$

Writing A for $2A$ and $\dfrac{A}{2}$ for A, these become

$$2 \cos^2 \frac{A}{2} = 1 + \cos A, \tag{99}$$

$$2 \sin^2 \frac{A}{2} = 1 - \cos A, \tag{100}$$

the quotient of which is

$$\tan^2 \frac{A}{2} = \frac{1 - \cos A}{1 + \cos A}. \tag{101}$$

Multiplying the numerator and denominator of the second member of (101) by $1 + \cos A$, we get

$$\tan^2 \frac{A}{2} = \frac{(1 - \cos^2 A)}{(1 + \cos A)^2}$$
$$= \frac{\sin^2 A}{(1 + \cos A)^2},$$

therefore
$$\tan\frac{A}{2} = \frac{\sin A}{1+\cos A}. \qquad (102)$$

Multiplying the numerator and denominator of the second member of (102) by $1-\cos A$, we get

$$= \frac{(1-\cos A)\sin A}{1-\cos^2 A}$$

$$= \frac{(1-\cos A)\sin A}{\sin^2 A}$$

$$= \frac{1-\cos A}{\sin A}. \qquad (102\ bis)$$

The formulæ of this Article have already been deduced for angles less than 90°, in Art. 28, by a less general process, and the student will now see that they are true for angles of any magnitude.

68. To find expressions for sin mA and cos mA.

From (51) and (53) we have

$$\sin mA + \sin(m-2)A = 2\sin(m-1)A \cos A$$
and
$$\cos mA + \cos(m-2)A = 2\cos(m-1)A \cos A,$$
hence
$$\left.\begin{array}{l}\sin mA = 2\sin(m-1)A \cos A - \sin(m-2)A \\ \cos mA = 2\cos(m-1)A \cos A - \cos(m-2)A.\end{array}\right\} \qquad (103)$$

If we make m successively 1, 2, 3, &c., these give

$$\left.\begin{array}{l}\sin A = \sin A \\ \sin 2A = 2\sin A \cos A \\ \sin 3A = 4\sin A \cos^2 A - \sin A, \\ \quad \&c., \qquad \&c. \\ \\ \cos A = \cos A \\ \cos 2A = \cos^2 A - \sin^2 A \\ \cos 3A = 4\cos^3 A - 3\cos A, \\ \quad \&c., \qquad \&c.\end{array}\right\} \qquad (104)$$

Other general formulæ for $\sin mA$ and $\cos A$ will be given in a subsequent chapter.

69. General Formulæ involving the Functions of three Angles.

Let A, B, C be any three angles, then by (45) and (46)

$$\sin(A+B+C) = \sin\{(A+B)+C\}$$
$$= \sin(A+B)\cos C + \cos(A+B)\sin C$$
$$= \sin A \cos B \cos C + \cos A \sin B \cos C$$
$$+ \cos A \cos B \sin C - \sin A \sin B \sin C. \quad (105)$$

$$\cos(A+B+C) = \cos\{(A+B)+C\}$$
$$= \cos(A+B)\cos C - \sin(A+B)\sin C$$
$$= \cos A \cos B \cos C - \sin A \sin B \cos C$$
$$- \sin A \cos B \sin C - \cos A \sin B \sin C. \quad (106)$$

In the same way we may develop the sine and cosine of $(A-B+C)$, &c.

If we divide (105) by (106) we get, after dividing the numerator and denominator by $\cos A \cos B \cos C$,

$$\tan(A+B+C) = \frac{\tan A + \tan B + \tan C - \tan A \tan B \tan C}{1 - \tan A \tan B - \tan A \tan C - \tan B \tan C}. \quad (107)$$

If $A+B+C = n\pi$, then $\tan(A+B+C) = 0$, and therefore from (107) we obtain

$$\tan A + \tan B + \tan C = \tan A \tan B \tan C. \quad (108)$$

If $A+B+C = (2n+1)\frac{\pi}{2}$, that is an *odd* multiple of $\frac{\pi}{2}$, then $\tan(A+B+C) = \infty$, and therefore the denominator of (107) must be zero,

hence $\quad \tan A \tan B + \tan A \tan C + \tan B \tan C = 1. \quad (109)$

If (109) be divided by $\tan A \tan B \tan C$, we get

$$\cot A + \cot B + \cot C = \cot A \cot B \cot C. \quad (110)$$

70.
Again, let $A+B+C = n\pi = 2n\frac{\pi}{2}$, an *even* multiple of $\frac{\pi}{2}$, then

$$\left.\begin{array}{l}\sin(A+B+C) = \sin n\pi = 0 \\ \sin(-A+B+C) = \sin(n\pi - 2A) = -(-1)^n \sin 2A \\ \sin(A-B+C) = \sin(n\pi - 2B) = -(-1)^n \sin 2B \\ \sin(A+B-C) = \sin(n\pi - 2C) = -(-1)^n \sin 2C\end{array}\right\} \quad (111)$$

The sum of this group is, by (55) and (58),

$4\sin A \sin B \sin C = -(-1)^n (\sin 2A + \sin 2B + \sin 2C).$ (112)

Again, we have

$$\begin{aligned}
\cos(A+B+C) &= \cos n\pi &&= (-1)^n. \\
\cos(-A+B+C) &= \cos(n\pi - 2A) &&= (-1)^n \cos 2A. \\
\cos(A-B+C) &= \cos(n\pi - 2B) &&= (-1)^n \cos 2B. \\
\cos(A+B-C) &= \cos(n\pi - 2C) &&= (-1)^n \cos 2C.
\end{aligned}$$ (113)

The sum of this group is, by (57),

$4\cos A \cos B \cos C = (-1)^n (\cos 2A + \cos 2B + \cos 2C + 1).$ (114)

71. If $A+B+C = (2n+1)\frac{\pi}{2}$, an *odd* multiple of $\frac{\pi}{2}$, we have

$$\begin{aligned}
\sin(A+B+C) &= \sin(2n+1)\frac{\pi}{2} &&= (-1)^n. \\
\sin(-A+B+C) &= \sin\left\{(2n+1)\frac{\pi}{2} - 2A\right\} &&= (-1)^n \cos 2A. \\
\sin(A-B+C) &= \sin\left\{(2n+1)\frac{\pi}{2} - 2B\right\} &&= (-1)^n \cos 2B. \\
\sin(A+B-C) &= \sin\left\{(2n+1)\frac{\pi}{2} - 2C\right\} &&= (-1)^n \cos 2C.
\end{aligned}$$ (111 *bis*)

Subtracting the difference of the first two of this group from the sum of the last two, we find

$4\sin A \sin B \sin C = (-1)^n (\cos 2A + \cos 2B + \cos 2C - 1).$ (112 *bis*)

Again,

$$\begin{aligned}
\cos(A+B+C) &= \cos(2n+1)\frac{\pi}{2} &&= 0. \\
\cos(-A+B+C) &= \cos\left\{(2n+1)\frac{\pi}{2} - 2A\right\} &&= (-1)^n \sin 2A. \\
\cos(A-B+C) &= \cos\left\{(2n+1)\frac{\pi}{2} - 2B\right\} &&= (-1)^n \sin 2B. \\
\cos(A+B-C) &= \cos\left\{(2n+1)\frac{\pi}{2} - 2C\right\} &&= (-1)^n \sin 2C.
\end{aligned}$$ (113 *bis*)

The sum of which is

$4\cos A \cos B \cos C = (-1)^n (\sin 2A + \sin 2B + \sin 2C).$ (114 *bis*)

GENERAL FORMULÆ.

By combining the equations of groups (111) and (113), many other important results are obtained. Thus, in (113), taking the difference between the sum of the first two and the sum of the second two, we find

$$4\cos A \sin B \sin C = (-1)^n (-\cos 2A + \cos 2B + \cos 2C - 1). \quad (115)$$

In a similar manner, from (111) we obtain

$$4\sin A \cos B \cos C = -(-1)^n (-\sin 2A + \sin 2B + \sin 2C). \quad (116)$$

Again, from groups (111 *bis*) and (113 *bis*) the following are easily obtained, which will serve as exercises for the student.

$$\left.\begin{array}{l} 4\sin A \cos B \cos C = (-1)^n (-\cos 2A + \cos 2B + \cos 2C + 1). \\ 4\cos A \sin B \cos C = (-1)^n (\cos 2A - \cos 2B + \cos 2C + 1). \\ 4\cos A \cos B \sin C = (-1)^n (\cos 2A + \cos 2B - \cos 2C + 1). \\ 4\cos A \sin B \sin C = (-1)^n (-\sin 2A + \sin 2B + \sin 2C). \\ 4\sin A \cos B \sin C = (-1)^n (\sin 2A - \sin 2B + \sin 2C). \\ 4\sin A \sin B \cos C = (-1)^n (\sin 2A + \sin 2B - \sin 2C). \end{array}\right\} \quad (117)$$

72. The following six formulæ are of great utility in computing the values of the trigonometrical functions, and in transforming a sum or a difference into a product.

$$\begin{aligned} \cot A + \tan A &= \frac{\cos A}{\sin A} + \frac{\sin A}{\cos A} \\ &= \frac{\cos^2 A + \sin^2 A}{\sin A \cos A} \\ &= \frac{2}{2 \sin A \cos A} = \frac{2}{\sin 2A} \\ &= 2 \operatorname{cosec} 2A. \end{aligned} \quad (118)$$

$$\begin{aligned} \cot A - \tan A &= \frac{\cos A}{\sin A} - \frac{\sin A}{\cos A} \\ &= \frac{2(\cos^2 A - \sin^2 A)}{2 \sin A \cos A} \\ &= \frac{2 \cos 2A}{\sin 2A} \\ &= 2 \cot 2A. \end{aligned} \quad (119)$$

$$\operatorname{cosec} A + \cot A = \frac{1}{\sin A} + \frac{\cos A}{\sin A}$$

$$= \frac{1 + \cos A}{\sin A}$$

$$= \frac{2 \cos^2 \frac{A}{2}}{2 \sin \frac{A}{2} \cos \frac{A}{2}}, \text{ by (99) and (94)}$$

$$= \cot \frac{A}{2}. \qquad (120)$$

$$\operatorname{cosec} A - \cot A = \frac{1}{\sin A} - \frac{\cos A}{\sin A}$$

$$= \frac{1 - \cos A}{\sin A}$$

$$= \frac{2 \sin^2 \frac{A}{2}}{2 \sin \frac{A}{2} \cos \frac{A}{2}}, \text{ by (100) and (94)}$$

$$= \tan \frac{A}{2}. \qquad (121)$$

From (81) we have

$$\tan(45° + A) = \frac{1 + \tan A}{1 - \tan A}$$

$$\tan(45° - A) = \frac{1 - \tan A}{1 + \tan A},$$

hence

$$\tan(45° + A) \div \tan(45° - A) = \frac{1 + \tan A}{1 - \tan A} + \frac{1 - \tan A}{1 + \tan A}$$

$$= \frac{2(1 + \tan^2 A)}{1 - \tan^2 A}$$

$$= \frac{2(\cos^2 A + \sin^2 A)}{\cos^2 A - \sin^2 A}, \text{ by (13)}$$

$$= \frac{2}{\cos 2A}$$

$$= 2 \sec 2A. \qquad (122)$$

$$\tan(45° + A) - \tan(45° - A) = \frac{4\tan A}{1 - \tan^2 A}$$

$$= \frac{4}{\cot A - \tan A}$$

$$= \frac{4}{2\cot 2A}, \text{ by (119)}$$

$$= 2\tan 2A. \quad (123)$$

Formulæ of Verification.

The following four formulæ are useful for testing the accuracy of the trigonometrical tables:

$$\sin(36° + A) - \sin(36° - A) = 2\cos 36° \sin A = \tfrac{1}{2}(\sqrt{5} + 1)\sin A.$$
$$\sin(72° + A) - \sin(72° - A) = 2\cos 72° \sin A = \tfrac{1}{2}(\sqrt{5} - 1)\sin A.$$

The difference of which is

$$\sin(36° + A) - \sin(72° + A) + \sin(72° - A) - \sin(36° - A)$$
$$= \sin A. \quad \text{(Euler's formula.)} \quad (124)$$

Substituting $(90° - A)$ for A in (124) we obtain

$$\sin(54° + A) - \sin(18° + A) + \sin(54° - A) - \sin(18° - A)$$
$$= \cos A. \quad \text{(Legendre's formula.)} \quad (125)$$

By (55) we find

$$\sin(30° - A) + \sin(30° + A) = \cos A. \quad (126)$$

Similarly, by (58) we find

$$\cos(30° - A) - \cos(30° + A) = \sin A. \quad (127)$$

Examples.

1. Prove that $\dfrac{1 + \sin A}{1 + \cos A} = \tfrac{1}{2}\left(1 + \tan\dfrac{A}{2}\right)^2.$

From (94) and (99) we have

$$\frac{1+\sin A}{1+\cos A} = \frac{\sin^2\frac{A}{2} + 2\sin\frac{A}{2}\cos\frac{A}{2} + \cos^2\frac{A}{2}}{2\cos^2\frac{A}{2}}$$

$$= \tfrac{1}{2}\left(\frac{\sin\frac{A}{2}+\cos\frac{A}{2}}{\cos\frac{A}{2}}\right)^2$$

$$= \tfrac{1}{2}\left(1+\tan\frac{A}{2}\right)^2.$$

2. Prove that $\tan\dfrac{\theta}{2} = \dfrac{\tan\theta}{1+\sec\theta}$.

From (102) we have

$$\tan\frac{\theta}{2} = \frac{\sin\theta}{1+\cos\theta} = \frac{\frac{\sin\theta}{\cos\theta}}{\frac{1}{\cos\theta}+1} = \frac{\tan\theta}{1+\sec\theta}.$$

3. Prove that $\tan\dfrac{\theta}{2} - \tan\dfrac{\theta}{3} - \tan\dfrac{\theta}{6} = \tan\dfrac{\theta}{2}\tan\dfrac{\theta}{3}\tan\dfrac{\theta}{6}$.

$$\tan\frac{\theta}{2} - \tan\frac{\theta}{3} - \tan\frac{\theta}{6} = \frac{\sin\frac{\theta}{2}}{\cos\frac{\theta}{2}} - \frac{\sin\frac{\theta}{2}}{\cos\frac{\theta}{3}\cos\frac{\theta}{6}}, \text{ by (13) and (65)}$$

$$= \frac{\sin\frac{\theta}{2}}{\cos\frac{\theta}{2}\cos\frac{\theta}{3}\cos\frac{\theta}{6}}(\cos\frac{\theta}{3}\cos\frac{\theta}{6} - \cos\frac{\theta}{2})$$

$$= \frac{\sin\frac{\theta}{2}}{\cos\frac{\theta}{2}\cos\frac{\theta}{3}\cos\frac{\theta}{6}}\left\{\tfrac{1}{2}(\cos\frac{\theta}{6} - \cos\frac{\theta}{2})\right\}, \text{by (53)}$$

$$= \tan\frac{\theta}{2}\tan\frac{\theta}{3}\tan\frac{\theta}{6}, \text{ by (58)}$$

EXAMPLES. 73

4. If $\cos\theta = \dfrac{\cos\alpha + \cos\beta}{1 + \cos\alpha\cos\beta}$, prove that $\tan\dfrac{\theta}{2} = \tan\dfrac{\alpha}{2}\tan\dfrac{\beta}{2}$.

From (101) we have

$$\tan^2\frac{\theta}{2} = \frac{1-\cos\theta}{1+\cos\theta} = \frac{1 - \dfrac{\cos\alpha + \cos\beta}{1+\cos\alpha\cos\beta}}{1 + \dfrac{\cos\alpha + \cos\beta}{1+\cos\alpha\cos\beta}}$$

$$= \frac{1 + \cos\alpha\cos\beta - \cos\alpha - \cos\beta}{1 + \cos\alpha\cos\beta + \cos\alpha + \cos\beta}$$

$$= \frac{1-\cos\alpha}{1+\cos\alpha} \cdot \frac{1-\cos\beta}{1+\cos\beta}$$

$$= \tan^2\frac{\alpha}{2}\tan^2\frac{\beta}{2}, \quad \text{by (101)},$$

therefore $\tan\dfrac{\theta}{2} = \tan\dfrac{\alpha}{2}\tan\dfrac{\beta}{2}$.

5. Prove that $\cos\left(30° - \dfrac{\theta}{2}\right) - \cos\left(30° + \dfrac{\theta}{2}\right) = \sin\dfrac{\theta}{2}$.

6. Prove that $\cot\left(30° + \dfrac{\theta}{2}\right) - \tan\left(30° + \dfrac{\theta}{2}\right) = 2\tan(30° - \theta)$.

7. Prove that $\tan\left(30° + \dfrac{\theta}{2}\right)\tan\left(30° - \dfrac{\theta}{2}\right) = \dfrac{2\cos\theta - 1}{2\cos\theta + 1}$.

8. Prove that $\tan^2(45° - \theta) = \dfrac{1 - \sin 2\theta}{1 + \sin 2\theta}$.

9. Prove that $\dfrac{\sin 2\theta + \sin\theta}{\cos 2\theta + \cos\theta} = \tan\dfrac{3\theta}{2}$.

10. Prove that $\dfrac{\tan\alpha + \sec\alpha}{\cot\alpha + \operatorname{cosec}\alpha} = \tan\dfrac{\alpha}{2}\tan\left(45° + \dfrac{\alpha}{2}\right)$.

11. Prove that $\dfrac{\sin\theta + \sin 3\theta + \sin 5\theta}{\cos\theta + \cos 3\theta + \cos 5\theta} = \tan 3\theta$.

12. Prove that $\tan(\alpha + \beta) = \dfrac{\sin^2\alpha - \sin^2\beta}{\sin\alpha\cos\alpha - \sin\beta\cos\beta}$.

13. Prove that $\dfrac{\cos\theta - \cos 2\theta}{\cos\theta + \cos 2\theta} = \tan\dfrac{\theta}{2}\tan\dfrac{3\theta}{2}$.

14. Prove that $\dfrac{\sin \frac{2}{3}\theta + \sin \frac{4}{3}\theta}{\cos \frac{2}{3}\theta + \cos \frac{4}{3}\theta} = \tan \theta$.

15. Express by means of a sum or difference
$$\cos 2\theta \cos \theta; \quad \sin 2\theta \cos 3\theta; \quad \sin 5\theta \sin \theta.$$
Ans. $\tfrac{1}{2}(\cos 3\theta + \cos \theta);\ \tfrac{1}{2}(\sin 5\theta - \sin \theta);\ \tfrac{1}{2}(\cos 4\theta - \cos 6\theta)$.

16. Reduce $\sin^2 3\theta - \sin^2 2\theta$ and
$$\cos^2\left(45° + \dfrac{\theta}{2}\right) - \sin^2\left(45° - \dfrac{5\theta}{2}\right) \text{ to products.}$$
Ans. $\sin \theta \sin 5\theta;\ \sin 2\theta \cos 3\theta$.

17. Prove that $(1 + \cot \theta + \operatorname{cosec} \theta)(1 + \cot \theta - \operatorname{cosec} \theta)$
$$= \cot \dfrac{\theta}{2} - \tan \dfrac{\theta}{2}.$$

18. Express by a product each of the following:
$\sin \theta + \sin(\theta - 2\phi);\ \sin^2 \theta - \sin^2(\theta - 2\phi);\ \sin \theta + \cos \theta;$
$\sin(\theta + \phi) + \cos(\theta - \phi);\ \cos^2(\theta - \phi) - \sin^2(\theta + \phi)$.
Ans. $2\sin(\theta - \phi)\cos \phi;\ \sin 2(\theta - \phi)\sin 2\phi;$
$\sqrt{2}\cos(45° - \theta);\ 2\cos(45° - \theta)\cos(45° - \phi);$
$\cos 2\theta \cos 2\phi$.

19. Prove that $\cos^2 A - \cos^2 3A = \sin 4A \sin 2A$.

20. If $\tan^2 \theta = \tan(a - \theta)\tan(a + \theta)$, shew that
$$\sin 2\theta = \sqrt{2} \sin a.$$

21. Shew that $\tan(30° + \theta)\tan(30° - \theta) = \tan \theta \cot 3\theta$.

22. If $\sec(\phi + a) + \sec(\phi - a) = 2\sec \phi$, shew that
$$\cos \phi = \sqrt{2} \cos \dfrac{a}{2}.$$

23. If $\tan(\phi + a) + \tan(\phi + \beta) = 2\tan \phi$, shew that
$$\tan \phi = \tfrac{1}{2}(\cot a + \cot \beta).$$

24. Shew that $\cos 25° - \cos 11° + \cos 47° - \cos 61° = \sin 7°$.

25. If a, β, γ be in arithmetical progression, prove that
$$(\sin a - \sin \gamma)\sin \beta = (\cos \gamma - \cos a)\cos \beta.$$

EXAMPLES.

26. Prove that
$$\tan^2\theta - \tan^2\phi = \sin(\theta+\phi)\sin(\theta-\phi)\sec^2\theta\sec^2\phi.$$

27. Shew that $\cos^2(\theta+\phi) - \sin^2\theta = \cos\phi\cos(2\theta+\phi)$.

28. Shew that $\sin(\theta+\phi)\cos\theta - \cos(\theta+\phi)\sin\theta = \sin\phi$.

29. Prove that $\sin\alpha + \sin\beta + \sin\gamma - \sin(\alpha+\beta+\gamma)$
$$= 4\sin\tfrac{1}{2}(\alpha+\beta)\sin\tfrac{1}{2}(\alpha+\gamma)\sin\tfrac{1}{2}(\beta+\gamma).$$

30. Prove that $\cos\alpha + \cos\beta + \cos\gamma + \cos(\alpha+\beta+\gamma)$
$$= 4\cos\tfrac{1}{2}(\alpha+\beta)\cos\tfrac{1}{2}(\alpha+\gamma)\cos\tfrac{1}{2}(\beta+\gamma).$$

31. Prove that $\tan\alpha + \tan\beta + \tan\gamma - \tan\alpha\tan\beta\tan\gamma$
$$= \frac{\sin(\alpha+\beta+\gamma)}{\cos\alpha\cos\beta\cos\gamma}.$$

32. Prove that $\cot\alpha + \cot\beta + \cot\gamma - \cot\alpha\cot\beta\cot\gamma$
$$= -\frac{\cos(\alpha+\beta+\gamma)}{\sin\alpha\sin\beta\sin\gamma}.$$

33. Prove that $\cos(\alpha+\beta+\gamma) + \cos(\alpha+\beta-\gamma) + \cos(\alpha-\beta+\gamma)$
$+ \cos(-\alpha+\beta+\gamma) = 4\cos\alpha\cos\beta\cos\gamma$.

34. If $\alpha+\beta+\gamma = 180°$, prove that
$$\sin(\alpha+\beta)\sin(\beta+\gamma) = \sin\alpha\sin\gamma;$$
$$\sin 2\alpha + \sin 2\beta + \sin 2\gamma = 4\sin\alpha\sin\beta\sin\gamma.$$

35. If $\alpha+\beta+\gamma = 90°$, prove that
$$\frac{\cos\alpha + \sin\gamma - \sin\beta}{\cos\beta + \sin\gamma - \sin\alpha} = \frac{1+\tan\frac{\alpha}{2}}{1+\tan\frac{\beta}{2}}.$$

36. If α, β, γ be in arithmetical progression, prove that
$$\sin\alpha - \sin\gamma = 2\sin(\alpha-\beta)\cos\beta;$$
$$\tan(\alpha-\beta) = \frac{\cos\alpha - \cos\gamma}{\sin\alpha + \sin\gamma} = -\frac{\sin\alpha - \sin\gamma}{\cos\alpha + \cos\gamma}.$$

37. If $\sin\phi = m\sin(2\theta+\phi)$, prove that
$$\tan(\theta+\phi) = \frac{1+m}{1-m}\tan\theta.$$

38. If $\tan \frac{1}{2}(\alpha+\beta) \tan \frac{1}{2}(\alpha-\beta) = \tan^2 \frac{\gamma}{2}$, prove that
$$\cos \alpha = \cos \beta \cos \gamma.$$

39. If $\tan \dfrac{\theta}{2} = \dfrac{1-\tan^3 \phi}{1+\tan^3 \phi}$, prove that
$$2 \cot 2\phi = \sqrt[4]{\sec \theta + \tan \theta} - \sqrt[4]{\sec \theta - \tan \theta}.$$

40. If $\cos \theta = \sqrt{\frac{2}{3}}$, and $\cos \phi = \dfrac{\sqrt{3}+\sqrt{2}}{2\sqrt{3}}$, shew that
$$\cos(\theta+\phi) = \tfrac{1}{2}.$$

41. If $\tan \theta = \frac{1}{3}$ and $\tan \phi = \frac{1}{7}$, prove that $\sin(2\theta+\phi) = \dfrac{\sqrt{2}}{2}$.

42. If $\cot 2\theta = -\tan \phi$, prove that $\tan(\theta-\phi) = \cot \theta$.

43. If $\tan \phi = \dfrac{2 \sin \theta \sin \psi}{\sin(\theta+\psi)}$, then $\tan \theta$, $\tan \phi$ and $\tan \psi$ are in harmonical progression.

44. If $\cos(\theta-\psi) \cos \phi = \cos(\theta-\phi+\psi)$, then $\tan \theta$, $\tan \phi$ and $\tan \psi$ are in harmonical progression.

45. If $\tan \alpha \tan \theta + \sec \alpha \sec \theta = \sec \beta$, shew that
$$\tan \theta = \pm \dfrac{\sin \beta \mp \sin \alpha}{\cos \alpha \cos \beta}.$$

46. Prove that $\sin^2 30° = \sin 18° \sin 54°$.

47. Prove that $\tan 50° + \cot 50° = 2 \sec 10°$.

48. Prove that $\tan 52\tfrac{1}{2}° = (\sqrt{3}+\sqrt{2})(\sqrt{2}-1)$.

49. If $A+B+C = 180°$, prove that
$$\dfrac{\sin A + \sin B - \sin C}{\sin A + \sin B + \sin C} = \tan \dfrac{A}{2} \tan \dfrac{B}{2};$$
$$\sin\left(A+\dfrac{B}{2}\right) + \sin\left(B+\dfrac{C}{2}\right) + \sin\left(C+\dfrac{A}{2}\right) + 1$$
$$= 4 \cos \tfrac{1}{4}(A-B) \cos \tfrac{1}{4}(B-C) \cos \tfrac{1}{4}(C-A).$$

50. If $\sin \theta = \sin \alpha \sin(\beta+\theta)$, prove that
$$\tan\left(\dfrac{\beta}{2}+\theta\right) = \tan \dfrac{\beta}{2} \tan^2\left(45° + \dfrac{\alpha}{2}\right).$$

EXAMPLES.

51. If $\sin(x+y)\cos z = 2\cos(y-z)\sin x$, prove that
$\cot x \div \cot y = 2 \tan z$.

52. Prove that $2 \operatorname{cosec} 4x + 2 \cot 4x = \cot x - \tan x$.

53. Prove that
$2 \sin \tfrac{1}{2}(A+B-90°) \cos \tfrac{1}{2}(A-B+90°) = \sin A - \cos B$.

54. Prove that $\left(\cot^2 \dfrac{\theta}{2} - \tan^2 \dfrac{\theta}{2}\right) \tan \theta = 4 \operatorname{cosec} \theta$.

55. Prove that $\sin(A-B) + \sin(B-C) + \sin(C-A)$
$= 4 \sin \tfrac{1}{2}(B-A) \sin \tfrac{1}{2}(C-B) \sin \tfrac{1}{2}(A-C)$.

56. Prove that
$\sin A \operatorname{cosec}(A-B) \operatorname{cosec}(A-C) + \sin B \operatorname{cosec}(B-C) \operatorname{cosec}(B-A)$
$+ \sin C \operatorname{cosec}(C-A) \operatorname{cosec}(C-B) = 0$.

57. If $\cos(A+B)\sin(C+D) = \cos(A-B)\sin(C-D)$, shew that $\tan D = \tan A \tan B \tan C$.

58. If $\sin(\alpha-\gamma)\cos\beta + \sin(\beta-\gamma)\cos\alpha = 0$, shew that
$\tan \alpha + \tan \beta = 2 \tan \gamma$.

59. If $\cot \alpha + \cos \beta = \sqrt{2} \cot \theta \sin \beta$,
and $\cot \gamma + \cos \theta = \sqrt{2} \cot \beta \sin \theta$, prove that
$\sin \theta = \sin \alpha \sin \beta \operatorname{cosec} \gamma$.

60. Prove that $\dfrac{\cos A \cot A}{\cos A + \cot A} = \dfrac{\cot A - \cos A}{\cos A \cot A}$.

61. Prove that $\operatorname{cosec} 4\theta + \cot 4\theta = \tfrac{1}{2}(\cot \theta - \tan \theta)$.

62. If $\cot \theta = 2 \tan \phi$, then $\cos(\theta-\phi) = 3\cos(\theta+\phi)$, and $\sec(\theta+\phi) = 2 \sec \theta \sec \phi$.

63. If $A+B+C = 180°$, shew that $\sin^2 A + \sin^2 B + \sin^2 C$
$= 2(\cos A \sin B \sin C + \cos B \sin A \sin C + \cos C \sin A \sin B)$,
and $\sin^2 \dfrac{A}{2} + \sin^2 \dfrac{B}{2} + \sin^2 \dfrac{C}{2} + 2 \sin \dfrac{A}{2} \sin \dfrac{B}{2} \sin \dfrac{C}{2} = 1$.

64. Prove that
$\sin(x+y)\sin(y+z) = \sin x \sin z + \sin y \sin(x+y+z)$.

65. If $A + B + C = 180°$, prove that
$$\frac{1 - \cos A + \cos B + \cos C}{1 + \cos A + \cos B - \cos C} = \tan \frac{A}{2} \cot \frac{C}{2},$$
and $\sin^2 A + \sin^2 B - \sin^2 C = 2 \sin A \sin B \cos C$.

66. Prove that $\cos (x + y) \sin y - \cos (x + z) \sin z$
$$= \sin (x + y) \cos y - \sin (x + z) \cos z.$$

67. Determine θ from the equation
$$\sin a + \sin (\theta - a) + \sin (2\theta + a) = \sin (\theta + a) + \sin (2\theta - a),$$
Ans. $\theta = (10n \pm 1)\frac{\pi}{5}$ or $\{5(2n + 1) \pm 1\}\frac{\pi}{5}$.

68. Prove that $\sin^2 (x + y) + \cos^2 (x - y) = 1 + \sin 2x \sin 2y$.

69. If $A + B + C = 180°$, prove that
$$\frac{\cot A + \cot B}{\tan A + \tan B} + \frac{\cot B + \cot C}{\tan B + \tan C} + \frac{\cot A + \cot C}{\tan A + \tan C} = 1,$$
and
$$\cot A + \cot B + \cot C = \cot A \cot B \cot C + \operatorname{cosec} A \operatorname{cosec} B \operatorname{cosec} C$$

70. If $\frac{1 + \tan \theta}{1 - \tan \theta} = \frac{3}{2} \sec 2\theta$, shew that $\theta = \frac{1}{2}(n\pi + (-1)^n \frac{\pi}{6})$

71. If $A + B + C = 180°$, shew that
$$\cos^2 A + \cos^2 B + \cos^2 C + 2 \cos A \cos B \cos C = 1.$$

72. Shew that $\dfrac{1 - \cos 2x \cos 2y}{1 + \cos 2x \cos 2y} = \dfrac{\tan^2 x + \tan^2 y}{1 + \tan^2 x \tan^2 y}$.

73. Shew that $1 + \cos (\alpha - \beta) + \cos (\beta - \gamma) + \cos (\gamma - \alpha)$
$$= 4 \cos \tfrac{1}{2}(\alpha - \beta) \cos \tfrac{1}{2}(\beta - \gamma) \cos \tfrac{1}{2}(\gamma - \alpha),$$
and $\sin (\alpha - \beta) \sin \gamma - \sin (\alpha - \gamma) \sin \beta = \sin (\gamma - \beta) \sin \alpha$.

74. If $2 \tan \theta + \tan (\alpha - \theta) = \tan (\beta + \theta)$, shew that
$$\tan \theta = \tfrac{1}{2} \sin (\alpha - \beta) \operatorname{cosec} \alpha \operatorname{cosec} \beta.$$

75. If $\tan \dfrac{A}{2}$, $\tan \dfrac{B}{2}$ and $\tan \dfrac{C}{2}$ be in arithmetical progression, so also are
$$\cos A, \cos B \text{ and } \cos C, \text{ where } A + B + C = 180°.$$

76. Given $\cos 2\theta - \cos 4\theta = \sin \theta$, find θ.
Ans. $\theta = n\pi$ or $\tfrac{1}{3}(n\pi + (-1)^n \frac{\pi}{6})$.

EXAMPLES.

77. Find the general value of θ in $\sin(a-\theta) = \cos(a+\theta)$.

$$\text{Ans. } \theta = (4n-1)\frac{\pi}{4}.$$

78. If $a + \beta + \gamma = 180°$, shew that

$$\frac{\tan a}{\tan \beta} + \frac{\tan \beta}{\tan \gamma} + \frac{\tan \gamma}{\tan a} + \frac{\tan a}{\tan \gamma} + \frac{\tan \gamma}{\tan \beta} + \frac{\tan \beta}{\tan a} + 2$$
$$= \sec a \sec \beta \sec \gamma.$$

79. Prove that $\cos(m+n+r)a + \cos(m+n-r)a$
$+ \cos(m-n+r)a + \cos(m-n-r)a = 4 \cos ma \cos na \cos ra$.

80. If $\tan(a+\beta) = 3 \tan a$, shew that
$$\sin 2(a+\beta) = 2 \sin 2\beta - \sin 2a.$$

81. If $A + B + C = 180°$, shew that

$$\tan\frac{A}{2} + \cos\frac{A}{2}\sec\frac{B}{2}\sec\frac{C}{2} = \tan\frac{B}{2} + \cos\frac{B}{2}\sec\frac{A}{2}\sec\frac{C}{2}$$

$$= \tan\frac{C}{2} + \cos\frac{C}{2}\sec\frac{A}{2}\sec\frac{B}{2},$$

and $(\sin A + \sin B + \sin C)\left(\tan\frac{A}{2} + \tan\frac{B}{2} + \tan\frac{C}{2}\right)$

$$= 4\left(1 + \sin\frac{A}{2}\sin\frac{B}{2}\sin\frac{C}{2}\right).$$

82. Prove that

$$\frac{\cos\theta - \sin\phi}{\cos\theta + \sin\phi} = \tan\left\{45° - \frac{\theta + \phi}{2}\right\}\tan\left\{45° - \frac{\theta - \phi}{2}\right\}.$$

83. If a, β, γ be in arithmetical progression, shew that

$$\frac{1}{\tan a + \tan \gamma} - \frac{1}{2 \tan \beta} = \frac{1}{\cot a + \cot \gamma} - \frac{1}{2 \cot \beta}.$$

84. If $A + B + C = 90°$, shew that
$\tan A + \tan B + \tan C = \tan A \tan B \tan C + \sec A \sec B \sec C.$

85. Given $\cos\theta + \cos 7\theta = \cos 4\theta$, find θ.

$$\text{Ans. } \theta = (2n+1)\frac{\pi}{8} \text{ or } \tfrac{1}{3}\left(2n\pi \pm \frac{\pi}{3}\right).$$

CHAPTER VI.

ON THE CALCULATION OF THE NUMERICAL VALUES OF THE TRIGONOMETRICAL FUNCTIONS.

73. The Circular Measure of an Angle is greater than the sine, but less than the tangent of the Angle.

Let PBQ be an arc of a circle whose centre is A; bisect the angle PAQ by AB and draw the chord PQ and the tangent DBE. Now it is evident that the arc PBQ is greater than PQ and less than DE, therefore PB is greater than PC and less than DB, or $\frac{PB}{AP}$ is greater than $\frac{PC}{AP}$, and less than $\frac{DB}{AB}$.

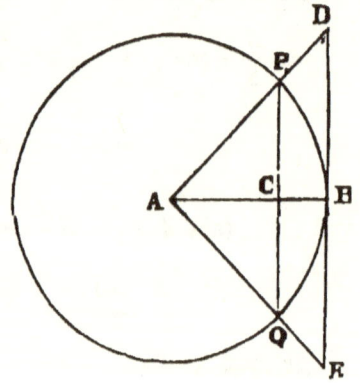

But $\frac{PB}{AP}$ is the circular measure of the angle BAP, (Art. 9.)

$\frac{PC}{AP}$ is the sine of the angle BAP

and $\frac{DB}{AB}$ is the tangent of the angle BAP,

or, if the radius be unity, we have PC the sine, BD the tangent and PB the circular measure of the angle BAP. Hence, representing the circular measure by θ, we have

θ greater than $\sin \theta$, and less than $\tan \theta$.

LIMITING VALUES OF $\dfrac{\sin \theta}{\theta}$ AND $\dfrac{\tan \theta}{\theta}$. 81

74. To find the limit of $\dfrac{\sin \theta}{\theta}$ and of $\dfrac{\tan \theta}{\theta}$, when θ is indefinitely diminished.

In the last Article we have shewn that

$$\sin \theta, \theta \text{ and } \tan \theta$$

are in ascending order of magnitude. Hence, $1, \dfrac{\theta}{\sin \theta}$ and $\dfrac{1}{\cos \theta}$ are also in ascending order of magnitude, that is, $\dfrac{\theta}{\sin \theta}$ lies between 1 and $\dfrac{1}{\cos \theta}$, whatever the value of θ may be.

But as θ is diminished, $\cos \theta$ tends to 1 and ultimately becomes 1 when $\theta = 0$; hence we have

$$\dfrac{\theta}{\sin \theta} = 1, \text{ when } \theta = 0,$$

and therefore also $\dfrac{\sin \theta}{\theta} = 1$, when $\theta = 0$.

Since $\dfrac{\tan \theta}{\theta} = \dfrac{\sin \theta}{\theta} \cdot \dfrac{1}{\cos \theta}$,

we have $\dfrac{\tan \theta}{\theta} = 1$, when $\theta = 0$.

From these results it follows that when θ is *very small*,

$$\sin \theta = \theta = \tan \theta, \text{ very nearly.}$$

75. To shew that $\sin \theta > \theta - \dfrac{\theta^3}{4}$, where θ is the Circular Measure of any Angle between 0° and 90°.

By Art. 73 we have

$$\tan \dfrac{\theta}{2} > \dfrac{\theta}{2}, \text{ or } \dfrac{\sin \dfrac{\theta}{2}}{\cos \dfrac{\theta}{2}} > \dfrac{\theta}{2},$$

hence $\quad 2 \sin \dfrac{\theta}{2} > \theta \cos \dfrac{\theta}{2}.$

Multiplying both members of this inequality by $\cos\frac{\theta}{2}$, we have by (94)

$$\sin\theta > \theta \cos^2\frac{\theta}{2}$$

$$> \theta\left(1 - \sin^2\frac{\theta}{2}\right)$$

$$> \theta\left(1 - \frac{\theta^2}{4}\right), \text{ since } \frac{\theta}{2} > \sin\frac{\theta}{2},$$

$$> \theta - \frac{\theta^3}{4}.$$

76. To find the Numerical Values of the sines and cosines of all Angles from 0° to 90° at intervals of 10″.

Let θ be the circular measure of 10″, then since the circular measure of 180° is π, or 3.141592653589793... as will be shewn in a subsequent chapter, we have

$$180° : 10'' :: \pi : \theta,$$

or

$$\theta = \frac{10\pi}{180 \times 60 \times 60}$$

$$= \frac{3.141592653589793}{64800}$$

$$= .000048481368.$$

But $\quad \sin\theta < \theta$ and $> \theta - \frac{\theta^3}{4}$, (Arts. 72 and 75.)

hence $\quad \sin\theta > .000048481368 - \frac{1}{4}(.000048481368)^3$
$\quad\quad\quad > .000048481368 - \frac{1}{4}(.00005)^3$
$\quad\quad\quad > .000048481368 - .00000000000003,$

from which it appears that the quantity to be subtracted from θ does not affect the first twelve places of decimals. Hence, the sine of 10″ coincides with the circular measure of 10″ to twelve places of decimals, and therefore

sin 10″ = circular measure of 10″ = .000048481368 ;
also sin 1″ = circular measure of 1″ = .0000048481368.

If a be the circular measure of any angle $A°$, then it is evident that

$$A'' = \frac{a}{\text{cir. meas. } 1''} = \frac{a}{\sin 1''} = a \times 206264''.806\ldots \quad (128)$$

which agrees with (6).

The cosine of 10″ is obtained from the formula

$$\cos 10'' = \sqrt{1 - \sin^2 10''} = \sqrt{(1 - \sin 10'')(1 + \sin 10'')}$$
$$= .999999998824.$$

77. From (51) and (53), we have

$$\sin(A + B) = 2 \sin A \cos B - \sin(A - B)$$
$$\cos(A + B) = 2 \cos A \cos B - \cos(A - B),$$

in which let B be constantly equal to 10″;

thus, $\sin(A + 10'') = 2 \cos 10'' \sin A - \sin(A - 10'')$
$\cos(A + 10'') = 2 \cos 10'' \cos A - \cos(A - 10'').$

Now, if $A = 10''$, 20″, 30″, &c., in succession, we find for the sines

$$\sin 20'' = 2 \cos 10'' \sin 10'' - \sin\ 0'',$$
$$\sin 30'' = 2 \cos 10'' \sin 20'' - \sin 10'',$$
$$\sin 40'' = 2 \cos 10'' \sin 30'' - \sin 20'',$$
$$\&c., \quad \&c.,$$

and for the cosines

$$\cos 20'' = 2 \cos 10'' \cos 10'' - \cos\ 0'',$$
$$\cos 30'' = 2 \cos 10'' \cos 20'' - \cos 10'',$$
$$\cos 40'' = 2 \cos 10'' \cos 30'' - \cos 20'',$$
$$\&c., \quad \&c.$$

These results may be further simplified as follows:

$$2 \cos 10'' = 1.999999997648,$$
$$= 2 - .000000002352,$$
$$= 2 - m, \text{ suppose.}$$

Making this substitution, we have

$$\sin 20'' = 2 \sin 10'' - \sin\ 0'' - m \sin 10'' = .0000969627,$$
$$\sin 30'' = 2 \sin 20'' - \sin 10'' - m \sin 20'' = .0001454441,$$
$$\sin 40'' = 2 \sin 30'' - \sin 20'' - m \sin 30'' = .0001939254,$$
$$\&c., \qquad \&c.$$

$$\cos 20'' = 2 \cos 10'' - \cos\ 0'' - m \cos 10'' = .9999999953,$$
$$\cos 30'' = 2 \cos 20'' - \cos 10'' - m \cos 20'' = .9999999894,$$
$$\cos 40'' = 2 \cos 30'' - \cos 20'' - m \cos 30'' = .9999999812,$$
$$\&c., \qquad \&c.$$

78. Having computed the sines and cosines by this process up to 30°, those for angles greater than 30° may be easily found by (126) and (127), thus,

$$\sin (30° + A) = \cos A - \sin (30° - A)$$
$$\cos (30° + A) = \cos (30° - A) - \sin A,$$

in which let $A = 10''$, $20''$, $30''$, &c., in succession, and we have

$$\sin 30° \ 0' \ 10'' = \cos 10'' - \sin 29° \ 59' \ 50'',$$
$$\sin 30° \ 0' \ 20'' = \cos 20'' - \sin 29° \ 59' \ 40'',$$
$$\&c., \qquad \&c.$$
$$\cos 30° \ 0' \ 10'' = \cos 29° \ 59' \ 50'' - \sin 10'',$$
$$\cos 30° \ 0' \ 20'' = \cos 20° \ 59' \ 40'' - \sin 20'',$$
$$\&c., \qquad \&c.$$

It is not necessary to continue this process beyond 45°, since the sines and cosines of angles above 45° are respectively the cosines and sines of their complements; thus, $\sin 46° = \cos 44°$; $\sin 50° = \cos 40°$; $\cos 46° = \sin 44°$; $\cos 50° = \sin 40°$, &c.

COMPUTATION OF TRIGONOMETRICAL FUNCTIONS.   85

79. The tangent is found from the formula $\tan \theta = \dfrac{\sin \theta}{\cos \theta}$; but when the tangents have been thus computed up to 45°, the rest may be found by the addition of those already computed; thus, from (123) we have

$$\tan(45° + A) = \tan(45° - A) + 2 \tan 2A$$

in which put $A = 10'', 20''$, &c., in succession, and we have

$$\tan 45° \ 0' \ 10'' = \tan 44° \ 59' \ 50'' + 2 \tan 20''$$

&c.,   &c.

The cotangent is already known from the tangent, thus,
$$\cot 1° = \tan 89°, \ \cot 10° = \tan 80°, \ \&c.$$

80. The secant is easily computed from the tangent by (122), thus,

$$\sec 2A = \tfrac{1}{2}\{\tan(45° + A) + \tan(45° - A)\}$$

in which put $A = 10'', 20''$, &c., in succession, and we have

$$\sec 20'' = \tfrac{1}{2}\{\tan 45° \ 0' \ 10'' + \tan 44° \ 59' \ 50''\}$$

&c.,   &c.

The cosecant may be found from (118), (120) or (121).
From (118) we have

$$\csc 2A = \tfrac{1}{2}(\tan A + \cot A),$$

therefore $\qquad \csc 20'' = \tfrac{1}{2}(\tan 10'' + \cot 10''),$

&c.,   &c.

81. The accuracy of the work should be verified from time to time by separate and independent computations. If, for example, the functions of 15°, 18°, 30°, 36° and 54°, agree with those found by the process of Chapter II., the work is correct. Formulæ (124) and (125) may also be used to verify the results. Thus, if we write 6° for A in (125), we have

$$\sin 60° - \sin 24° + \sin 48° - \sin 12° = \cos 6° = \sin 84°,$$

a relation involving the sines of five angles, which will be satisfied if the computed values are correct.

62. The tabulated numerical values of the sines, cosines, &c., for each given angle constitute what is called "A table of natural sines, &c.," to distinguish it from "A table of logarithmic sines, &c.," to be described in the next chapter. The latter table is by far the more useful, as nearly all computations in trigonometry are carried on by logarithms.

The arrangement of both tables is most easily understood from a simple inspection of them. As ample explanations are always prefixed to every set of trigonometrical tables with regard to their arrangement and to the manner of using them, we shall here refrain from giving the usual mechanical directions for their use. We shall, however, give a few illustrations shewing how to use them in accordance with the principles on which they are computed.

It will be necessary, too, for the student to make himself perfectly familiar with the arrangement of the tables he uses, in order to understand any peculiarities which belong to them, and to use them intelligently and not mechanically.

83. Since all the trigonometrical functions pass through all their possible numerical values, as the angle varies from 0° to 90°, the tables are not extended beyond the first quadrant. The functions of angles greater than 90° are found by the principles of Chapter V., Arts. 53–59, thus,

$$\sin 153° \ 10' = \quad \sin 26° \ 50' = .4513967,$$
$$\cos 124° \ 30' = - \cos 55° \ 30' = -.5664072,$$
$$\tan 170° \ 42' = - \tan 9° \ 18' = -.1637563,$$
$$\&c., \qquad \&c.$$

For angles greater than 180° and less than 270° we deduct 180° and take the same function of the remainder, being careful to prefix the signs that belong to that quadrant. Thus, by Art. 55, we have

USE OF TRIGONOMETRICAL TABLES. 87

$$\sin 225° \ 15' = -\sin 45° \ 15' = -.7101854,$$
$$\cos 250° \ \ 2' = -\cos 70° \ \ 2' = -.3414734,$$
$$\tan 192° \ 50' = +\tan 12° \ 50' = +.2278063,$$
$$\&c., \qquad \&c.$$

Again, by Art. 57, we have

$$\sin 300° \ 10' = -\cos 30° \ 10',$$
$$\cos 350° \ \ 0' = +\sin 80° \ \ 0',$$
$$\tan 354° \ 40' = -\cot 84° \ 40',$$
$$\&c., \qquad \&c.$$

84. The tables are computed for intervals of 10″ or 1′, and to five, six or seven places of decimals. The sine, cosine, &c., of any angle not given in the tables, can be found with sufficient accuracy for most practical purposes by means of simple proportion, except near the limits of the quadrant where the disparity between the variation of the angle and that of the function, changes too rapidly to admit of the application of this principle; and conversely, an angle whose sine, cosine, &c., is not found exactly in the tables, can be obtained in a similar manner.

Ex. 1. Required the sine and cosine of 42° 17′ 24″.

From the tables, we have

$$\sin 42° \ 17' = .6727973$$
$$\sin 42° \ 18' = .6730125$$

Difference for 1′ or 60″ = .0002152

Hence the proportional part for 24″ is $\frac{24}{60} \times .0002152 = .0000861$, which must be *added* to the sine of 42° 17′, since the sine increases as the angle increases, (Art. 34)

therefore $\qquad \sin 42° \ 17' \ 24'' = .6728834.$

Again, from the tables, we have

$$\cos 42° \ 17' = .7398268$$
$$\cos 42° \ 18' = .7396311$$

Difference for 1' or 60" = .0001957

Hence the proportional part for 24" is $\frac{24}{60} \times .0001957 = .0000783$, which must be *subtracted* from the cosine of 42° 17', since the cosine decreases as the angle increases, (Art. 34)

therefore $\quad \cos 42° \ 17' \ 42'' = .7397485$.

Ex. 2 Given $\cos A = .7216446$, to find A.

From the tables, we have

$$\cos 43° \ 48' = .7217602$$
$$\cos 43° \ 49' = .7215589$$

Difference for 1' or 60" = .0002013

$$\cos A = .7216446, \text{ the given cosine.}$$
$$\cos 43° \ 49' = .7215589, \text{ the next less in the table.}$$

Difference = .0000857

then $\quad .0002013 : .0000857 :: 60'' : 25''.5$,

which must be *subtracted* from 43° 49',

therefore $\quad A = 43° \ 48' \ 34''.5$.

Ex. 3. Given $\tan \theta = 1.1726470$, to find θ.

From the tables, we find

$$\tan 49° \ 32' = 1.1722298$$
$$\tan 49° \ 33' = 1.1729207$$

Difference for 1' or 60" = .0006909

INCREMENTS OF TRIGONOMETRICAL FUNCTIONS. 89

$\tan \theta = 1.1726470$, the given tangent.
$\tan 49° 32' = 1.1722298$, the next less in the table.

Difference = .0004172

then .0006909 : .0004172 :: 60'' : 36''.2,

which must be *added* to 49° 32',

therefore $\theta = 49° 32' 36''.2$.

The student must not fail to observe that the proportional part must be *added* for the sine, tangent and secant, since these functions *increase* as the angle *increases*; and *subtracted* for the cosine, cotangent and cosecant, because they *decrease* as the angle *increases*.

85. We will now deduce general formulæ for the increments of the trigonometrical functions of an angle corresponding to a given increment of the angle.

Let any angle θ be increased by a very small increment $\Delta\theta$, *expressed in circular measure*, and let the corresponding increment of the sine, &c., be denoted by $\Delta \sin \theta$, $\Delta \tan \theta$, &c.; then, we have

$$\Delta \sin \theta = \sin(\theta + \Delta\theta) - \sin \theta$$
$$= \cos \theta \sin \Delta\theta - \sin \theta (1 - \cos \Delta\theta)$$
$$= \cos \theta \sin \Delta\theta \left(1 - \tan \theta \frac{1 - \cos \Delta\theta}{\sin \Delta\theta}\right)$$
$$= \cos \theta \sin \Delta\theta \left(1 - \tan \theta \tan \frac{\Delta\theta}{2}\right), \quad \text{by (102 } bis\text{)}$$
$$= \cos \theta \Delta\theta \left(1 - \frac{\Delta\theta}{2} \tan \theta\right). \quad \text{Art. 74.} \quad (129)$$

Now, if $\tan \theta$ is not very large, that is, if θ is not near 90°, $\frac{\Delta\theta}{2} \tan \theta$ may be neglected, and we have approximately

$$\Delta \sin \theta = \cos \theta \Delta\theta. \quad (130)$$

$$\Delta \cos \theta = \cos (\theta + \Delta\theta) - \cos \theta$$
$$= -\sin \theta \sin \Delta\theta - \cos \theta (1 - \cos \Delta\theta)$$
$$= -\sin \theta \sin \Delta\theta \left(1 + \cot \theta \, \frac{1 - \cos \Delta\theta}{\sin \Delta\theta}\right)$$
$$= -\sin \theta \sin \Delta\theta \left(1 + \cot \theta \tan \frac{\Delta\theta}{2}\right)$$
$$= -\sin \theta \, \Delta\theta \left(1 + \frac{\Delta\theta}{2} \cot \theta\right). \tag{131}$$

When θ is near $90°$, $\dfrac{\Delta\theta}{2} \cot \theta$ may be neglected, and we have approximately

$$\Delta \cos \theta = -\sin \theta \, \Delta\theta. \tag{132}$$

$$\Delta \tan \theta = \tan (\theta + \Delta\theta) - \tan \theta$$
$$= \frac{\sin (\theta + \Delta\theta)}{\cos (\theta + \Delta\theta)} - \frac{\sin \theta}{\cos \theta}$$
$$= \frac{\sin \Delta\theta}{\cos^2 \theta \cos \Delta\theta (1 - \tan \theta \tan \Delta\theta)}$$
$$= \frac{\sec^2 \theta \tan \Delta\theta}{1 - \tan \theta \tan \Delta\theta}$$
$$= \frac{\sec^2 \theta \, \Delta\theta}{1 - \Delta\theta \tan \theta}. \tag{133}$$

When θ is not nearly equal to $90°$, $\Delta\theta \tan \theta$ may be neglected, so that we have approximately

$$\Delta \tan \theta = \sec^2 \theta \, \Delta\theta. \tag{134}$$

$$\Delta \cot \theta = \cot (\theta + \Delta\theta) - \cot \theta$$
$$= \frac{\cos (\theta + \Delta\theta)}{\sin (\theta + \Delta\theta)} - \frac{\cos \theta}{\sin \theta}$$
$$= \frac{-\sin \Delta\theta}{\sin^2 \theta \cos \Delta\theta (1 + \cot \theta \tan \Delta\theta)}$$
$$= \frac{-\csc^2 \theta \tan \Delta\theta}{1 + \cot \theta \tan \Delta\theta}$$
$$= \frac{-\csc^2 \theta \, \Delta\theta}{1 + \Delta\theta \cot \theta}. \tag{135}$$

INCREMENTS OF TRIGONOMETRICAL FUNCTIONS. 91

When θ is not nearly equal to $0°$, $\Delta\theta \cot \theta$ may be neglected, and we have approximately

$$\Delta \cot \theta = -\csc^2 \theta \, \Delta a. \qquad (136)$$

$$\Delta \sec \theta = \sec(\theta + \Delta\theta) - \sec \theta$$

$$= \frac{1}{\cos(\theta + \Delta\theta)} - \frac{1}{\cos \theta}$$

$$= \frac{\cos \theta - \cos(\theta + \Delta\theta)}{\cos \theta \cos(\theta + \Delta\theta)}$$

$$= \frac{\sec \theta \left(\tan \theta + \tan \frac{\Delta\theta}{2}\right) \tan \Delta\theta}{1 - \tan \theta \tan \Delta\theta}$$

$$= \frac{\sec \theta \left(\tan \theta + \frac{\Delta\theta}{2}\right) \Delta\theta}{1 - \Delta\theta \tan \theta}. \qquad (137)$$

When θ is neither very small nor very nearly equal to $90°$, $\dfrac{\Delta\theta^2}{2} \sec \theta$ and $\Delta\theta \tan \theta$ may be neglected, and we have approximately

$$\Delta \sec \theta = \sin \theta \sec^2 \theta \, \Delta\theta. \qquad (138)$$

$$\Delta \csc \theta = \csc(\theta + \Delta\theta) - \csc \theta$$

$$= \frac{1}{\sin(\theta + \Delta\theta)} - \frac{1}{\sin \theta}$$

$$= \frac{\sin \theta - \sin(\theta + \Delta\theta)}{\sin \theta \sin(\theta + \Delta\theta)}$$

$$= \frac{\csc \theta \left(\tan \frac{\Delta\theta}{2} - \cot \theta\right) \tan \Delta\theta}{1 - \cot \theta \tan \Delta\theta}$$

$$= \frac{\csc \theta \left(\frac{\Delta\theta}{2} - \cot \theta\right) \Delta\theta}{1 - \Delta\theta \cot \theta}. \qquad (139)$$

When θ is neither very small nor very nearly equal to $90°$,

$\frac{\Delta\theta^2}{2}$ cosec θ and $\Delta\theta$ cot θ may be neglected, and we have approximately
$$\Delta \text{ cosec } \theta = -\cos\theta \text{ cosec}^2\theta\, \Delta\theta. \qquad (140)$$

From these formulæ we observe, (1) that a *small* increment of the angle causes a small *increment* of the sine, tangent and secant, and a small *decrement* of the cosine, cotangent and cosecant; (2) that for all the trigonometrical functions, a *small* increment of the angle produces a small *proportional* increment or decrement of the function, with the exception of the particular cases noticed above.

Hence, if an angle is near 90°, it cannot be found from the ordinary tables with *great* accuracy from its sine, tangent or secant; nor if near 0° or 180°, from its cosine, cotangent or cosecant.

In the next chapter, however, we will explain the construction of a special table by which the trigonometrical functions of an angle near the limits of the quadrant, can be found with great accuracy.

86. The results of this chapter enable us to solve many interesting and useful problems in surveying and astronomy without using the trigonometrical tables.

Examples.

1. An object standing on a horizontal plane subtends an angle of 3′ 20″ at the distance of two miles, find its height.

Let BC be the object whose height is required. The angle $A = 3'\ 20'' = \theta$, and $AC = 2$ miles $= a$.

Then
$$BC = AC \tan\theta, \quad \text{by (32)}$$
$$= a\theta, \quad \text{by Art. 74,}$$
$$= a \times \text{circular measure of } \theta$$
$$= a \times \frac{\theta''}{206264''.8}, \quad \text{by (128)}$$
$$= \frac{2 \times 200}{206264.8}$$
$$= 10\tfrac{1}{5} \text{ feet, very nearly.}$$

EXAMPLES. 93

2. The equatorial radius of the earth is 3962.8 miles, which is found by astronomical observations and calculations to subtend at the sun's centre an angle of $8''.95$. Find the sun's distance from the earth.

Let S and E represent the centres of the sun and earth. The angle $ASE = 8''.95 = \theta$, $AE = 3962.8$ miles $= r$, and $SE = x$.

Then
$$x \sin \theta = r$$

and
$$x = \frac{r}{\sin \theta}, \quad \text{by (35)}$$

$$= \frac{r}{\text{cir. meas. } \theta}, \quad \text{by Art. 74}$$

$$= r \times \frac{206264''.8}{8''.95}, \quad \text{by (128)}$$

$$= 91328000 \text{ miles.}$$

3. The apparent semi-diameter of the sun is $16'\ 1''.82$ at the earth's mean distance, as given in the last problem. Find the sun's radius.

In the figure of the last problem, let S be the earth's centre and E the sun's. The angle $ASE = 16'\ 1''.82 = \theta$, $SE = 91328000$ miles $= d$ and AE the sun's radius $= r$.

Then
$$r = d \sin \theta, \quad \text{by (27)}$$

$$= d \times \text{cir. meas. of } \theta, \quad \text{by Art. 74}$$

$$= d \times \frac{\theta''}{206264''.8}, \quad \text{by (128)}$$

$$= d \times \frac{961''.82}{206264''.8}$$

$$= 425860 \text{ miles.}$$

Hence the sun's radius = 107.4 times the earth's radius and as the volumes of spheres are as the cubes of their radii, we have

$$\text{volume of sun} : \text{volume of earth} :: (107.4)^3 : 1$$
$$:: 1238833 : 1,$$

so that the sun is more than a million times as large as the earth.

Dip of the Horizon.

87. The dip of the horizon is the angle of depression of the visible horizon below the true horizon, arising from the elevation of the observer's eye above the level of the sea. Let BDG be a section of the earth regarded as a sphere, and A the position of an observer above the surface. Join AC and produce it to G, also draw AE at right angles to the vertical line AG, and AF touching the surface at D, then neglecting the effect of atmospheric refraction, D is the most distant point visible from A. If we conceive AE and AF to revolve about AG as an axis, AE will describe the plane of the true or celestial horizon and AF will describe the surface of a cone touching the earth in the small circle called the apparent or visible horizon, and the angle EAD is called the dip of the horizon.

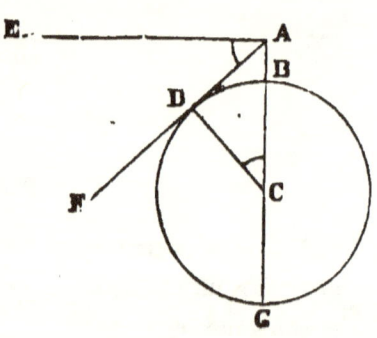

Let $h = AB$, the height of the observer's eye,

$r = DC$, the radius of the earth,

$D = $ the dip of the horizon.

The angle $EAD = $ the angle $ACD = D$.

Then
$$\tan D = \frac{AD}{DC} = \frac{\sqrt{AG \times AB}}{DC}, \quad (Euc.\ III.,\ 36)$$

$$= \frac{\sqrt{(2r+h)h}}{r}$$

$$= \sqrt{\frac{2h}{r} + \left(\frac{h}{r}\right)^2}.$$

As h is always very small compared with r, the square of the fraction $\frac{h}{r}$ is quite inappreciable, and therefore may be neglected, so that we have

$$\tan D = \sqrt{\frac{2h}{r}}. \tag{141}$$

Again, as the angle D is always very small for all accessible heights above the earth's surface, we may write the circular measure of the dip for its tangent (Art. 74), and expressing it in seconds by (128) we obtain

$$D'' = 206264''.8 \sqrt{\frac{2h}{r}}. \tag{142}$$

The distance AD of the visible horizon = the arc BD very nearly; let $AD = DB = d$, then

$$d = DC \times \text{the circular measure of } D$$

$$= r\sqrt{\frac{2h}{r}} = \sqrt{2rh}. \tag{143}$$

The mean value of r is 20888628 feet. Substituting this in (142) and reducing the *constant* coefficient of $\sqrt{h}$ we have

$$D'' = 63''.82 \sqrt{h}, \tag{142 bis}$$

where h is expressed in feet.

Owing to the effect of refraction, this coefficient is diminished by about $\frac{1}{11}$ part. Deducting $\frac{1}{11}$ of 63.82 we have finally

$$D'' = 58''.02 \sqrt{h}. \tag{144}$$

By this formula, the dip is computed and tabulated for heights varying from 1 to 100 feet, in works on navigation and astronomy.

Expressing d and r in *miles* and h in *feet* we have, by computing the constant coefficient in (143),

$$d = 1.224 \sqrt{h}. \qquad (145)$$

which gives the distance, unaffected by refraction. The amount of refraction however varies very much with the temperature of the atmosphere, in ordinary states of which, this distance is increased by about $\frac{1}{13}$ part. Hence, increasing the coefficient 1.224 by $\frac{1}{13}$ of itself, we have

$$d = 1.317 \sqrt{h}$$
$$= \tfrac{33}{25} \sqrt{h}, \text{ very nearly.} \qquad (146)$$

In (145) let $d = 1$ mile, then we have

$$h = .6675 \text{ feet}$$
$$= 8 \text{ inches, very nearly.}$$

Hence, if the tangent DA in the figure, be one mile, AB is 8 inches, which is expressed by saying that on the surface of still water the depression of the horizon is 8 inches for a horizontal distance of one mile; therefore in levelling for canals, aquaducts, &c., a reduction of 8 inches per mile must be made for the curvature of the earth's surface.

Since h varies as the square of the distance, the depression for 2 miles is 32 inches; for 3 miles, 6 feet, &c.

Examples.

1. Given $\cos A = .6400566$, to find A.

From the tables, we have

EXAMPLES.

$$\cos 50° \, 13' = .6398862$$
$$\cos 50° \, 12' = .6401097$$

Difference for 1' or 60" = .0002235

$$\cos A = .6400566, \text{ the given cosine,}$$
$$\cos 50° \, 13' = .6398862, \text{ next less in the table.}$$

Difference = .0001704

Hence we have

.0002235 : .0001704 :: 60" : 45".7,

Therefore $\quad A = 50° \, 12' \, 14''.3$.

2. Find in a similar manner the angle whose cotangent is 2.1453675.

Ans. 24° 59' 28".2.

3. Given tan 45° 1' = 1.0005819, find tan 45° 0' 40".

Ans. 1.0003879.

4. Find the angle whose sine is .7126666.

Ans. 45° 27' 8".2.

5. Find the angle whose tangent is .8952524.

Ans. 41° 50' 11".6.

6. Given $\sin 45° = \dfrac{\sqrt{2}}{2}$, show that tan 22° 30' = .4142136, and sec 22° 30' = 1.0823922.

7. The circular measure of an angle is .1047683; find the number of degrees in it.

Ans. 6° 0' 10".

8. Given cosec A + cot $A = \sqrt{3}$, find A.

Ans. $A = 60°$.

9. Find the dip of the horizon for a height of 25 feet.

Ans. 4' 50".

10. How far can the top of a lighthouse 200 feet high be seen at sea?

Ans $18\frac{2}{3}$ miles.

11. From what height will the horizon be 44 miles distant?

Ans. $1111\frac{1}{3}$ feet.

12. From the top of a ship's mast whose height is 81 feet, the top of a lighthouse 100 feet above the level of the sea, was just visible; required their distance, taking the effect of refraction into account.

Ans. $25\frac{2}{15}$ miles.

13. What is the difference between the true and the apparent level, at the distance of 1000 feet?

Ans. .2873 inch.

14. A ship whose mast is 90 feet high, is sailing directly towards an observer at the rate of 10 miles per hour, and the time from its first appearance till its arrival at the observer, is 1 hour $9\frac{2}{3}$ minutes; find approximately the earth's radius, supposing the observer's eye to be on a level with the surface of the sea and no allowance made for refraction.

Ans. 3958 miles.

15. When the moon's distance from the earth is 238824 miles, her radius subtends at the earth an angle of $15'\ 33''.5$; find her diameter.

Ans. 2161.6 miles.

16. Find the distance and the dip of the horizon for a height of 6 feet, taking the effect of refraction into account.

Ans. 3.226 miles; $2'\ 22''$.

CHAPTER VII.

ON LOGARITHMS AND LOGARITHMIC TABLES OF THE TRIGONOMETRICAL FUNCTIONS.

88. The logarithm of a number is the index of the power to which a fixed number, called the *base*, must be raised, in order to produce the given number. Thus, if

$$a^x = N,$$

then x is the logarithm of the number N, to the base a, and is expressed thus,

$$x = \log_a N;$$

or, if there be no occasion for mentioning the base, simply by

$$x = \log N.$$

Again, since $10^0 = 1$; $10^1 = 10$; $10^2 = 100$, &c.,
$\log_{10} 1 = 0$; $\log_{10} 10 = 1$; $\log_{10} 100 = 2$, &c.

It is evident that any number except unity may be taken as the base; and if for any base as 10, the logarithms of all numbers be computed and registered in a table, the table thus formed constitutes what is called "A table of a system of logarithms to base 10." There may therefore be an infinite number of systems of logarithms.

89. Although in theory any number except unity may be used as the base, yet in actual practice it has been found most convenient to use only two systems, viz.:

I.—*Logarithms to base 2.71828*... (denoted by e) which are called Napierian logarithms, from the name of the inventor,

Lord Napier, Baron of Merchiston in Scotland, who published the first table of logarithms in the year 1614; and·

II.—*Logarithms to base 10*, the radix of the ordinary scale of arithmetical notation, which are called common logarithms. These possess some peculiar advantages, and all the tables in common use are calculated to this base. Hence when we speak of logarithms, we mean logarithms to base 10, unless the contrary is stated.

90. Since $a^0 = 1$, and $a^1 = a$, we have

$$\log_a 1 = 0, \text{ and } \log_a a = 1. \tag{147}$$

whatever a may be.

If a be greater than unity, then

$$a^{-\infty} = 0 \,;\; a^0 = 1 \text{ and } a^{+\infty} = \infty,$$

therefore $\quad \log_a 0 = -\infty \,;\; \log_a 1 = 0 \,;\; \log_a \infty = \infty. \qquad (148)$

Hence, the logarithm of any number will be between 0 and $-\infty$, or between 0 and $+\infty$, that is, will be *negative* or *positive* according as the number is *less* or *greater* than unity.

Again, if a be *less* than unity, it may be shewn in a similar manner that the logarithm of any number will be *positive* or *negative* according as the number is less or greater than one.

Since a^x and a^{-x} are always positive quantities, negative numbers have no *real* logarithms.

91. In any System of Logarithms, the Logarithm of the product of two numbers is equal to the sum of the Logarithms of the numbers.

Let a be the base, m and n any two numbers, and x and y their logarithms respectively. Then by the definition of a logarithm, we have

and
and therefore
$$m = a^x \text{ or } x = \log_a m$$
$$n = a^y \text{ or } y = \log_a n.$$
$$mn = a^x \cdot a^y$$
$$= a^{x+y};$$

but by the definition, $x + y$ is the logarithm of mn to base a,

therefore
$$\log_a mn = x + y$$
$$= \log_a m + \log_a n.$$

92. In any System, the Logarithm of the quotient of two numbers is equal to the difference of their Logarithms.

Using the same notation as in the last Article, we have

$$\frac{m}{n} = \frac{a^x}{a^y} = a^{x-y};$$

but $x - y$ is the logarithm of $\frac{m}{n}$ to base a,

therefore
$$\log_a \frac{m}{n} = x - y$$
$$= \log_a m - \log_a n.$$

Hence it follows that
$$\log_a \frac{1}{m} = \log_a 1 - \log_a m$$
$$= -\log_a m. \quad \text{by Art. 90.}$$

93. In any System, the Logarithm of the nth power of any number is n times the Logarithm of the number.

Let $m = a^x$ as before, that is, $x = \log_a m$.

Then
$$m^n = (a^x)^n$$
$$= a^{nx};$$

but nx is the logarithm of m^n to base a,

therefore $\quad \log_a m^n = nx$
$$= n \log_a m,$$

which is true whether n be an integer or a fraction.

94. We thus see from the last three Articles that, by means of logarithms, the processes of multiplication and division are reduced to those of addition and subtraction; and those of involution and evolution to multiplication and division. The arithmetical operations of addition and subtraction cannot be performed by logarithms.

·The following examples shew the application of the preceding principles:

$$\log \frac{189}{\sqrt{7}} = \log \frac{3^3 \times 7}{\sqrt{7}} = \log 3^3 \times 7^{\frac{1}{2}}$$
$$= 3 \log 3 + \tfrac{1}{2} \log 7.$$

$$\log \frac{a^n c^{\frac{1}{m}}}{b^p} = n \log a + \frac{1}{m} \log c - p \log b.$$

$$\log \left(\frac{3}{17}\right)^{\frac{6}{7}} = \frac{6}{7} \log \frac{3}{17} = \frac{6}{7} \{\log 3 - \log 17\}.$$

$$\log \sqrt{a^2 - b^2} = \tfrac{1}{2} \log (a^2 - b^2)$$
$$= \tfrac{1}{2} \log \{(a+b)(a-b)\}$$
$$= \tfrac{1}{2} \log (a+b) + \tfrac{1}{2} \log (a-b).$$

95. Having given the logarithm of any number to a given base as e (where e denotes a certain number 2.71828... of which more hereafter), to find the logarithm of the same number to any other base as a.

Let x and y denote the logarithms of any number N to the bases a and e respectively, that is,

PROPERTIES OF LOGARITHMS.   103

let $\quad\quad\quad x = \log_a N$, and $y = \log_e N$,
hence $\quad\quad\quad a^x = N$ and $e^y = N$,
therefore $\quad\quad\quad a^x = e^y$.

Expressing this equation in logarithms to base e, we have

$$x \log_e a = y \log_e e$$
$$= y. \quad \text{by (147)}$$

Substituting the values of x and y, we have

$$\log_a N \log_e a = \log_e N,$$

therefore $\quad\quad \log_a N = \dfrac{1}{\log_e a} \cdot \log_e N. \quad\quad (149)$

If $a = 10$, the base of the common system,

then $\quad\quad \log_{10} N = \dfrac{1}{\log_e 10} \log_e N. \quad\quad (150)$

From (150) it appears that every common logarithm may be resolved into two factors, one of which is *constant* and depends upon the base of the system employed, the other is *variable* and depends upon the number itself. The constant factor or multiplier which thus connects two systems of logarithms, is called the *modulus*, and is denoted by M. Thus, the constant $\dfrac{1}{\log_e 10}$ is the modulus of the common system taken relatively to the system whose base is e.

The modulus of the common system may, therefore, be defined as the *constant* factor or multiplier, by which it is necessary to multiply the Napierian logarithms in order to convert them into common logarithms, and is equal to the reciprocal of the Napierian logarithm of the base of the common system.

It will be shewn hereafter, in the chapter on the computation of logarithms, that the Napierian logarithm of 10 is 2.302585..., therefore the modulus of the common system

$$= \dfrac{1}{\log_e 10} = \dfrac{1}{2.302585...} = .4342944...,$$

Denoting the *common logarithm* of a number by log, (150) may be written

$$\log N = .4342944 \times \log_e N. \qquad (151)$$

Hence we also find

$$\log_e N = \frac{1}{.4342944} \log N$$
$$= 2.302585 \times \log N, \qquad (152)$$

which enables us to find the Napierian logarithm when the common logarithm of any number is given.

96. Since $\log_a N = M \log_e N$, and $\log_b N = M' \log_e N$, where M and M' are the moduli, we have by division

$$\frac{\log_a N}{\log_b N} = \frac{M}{M'},$$

therefore the logarithms of the same number in different systems are proportional to the moduli of those systems.

97. Relation between the bases of two Systems.

Let x and y denote the logarithms of any number N to the bases a and b respectively, that is,

let $\quad x = \log_a N$, or $a^x = N$,
and $\quad y = \log_b N$, or $b^y = N$,
therefore $\quad a^x = b^y$.

Expressing this equation in logarithms to base a, we have

$$x = y \log_a b,$$

and expressing it in logarithms to base b, we have

$$x \log_b a = y.$$

Eliminating x and y from the last two equations, we obtain

$$\log_b a \, \log_a b = 1, \qquad (153)$$

which expresses the relation between the bases of any two systems.

COMMON LOGARITHMS. 105

98. Properties of Common Logarithms.

In the common system, the logarithms of all numbers which are integral powers of 10, are immediately known. Thus

$$10^0 = 1, \quad \text{therefore} \quad \log 1 = 0,$$
$$10^1 = 10, \quad \text{``} \quad \log 10 = 1,$$
$$10^2 = 100, \quad \text{``} \quad \log 100 = 2,$$
$$10^3 = 1000, \quad \text{``} \quad \log 1000 = 3,$$
$$\&c., \quad\quad\quad \&c.$$

Hence it appears that the logarithms of all numbers which are not integral powers of 10 consist of either a fraction or an integer and a fraction.

Thus, the logarithm of every number between 1 and 10, lies between 0 and 1, or is a proper fraction; the logarithm of every number between 10 and 100, lies between 1 and 2, or is 1 *plus* a fraction; the logarithm of every number between 100 and 1000, lies between 2 and 3, or is 2 *plus* a fraction, and so on.

The integral part of a logarithm is called the *characteristic*, and the decimal or fractional part the *mantissa*. Hence, if a number is between

1	and	10,	the characteristic of its log				$= 0,$
10	``	100,	``	``	``	``	$= 1,$
100	``	1000,	``	``	``	``	$= 2,$
1000	``	10000,	``	``	``	``	$= 3,$
....		....					....
10^{n-1}	``	$10^n,$	``	``	``	``	$= n-1.$

Therefore the characteristic of the logarithm of a number which has n integral places, is $n - 1$, or is less by unity than the number of integral places in the number.

Again, since

$$10^0 = 1 = 1, \text{ therefore } \log 1 = 0,$$
$$10^{-1} = \tfrac{1}{10} = .1, \text{ " } \log .1 = -1,$$
$$10^{-2} = \tfrac{1}{100} = .01, \text{ " } \log .01 = -2,$$
$$10^{-3} = \tfrac{1}{1000} = .001, \text{ " } \log .001 = -3,$$

&c., &c., &c.

Therefore the logarithm of every number between 1 and .1 lies between 0 and -1, or is -1 *plus* a fraction; the logarithm of every number between .1 and .01, lies between -1 and -2, or is -2 *plus* a fraction; the logarithm of every number between .01 and .001 lies between -2 and -3, or is -3 *plus* a fraction, and so on.

Hence if a number is between

1 and .1,	the characteristic of its log	= $\overline{1}$,
.1 " .01,	" " " "	= $\overline{2}$,
.01 " .001,	" " " "	= $\overline{3}$,
.001 " .0001,	" " " "	= $\overline{4}$,
....		....
10^{-n} $10^{-(n+1)}$ "	" " " "	= $\overline{(n+1)}$.

Therefore generally if a number has n ciphers after the decimal point, the characteristic of its logarithm is $\overline{(n+1)}$. The negative sign is written over the characteristic to shew that it *alone* is negative, while the mantissa, upon this supposition, is *always* positive.

Hence, we have the following general rules for finding the characteristic of the logarithm of any number.

Rule I.—The characteristic of the logarithm of any number *greater* than unity, is *positive* and less by unity than the number of integral places in the given number.

Rule II.—The characteristic of the logarithm of a decimal fraction, is *negative*, and numerically greater by unity than the number of ciphers after the decimal point.

Thus, for the following numbers, 14067, 521.64, 6.721, .364, .04271, .0027, .0001672, the characteristics are respectively 4, 2, 0, $\bar{1}$, $\bar{2}$, $\bar{3}$, $\bar{4}$.

And conversely, if logarithms be given, having characteristics 0, 1, 2, 3 .... n, there are in the numbers to which these respectively belong 1, 2, 3, 4 .... $(n+1)$ digits; and if the characteristics are $\bar{1}$, $\bar{2}$, $\bar{3}$ .... $\bar{n}$, there are in the corresponding numbers 0, 1, 2 .... $(n-1)$ ciphers respectively after the decimal point.

99. When the logarithm of any number is known, we can at once write down the logarithm of any other number having the same significant digits, but differing only from the given number in the position of the decimal point.

Let N be any number whose logarithm is known, then $10^n \times N$ is a number which has the same significant digits as N, but with the decimal point moved n places to the *right;* again, $\dfrac{N}{10^n}$ or $10^{-n} \times N$ is a number which has the same significant digits as N, but with the decimal point moved n places to the *left*.

Now, $\log(10^n \times N) = \log 10^n + \log N$,
$\qquad = n \log 10 + \log N$,
$\qquad = n + \log N$;

and $\log(10^{-n} \times N) = \log 10^{-n} + \log N$,
$\qquad = -n \log 10 + \log N$,
$\qquad = -n + \log N$.

Hence, as n is an integer, the logarithms of $10^n \times N$ and $10^{-n} \times N$ differ from the logarithm of N in the characteristic only; therefore by giving the logarithm of N its proper characteristic in accordance with the rules investigated in the last Article, we can at once deduce those of $10^n \times N$ and $10^{-n} \times N$.

Thus,
$$\log 2436000 = 6.386677,$$
$$\log 24360 = 4.386677,$$
$$\log 24.36 = 1.386677,$$
$$\log .2436 = \overline{1}.386677,$$
$$\log .002436 = \overline{3}.386677.$$

10c. In this system it is only necessary to register the mantissæ in the tables, for the characteristics can be determined by the preceding rules.

These properties render the common tables less bulky and more comprehensive than those computed for any other base, and give them an advantage over all others.

101. The arrangement and use of the common tables are easily understood from the following portion of a table taken from Chambers's Logarithms:

No.	0	1	2	3	4	5	6	7	8	9	D
4829	6838572	8662	8752	8842	8932	9022	9112	9202	9291	9381	
4830	9471	9561	9651	9741	9831	9921	.011	.101	.191	.280	90
4831	6840370	0460	0550	0640	0730	0820	0910	1000	1089	1179	
4832	1269	1359	1449	1539	1629	1719	1808	1898	1988	2078	
Diff.	90	9	18	27	36	45	54	63	72	81	

In this table we have the natural numbers between 48290 and 48329, and the mantissæ are between .6838572 and .6842078. The four left hand figures of the natural numbers are placed in the column marked No., and the fifth figure at the top of one of the other columns. As the first three figures of the mantissæ are the same for several consecutive numbers in the table, they are not repeated for each logarithm, but registered once only. Thus, the mantissæ of the logarithms of all numbers from 48290 to 48305 inclusive, contain the same three initial figures, viz.: 683, which are registered opposite the number 4829 and in the column under 0, while the last four figures of each mantissa, are registered in the columns 1, 2, 3, &c., to which they belong.

Thus,   the mantissa of log $48296 = .6839112$
    "   log $48303 = .6839741$
            &c.,   &c.

In the second horizontal line and in the column under 6, a dot is found in place of a figure. This indicates that the value of the three initial figures has changed, and therefore both here and all along the remainder of the same line, the three initial figures taken from the horizontal line next below, are to be used. The places of the dots are to be supplied with ciphers. Thus the mantissa of log $48308 = .6840191$.

In this manner we obtain from the tables the mantissæ of the logarithms of all numbers which contain, three, four or five digits. The logarithms of all numbers between 1 and 100 are generally tabulated separately. At the bottom of each page there is a "table of proportional parts" by which the mantissæ of the logarithms of all numbers containing six or seven digits may be calculated approximately; and conversely, the number corresponding to a given logarithm whose mantissa cannot be exactly found in the tables, may be found approximately to six or seven places of figures.

In the construction of the "table of proportional parts," it is assumed that the increase in a logarithm is proportional to the increase in the number. The proof of this principle will be given in the chapter on the computation of logarithms. The results given by it are only approximate, but the error is generally so small that it may be safely neglected, especially if the numbers are large.

The last column marked D contains the difference between two consecutive mantissæ; thus in the above table,

$$\text{the mantissa of } 48304 = .6839831$$
$$\text{`` } 48305 = .6839921$$

Difference corresponding to $1 = .0000090$

The significant digits only of this difference, are registered in column D.

Now, suppose the logarithms of 48304.1, 48304.2, 48304.3 &c., are required, we have by the principle just stated

$$1 : .1 :: .0000090 : .0000009$$
$$1 : .2 :: .0000090 : .0000018$$
$$1 : .3 :: .0000090 : .0000027$$
&c., &c., &c.

Hence, the mantissa of log $48304.1 = .6839840$
" log $48304.2 = .6839849$
" log $48304.3 = .6839858$
&c., &c., &c.

The significant digits, 9, 18, 27, 36, &c., of the proportional parts found above, are registered at the foot of the page, opposite to the whole difference 90, and in the columns 1, 2, 3, 4, &c., to which they respectively belong. The manner of using this table will be easily understood from the following examples:

Ex. 1.—Find the logarithm of 483136.

In the table we find   log $483130 = 5.6840640$
Proportional part for $6 =$ 54

Therefore   log $483136 = 5.6840694$

LOGARITHMIC TABLES.   111

Ex. 2.—Find the logarithm of 48.29689.

In the table we find   log 48.29600 = 1.6839112
Proportional part for   8 =   72
      "   "   9 =   81
                                      ─────────
Therefore   log 48.29689 = 1.6839192

Here we observe that since the *local* value of the seventh digit (9) is one-tenth of that of the preceding digit (8), the proportional part for the seventh digit is one-tenth of what it would be if it occupied the place of the sixth digit, and therefore must be moved one place to the right, as shewn above.

For the same reason the proportional part for the eighth digit must be moved two places to the right, as is seen in the following example:

Ex. 3.—Find the logarithm of .48306357.

From the table, we have   log .48306000 = $\overline{1}$.6840101
Proportional part for   3 =   27
      "   "   5 =   45
      "   "   7 =   63
                                      ─────────
Therefore   log .48306357 = $\overline{1}$.6840133

102. To find the Natural Number corresponding to a given Logarithm.

If the mantissa is exactly found in the table, the first four figures of the natural number will be found opposite to them in the column headed No., and the fifth figure at the top of the page in the column from which the last four figures of the mantissa were taken. The position of the decimal point is determined by the characteristic, according to Art. 98.

If the mantissa is not exactly found in the table, find the next *less* mantissa, and take out the first five figures of the natural number as before. The additional figures may be found approximately by the proportional parts, as in the following examples:

Ex. 1.—Find the number whose logarithm is 2.6841161.

The given mantissa is .6841161
Next less in the table is .6841089, and nat. No. = 48318

Difference 72, and nat. No. = 8

Therefore all the digits are 483188, and since the characteristic is 2, the required number is 483.188.

Ex. 2.—Find the number whose logarithm is $\bar{2}.6840700$.

The given mantissa is .6840700
Next less in table .6840640, and nat. No. = 48313

Difference 60
Proportional part next less is 54, and nat. No. = 6

Residual difference 6, and nat. No. = 7

Therefore the digits of the number are 4831367, and since the characteristic is $\bar{2}$, the required No. = .04831367.

103. Multiplication is performed by the principle of Art. 91.

Ex.—Find the product of .00098, .0761 and 41.

$$\begin{aligned}
\text{Let} \quad & x = .00098 \times .0761 \times 41, \\
\text{then} \quad & \log x = \log .00098 + \log .0761 + \log 41. \\
& \log .00098 = \bar{4}.991226 \\
& \log .0761 = \bar{2}.881385 \\
& \log 41 = 1.612784 \\
& \log x = \bar{3}.485395 \\
\text{therefore} \quad & x = .003057698
\end{aligned}$$

The student will observe that the characteristics are added *algebraically*; thus, in the above example we have 2 " to carry " to the characteristics,

then $\qquad 2 + 1 + (-2) + (-4) = -3.$

104. Division is performed by the principle of Art. 92.

Ex.—Divide 18.792 by .0007834.

Let $\qquad x \qquad = 18.792 \div .0007834,$
then $\qquad \log x \qquad = \log 18.792 - \log .0007834.$
$\log 18.792 = 1.2739730$
$\log .0007834 = \overline{4}.8939836$

$\log x \qquad = 4.3799894$
therefore $\qquad x \qquad = 23987.74$

Here we have 1 "to carry" to the characteristic -4, which makes -3. This subtracted from $+1$ gives $+4$, that is, the characteristics are subtracted *algebraically*.

105. Although negative numbers have no real logarithms, yet they may be multiplied or divided by means of logarithms, if we consider the quantities as positive, and prefix the proper sign to the result, according to the rules of algebra. To indicate that the natural number is negative, we append the letter n to its logarithm; thus the logarithm of -43 is written $1.6334685n$.

Ex. Divide 5 by $-.029$.

$\log 5 = 0.6989700$
$\log -.029 = \overline{2}.4623980n$

$\log -172.4131 = 2.2365720n$

therefore the quotient is -172.4131.

9

106. If a Logarithm be subtracted from 10, the remainder is called its Arithmetical Complement.

Let $x = \dfrac{M}{N}$,

then $\log x = \log M - \log N$
$= \log M + (10 - \log N) - 10$,

but, by the definition, $(10 - \log N)$ is the arithmetical complement of the logarithm of N, and is written ar. co-log N,

therefore $\log x = \log M + $ ar. co-log $N - 10$.

Hence, by means of the arithmetical complement the process of subtraction of logarithms may be converted into that of addition.

Ex. Divide .00815 by .000256.

$$\log .00815 = \overline{3}.9111576$$
$$\text{ar. co-log } .000256 = 13.5917600$$
$$\overline{}$$
$$11.5029176$$
$$\text{subtract } 10$$
$$\overline{}$$
$$\log 31.83593 = 1.5029176$$

therefore the quotient is 31.83593.

107. Involution and Evolution are performed by the principle of Art. 93.

Ex. 1.—Find the 7th power of .091.

Let $x = (.091)^7$,

then $\log x = 7 \times \log .091$
$= 7 \times \overline{2}.959041$
$= \overline{8}.713287$,

therefore $x = .000000051676$.

Ex. 2.—Find the 7th root of .00074.

Let $x = (.00074)^{\frac{1}{7}}$,

then $\log x = \frac{1}{7} \log .00074$

$$= \frac{\overline{4}.8692317}{7}$$

$$= \frac{\overline{7} + 3.8692317}{7}$$

$$= \overline{1}.5527474$$

therefore $x = .357065$.

Here we add -3 to the characteristic -4 and $+3$ to the mantissa, in order to make the characteristic divisible by 7.

Ex. 3.—Find the value of $\left(\frac{2}{7}\right)^{\frac{3}{5}}$

Let $x =$ the value of it,

then $\log x = \frac{3}{5} \log \frac{2}{7} = \frac{3}{5} (\log 2 - \log 7)$

$$= \frac{\overline{4}.367796}{5}$$

$$= \frac{\overline{5} + 1.367796}{5}$$

$$= \overline{1}.273559,$$

therefore $x = .18774$.

108. Table of Logarithmic sines, etc.

This table contains the logarithms of the natural sines, cosines, &c., computed at intervals of 1′ or 10″, and is arranged in precisely the same manner as the table of natural sines, &c., described in the last chapter.

PLANE TRIGONOMETRY.

As the sines and cosines of all angles, the tangents of angles *less* than 45°, and the cotangents of angles *greater* than 45°, are less than 1, their logarithms have negative characteristics (Arts. 90 and 98). In order, then, to avoid the use of negative characteristics and to secure uniformity in the entire table, the logarithms of all the trigonometrical functions are increased by 10, before being registered, and the logarithm so increased is called the *Tabular Logarithm* of the function, which is usually denoted by Log.

Thus,
$$\sin 60° = \frac{\sqrt{3}}{2} = .8660254.$$

$$\log \sin 60° = \overline{1}.9375306.$$

$$\text{(Tabular) Log} \sin 60° = 9.9375306.$$

Of course the real logarithm of any function is found from the tabular logarithm by subtracting 10.

109. Since
$$\operatorname{cosec} A = \frac{1}{\sin A},$$
we have $\quad \operatorname{cosec} A \sin A = 1$

and $\quad \log \operatorname{cosec} A + \log \sin A = 0;$

adding 10 to each of these, we have

$$(\log \operatorname{cosec} A + 10) + (\log \sin A + 10) = 20,$$

or $\quad \text{Log} \operatorname{cosec} A + \text{Log} \sin A = 20,$

hence $\quad \text{Log} \operatorname{cosec} A = 20 - \text{Log} \sin A.$ \qquad (154)

In a similar manner we find

$$\text{Log sec } A = 20 - \text{Log cos } A. \qquad (155)$$

and $\quad \text{Log cot } A = 20 - \text{Log tan } A. \qquad (156)$

LOGARITHMIC TABLES. 117

Again, since
$$\tan A = \frac{\sin A}{\cos A}$$

we have $\tan A \cos A = \sin A$

and $\log \tan A + \log \cos A = \log \sin A$

$(\log \tan A + 10) + (\log \cos A + 10) = 10 + (\log \sin A + 10)$

or $\quad$ Log tan A + Log cos A = 10 + Log sin A,

hence $\quad$ Log tan A = 10 + Log sin A - Log cos A. $\quad$ (157)

110. To find the Tabular Logarithmic Function of a given Angle.

If the given angle is found exactly in the table, the required quantity is at once obtained. When the angle is not exactly found in the table, the logarithm of the function is found approximately by the method of proportional parts, as in the case of the natural functions. (See Art. 84.)

Except near the limits of the quadrant, the increase or decrease of the logarithm of a trigonometrical function, is very nearly proportional to the increase of the angle. This statement will be proved in a subsequent chapter.

The results given by the method of proportional parts, are only approximate, but the error is in general so small that it may be safely neglected.

The application of this method will be easily understood from the following examples:

Ex. 1.—Find Log sin 17° 18′ 24″.5.

From the table we have

$\quad\quad\quad$ Log sin 17° 18′ 20″ = 9.473439
$\quad\quad\quad$ Log sin 17° 18′ 30″ = 9.473507
$\quad\quad\quad\quad\quad$ Difference for 10″ = .000068

Then we have

$$.10'' : 4''.5 :: .000068 : .000030$$

hence Log sin $17° \ 18' \ 24''.5 = 9.473469$.

Ex. 2.—Find Log cot $42° \ 17' \ 53''$.

From the table we find that the difference for $10''$ is $.000042$. Therefore we have

$$10'' : 3'' :: .000042 : .000013$$

hence Log cot $42° \ 17' \ 53''$ = Log cot $42° \ 17' \ 50'' - .000013$
$$= 10.041034 - .000013$$
$$= 10.041021$$

III. To find the Angle when the Logarithm of the Function is given.

Ex. 1.—Given Log sin $A = 9.473469$, find A.

From the table we find

Log sin $17° \ 18' \ 20'' = 9.473439$
and Log sin $17° \ 18' \ 30'' = 9.473507$

Difference for $10'' = .000068$

Hence it is evident that the required angle must lie between $17° \ 18' \ 20''$ and $17° \ 18' \ 30''$.

The given Log sin $A = 9.473469$
Log sin $17° \ 18' \ 20'' = 9.473439$

Difference $= .000030$

Then we have

$$.000068 : .000030 :: 10'' : 4''.5,$$

therefore $A = 17° \ 18' \ 20'' + 4''.5 = 17° \ 18' \ 24''.5$.

Ex. 2.—Given Log cos $A = 9.893586$, find A.

From the table we find

Log cos $38° \ 29' \ 30'' = 9.893595$

and Log cos $38° \ 29' \ 40'' = 9.893578$

Difference for $10'' = \ .000017$

The given Log cos $= 9.893586$

Log cos $38° \ 29' \ 40'' = 9.893578$

Difference $= \ .000008$

Then we have

$.000017 : .000008 :: 10'' : 4''.7$,

therefore $A = 38° \ 29' \ 40'' - 4''.7 = 38° \ 29' \ 35''.3$.

The chief point for the student to bear in mind here is, that the proportional part for the sine, tangent and secant must be *added*, while that for the cosine, cotangent and cosecant must be *subtracted* for the reason already given in Article 84.

112. The method of proportional parts which has just been used for finding the logarithmic functions of angles involving seconds and fractions of a second, is sufficiently accurate in all cases, except near the limits of the quadrant where, from an inspection of the table, it is seen that the differences for the Logarithmic sines, tangents and cotangents of angles less than 2° or 3°, and for the Logarithmic cosines, tangents and cotangents for the *last* 2° or 3° of the quadrant, are very variable, and therefore the proportional part cannot be accurately found by this method, since it is computed on the supposition that the differences are constant for a difference of 1' or 10'' in the angle.

The logarithmic functions of angles near the limit of the

quadrant, are computed with extreme accuracy by the following process. A special table is formed containing for every minute from 0° to 2°, the Logarithms of

$$\frac{\sin A}{A''} \text{ and } \frac{\tan A}{A''}$$

which vary quite slowly and uniformly for the first 2°, and therefore may be accurately found from the table for any intermediate value.

Thus, giving A the values 30', 31', 32' &c., in succession we have from the former of the above expressions,

$$\text{Log sin } 30' = 7.9408419$$
$$\text{Log } 1800'' = 3.2552725$$
$$\overline{\text{Log sin } 30' - \log 1800'' = 4.6855694}$$

In a similar manner we find

$$\text{Log sin } 31' - \log 1860'' = 4.6855690$$
$$\text{Log sin } 32' - \log 1920'' = 4.6855686$$
$$\&c., \quad \&c., \quad \&c.$$

These numbers are tabulated under the heading Log sin A − log A''; and in a similar manner is formed a column under the heading Log tan A − log A''. The cotangent being the reciprocal of the tangent, a column is also formed under the heading Log $\frac{A''}{\tan A}$ or Log cot A + log A''.

Since the cosine of an angle is the sine of its complement, the column headed Log sin A − log A'' at the top, is marked Log cos A − log. comp. A'' at the bottom, and when read upwards answers for the cosines of angles near the close of the quadrant.

SPECIAL LOGARITHMIC TABLE. 121

Again, to find the tangent of an angle near 90°, we find the tangent of its complement and then take its reciprocal, so that the column headed Log tan A + log comp. A'', is formed by subtracting the numbers found in the column headed Log tan A − log A'', from 20, according to (156).

The use of these special tables is easily learned from the following examples:

Ex. 1.—If $A = 31'\ 17''.4$, find Log sin A.

Here $\qquad A = 1877''.4.$

From the special table we find

$\qquad$ Log sin A − log $A'' = 4.6855689$
Add $\qquad\qquad\qquad$ log $A'' = 3.2735568$
$\qquad\qquad\qquad$ Log sin $A = 7.9591257$

Ex. 2.—Given $A = 17'\ 5''.3$, find Log tan A.

Here $\qquad A = 1025''.3.$

From the special table we have

$\qquad$ Log tan A − log $A'' = 4.6855784$
Add $\qquad\qquad\qquad$ log $A'' = 3.0108510$
$\qquad\qquad\qquad$ Log tan $17'\ 5''.3 = 7.6964294$

Ex. 3.—Find the Log tan $89°\ 47'\ 12''.8$.

The complement of this angle is $12'\ 47''.2 = 767''.2$.

From the table we get

$\qquad$ Log tan $89°\ 47'\ 12''.8$ + log $767''.2 = 15.3144232$
Subtract $\qquad\qquad$ log $767''.2 = 2.8849086$
$\qquad\qquad$ Log tan $89°\ 47'\ 12''.8 = 12.4295146$

Ex. 4.—Given Log sin $A = 8.0794466$, find A.

From the ordinary table we find that $A = 41'\ 16''$ nearly, therefore from the special table, we have

$$\text{Log sin } A - \log A'' = 4.6855645$$
but $\quad\quad\quad\quad\quad\text{Log sin } A = 8.0794466$

therefore $\quad\quad\quad\log A'' = 3.3938821$
and $\quad\quad\quad\quad\quad A = 2476''.75$
$\quad\quad\quad\quad\quad\quad\quad = 41'\ 16''.75.$

Ex. 5.—Given Log tan $A = 12.7478654$, find A.

From the ordinary table we find that $A = 89°\ 54'$ nearly, therefore from the special table, we have

$$\text{Log tan } A + \log \text{comp. } A'' = 15.3144251$$
but $\quad\quad\quad\quad\text{Log tan } A = 12.7478654$

therefore $\quad\log \text{comp. } A'' = 2.5665597$
and $\quad\quad\quad\text{comp. } A = 368''.604$
$\quad\quad\quad\quad\quad\quad = 6'\ 8''.604$
hence $\quad\quad\quad A = 89°\ 53'\ 51''.396.$

113. The best seven-figure tables hitherto published are those by Dr. L. Schrön, with an introduction by the late Professor De Morgan. They contain the logarithms of numbers from 1 to 108000, and of sines, cosines, tangents and cotangents, to every $10''$ of the quadrant, with a table of proportional parts.

Of the other seven-figure tables published in England, may be mentioned Hutton's, Chambers's, Babbage's and Shortrede's, the last of which gives the log. functions to every second of the quadrant.

The best American seven-figure tables are those of Stanley. They give the log. functions to every $10''$ for the first $15°$.

The best American six-figure tables are those of Loomis and Olney. The former gives the log. functions to every 10″ of the quadrant, with a table of proportional parts very conveniently arranged; the latter gives the log. functions to every minute of the quadrant.

Loomis's tables are sufficiently extensive for all ordinary purposes, and will be used hereafter in the examples of this work.

Examples.

1. Find the logarithm of 256 to the base $\sqrt{8}$.

Let $x =$ its logarithm,

then $$8^{\frac{x}{2}} = 256$$

or $$2^{\frac{3x}{2}} = 2^8$$

hence $$\frac{3x}{2} = 8 \text{ and } x = 5\tfrac{1}{3}.$$

2. Find the logarithm of 10 to the base $\frac{1}{3}$.

Let $x =$ its logarithm,

then $$\left(\frac{1}{3}\right)^x = 10$$

and $$x \log \frac{1}{3} = \log 10 = 1, \quad \text{(Art. 93)}$$

therefore $$x = \frac{1}{\log \frac{1}{3}} = -\frac{1}{\log 3}$$

$$= -2.0959.$$

3. The logarithm of a certain number to base 5 is n times the logarithm of the same number to base 3; find n.

If N denote the number and x and y the logarithms, we have
$$5^x = N \text{ and } 3^y = N,$$
therefore $\quad 5^x = 3^y$

and $\quad x \log 5 = y \log 3.$ (Art. 93)

But by the question $x = ny$, since $x = \log_5 N$ and $y = \log_3 N$, therefore
$$ny \log 5 = y \log 3,$$
whence $\quad n = \dfrac{\log 3}{\log 5}.$

4. Given $\log \dfrac{1}{2} = \overline{1}.698970$, and $\log \dfrac{1}{3} = \overline{1}.522878$, find the logarithms of $\sqrt{3}$, $\sqrt[5]{2}$ and $1440^{\frac{2}{5}}$.

$$\text{Log } \dfrac{1}{3} = \log 1 - \log 3 = \overline{1}.522878$$
$$= -\log 3 = \overline{1}.522878$$

or $\quad \log 3 = 1 - .522878$
$$= 0.477121$$

therefore $\quad \log \sqrt{3} = \dfrac{1}{2} \log 3.$
$$= 0.238560$$

Similarly we find
$$\log \sqrt[5]{2} = 0.100343$$
$$\log 1440^{\frac{2}{5}} = \dfrac{2}{5} \log 1440$$
$$= \dfrac{2}{5} \log (10 \times 2^4 \times 3^2)$$
$$= \dfrac{2}{5} (1 + 4 \log 2 + 2 \log 3)$$
$$= 1.263345.$$

5. Find $\log 243$ to the base $3^{\frac{2}{5}}$. *Ans.* $7\frac{1}{2}$.

EXAMPLES.

6. If the log of 9 is $\frac{2}{3}$, what is the base?   *Ans.* 27.

7. Given $\log 98 = 1.991226$, and $\log 112 = 2.049218$, find the logarithms of 50, .7 and 1750.

$$\textit{Ans. } 1.698970,\ \overline{1}.845098,\ 3.243038.$$

8. Given $\log 1\tfrac{1}{4} = .096910$, and $\log \frac{1}{9} = \overline{1}.045757$, find the logarithms of $2\tfrac{1}{2}$ and $2\tfrac{1}{4}$.

$$\textit{Ans. } 0.397940,\ 0.352183.$$

9. Shew that $5 \log \dfrac{8}{9} + 3 \log \dfrac{27}{16} + \log \dfrac{3}{4} = \log 2$.

10. Find the 50th root of 10.   *Ans.* 1.047126.

11. Find the value of $\left(\dfrac{3}{7}\right)^{1\frac{1}{6}}$.   *Ans.* .612295.

12. Find the 17th root of .071852.   *Ans.* .8565.

13. If a, b, c be in geometrical progression, prove that $\log_a n$, $\log_b n$, $\log_c n$ are in harmonical progression.

14. Shew that $\log_{\sin A} \operatorname{cosec} A = \log_{\cos A} \sec A$.

15. Find x from the equation $8^x = 1000$, having given $\log 2 = .301030$.   *Ans.* $x = 3.3219$.

16. Given $\log 1.4 = .146128$, $\log 144 = 2.158362$, and $\log 441 = 2.644438$, find the logarithms of the nine digits.

17. Given $\log 2 = .301030$, and $\log 3 = .477121$, find the logarithms of 135, 405, 3.24 and $\dfrac{1}{9}$.

$$\textit{Ans. } 2.130334,\ 2.607455,\ 0.510545,\ \overline{1}.045757.$$

18. Find the value of $\dfrac{(.327)^7 \times (191)^{\frac{1}{4}}}{(19)^{\frac{1}{4}} \times (.061)^2}$.   *Ans.* .340653.

19. Find $\log_2 6$.   *Ans.* 2.584.

20. If $3^{2x} \cdot 5^{x-1} = 2^{3x+1}$, shew that $x = \dfrac{1}{\log 45 - \log 8}$.

21. Given $x^y = y^x$, and $x^3 = y^2$, find x and y.

$$Ans. \ x = \frac{9}{4}, \ y = \frac{27}{8}.$$

22. Given $2^{3x} \cdot 5^{2x-1} = 4^{5x} \cdot 3^{x+1}$, find x.

$$Ans. \ -.991.$$

23. Given Log cot $68°\ 21'\ 10'' = 9.598661$,
Log cot $68°\ 21'\ 20'' = 9.598600$,
find Log cot $68°\ 21'\ 16''.4$.

$$Ans. \ 9.598622.$$

24. Given Log sin $37°\ 4'\ 40'' = 9.780244$,
Log sin $37°\ 4'\ 50'' = 9.780272$,
find Log cosec $37°\ 4'\ 47''$.

$$Ans. \ 10.219736.$$

25. Given Log cos $75°\ 13' = 9.406820$,
Log cos $75°\ 14' = 9.406341$,
find Log cos $75°\ 13'\ 22''.4$.

$$Ans. \ 9.406641.$$

26. Given Log tan $A = 10.572676$,
find A, having given
Log tan $75°\ 1'\ 20'' = 10.572622$,
Log tan $75°\ 1'\ 30'' = 10.572706$,

$$Ans. \ A = 75°\ 1'\ 26''.2.$$

CHAPTER VIII.

FORMULÆ FOR THE SOLUTION OF TRIANGLES.

RIGHT-ANGLED TRIANGLES.

114. The formulæ for right-angled triangles have been already given in Chapter III. They are immediately derived from the definition of the sine, cosine and tangent. Thus, from the right-angled triangle ABC, we have

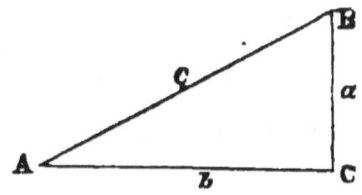

$$\sin A = \frac{a}{c}, \quad \cos A = \frac{b}{c}, \quad \tan A = \frac{a}{b}. \tag{158}$$

Special Formulæ for Right-angled Triangles.

115. When A is nearly 90°, the first of this group cannot be accurately computed from the tables, since the sines then differ very little from each other; and for a similar reason, the second of the group cannot be accurately found when A is very small, or when b is nearly equal to c. Hence, an angle near 90° should be determined by its cosine, and a small angle by its sine or tangent. Special formulæ are therefore frequently necessary, especially in surveying and astronomy, where great accuracy is required.

From (100) we have

$$2 \sin^2 \frac{A}{2} = 1 - \cos A.$$

Writing $90° - A$ for A, this becomes

$$2 \sin^2 (45° - \frac{A}{2}) = 1 - \sin A$$

$$= 1 - \frac{a}{c}, \text{ by the first of (158)}$$

$$= \frac{c - a}{c}.$$

whence
$$\sin(45° - \tfrac{A}{2}) = \sqrt{\tfrac{c-a}{2c}},\qquad(159)$$

which may be used with advantage when A is near 90°.

116. From the second of (158), we have
$$\sin^2 A = 1 - \cos^2 A = 1 - \tfrac{b^2}{c^2},$$
whence
$$\sin A = \sqrt{\tfrac{c^2 - b^2}{c^2}}$$
$$= \tfrac{\sqrt{(c+b)(c-b)}}{c},\qquad(160)$$

which may be used instead of $\cos A = \tfrac{b}{c}$, when A is small.

From (100) we have
$$2\sin^2 \tfrac{A}{2} = 1 - \cos A$$
$$= 1 - \tfrac{b}{c} = \tfrac{c-b}{c}$$
whence
$$c - b = 2c \sin^2 \tfrac{A}{2},\qquad(161)$$

by which $c-b$ may be accurately found when A is small.

117. From (83) we have
$$\tan(A - 45°) = \tfrac{\tan A - 1}{\tan A + 1}$$
$$= \tfrac{\tfrac{a}{b} - 1}{\tfrac{a}{b} + 1};\quad \text{by the third of (158)}$$
$$= \tfrac{a - b}{a + b},\qquad(162)$$

which may be used instead of $\tan A = \tfrac{a}{b}$, when A is near 90°.

118. From (101) we have
$$\tan \tfrac{A}{2} = \sqrt{\tfrac{1 - \cos A}{1 + \cos A}}$$

$$= \sqrt{\frac{1-\frac{b}{c}}{1+\frac{b}{c}}}, \quad \text{by the second of (158)}$$

$$= \sqrt{\frac{c-b}{c+b}} = \frac{c-b}{a},$$

whence $\quad c - b = a \tan \frac{A}{2},$ \hfill (163)

by which $c - b$ may be accurately found when A is near 90°.

Since $c^2 - b^2 = a^2$, we have from (163)

$$\frac{c^2 - b^2}{c-b} = \frac{a^2}{a \tan \frac{A}{2}}$$

or $\quad c + b = a \cot \frac{A}{2},$ \hfill (164)

by which $c+b$ may be found when A is near 90°.

OBLIQUE-ANGLED TRIANGLES.

119. The sides of a Triangle are proportional to the sines of the opposite Angles.

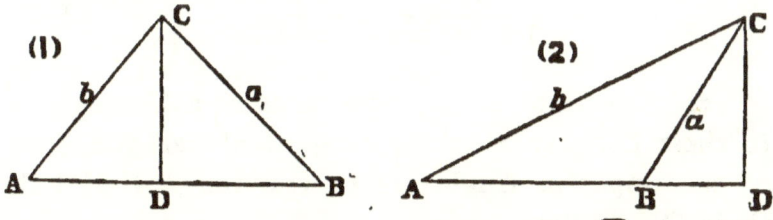

In the triangle ABC, let the sides opposite to the angles A, B and C be denoted by a, b, c respectively. Draw CD perpendicular to AB, produced if necessary.

Then, in the right-angled triangles ACD, BCD, we have by the first of (158)

(Fig. 1) $\quad CD = b \sin A$, and $CD = a \sin B$,
whence $\quad b \sin A = a \sin B;$

and (Fig. 2) $CD = b \sin A$, and $CD = a \sin CBD = a \sin B$, since CBD and ABC are supplementary angles, whence
$$b \sin A = a \sin B.$$

Therefore, in both cases we have
$$b \sin A = a \sin B,$$
which may be converted into a proportion, thus,
$$a : b :: \sin A : \sin B.$$

In the same way we may shew that
$$a : c :: \sin A : \sin C$$
and
$$b : c :: \sin B : \sin C.$$

These three proportions may be written as one, thus,
$$a : b : c :: \sin A : \sin B : \sin C,$$
or better thus,
$$\frac{a}{\sin A} = \frac{b}{\sin B} = \frac{c}{\sin C}. \qquad (165)$$

120. The Sum of any two sides of a Triangle is to their Difference as the tangent of half the Sum of the opposite Angles is to the tangent of half their Difference.

From (165) we have
$$\frac{a}{b} = \frac{\sin A}{\sin B},$$
whence, by composition and division,
$$\frac{a+b}{a-b} = \frac{\sin A + \sin B}{\sin A - \sin B}$$
$$= \frac{\tan \tfrac{1}{2}(A+B)}{\tan \tfrac{1}{2}(A-B)}, \quad \text{by (59)}. \qquad (166)$$

Since $A + B + C = 180°$, $\tfrac{1}{2}(A + B) = 90° - \dfrac{C}{2}$

and $\tan \tfrac{1}{2}(A + B) = \cot \dfrac{C}{2}$.

Hence (166) may be written

$$\tan \tfrac{1}{2}(A - B) = \dfrac{a - b}{a + b} \cot \dfrac{C}{2}. \qquad (167)$$

Similar relations may at once be inferred between b, c, B, C and a, c, A, C.

The last equation may be written thus

$$\tan \tfrac{1}{2}(A - B) = \dfrac{1 - \dfrac{b}{a}}{1 + \dfrac{b}{a}} \cot \dfrac{C}{2}.$$

Let $\dfrac{b}{a} = \tan \theta,$

an assumption always possible, since a tangent may have any value between 0 and ∞; then we have

$$\tan \tfrac{1}{2}(A - B) = \dfrac{1 - \tan \theta}{1 + \tan \theta} \cdot \cot \dfrac{C}{2}$$

$$= \tan (45° - \theta) \cot \dfrac{C}{2}. \qquad (168)$$

In the last three formulæ $a > b$, therefore we may put

$$\dfrac{b}{a} = \cos \phi,$$

then $\tan \tfrac{1}{2}(A - B) = \dfrac{1 - \cos \phi}{1 + \cos \phi} \cot \dfrac{C}{2}$

$$= \tan^2 \dfrac{\phi}{2} \cot \dfrac{C}{2}. \qquad (168 \text{ bis})$$

121. Geometrical proof of (166).

Let ABC be any triangle of which the side BC is greater than AC. With C as a centre and AC, the shorter of the two sides, as radius, describe a circle cutting AB in F, BC in E, and BC produced in D. Join AD, CF, AE, and draw EH at right-angles to AB.

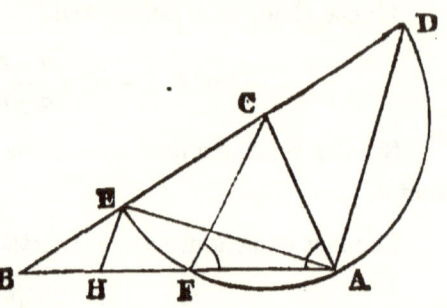

The angle EAD is a right angle; the angle AFC is equal to the angle CAF; the exterior angle ACD is equal to the angles CAB, ABC, that is, $A+B$. But the angle AEC is half of the angle ACD (*Euc.* III., 20), therefore the angle AEC is equal to $\frac{1}{2}(A+B)$. Again, the angle BCF is equal to the difference between the angles CFA and CBF (*Euc.* I., 32), that is, the angle BCF is equal to $A-B$; but the angle EAF is half of the angle ECF, therefore the angle EAF is equal to $\frac{1}{2}(A-B)$.

Now, $\quad AD = AE \tan AED = AE \tan \frac{1}{2}(A+B)$

and $\quad\quad EH = AE \tan EAF = AE \tan \frac{1}{2}(A-B)$,

and since EH is parallel to AD, we have

$$BD : BE :: AD : EH$$

that is, $a+b : a-b :: AE \tan \frac{1}{2}(A+B) : AE \tan \frac{1}{2}(A-B)$

$$:: \quad \tan \frac{1}{2}(A+B) : \quad \tan \frac{1}{2}(A-B)$$

therefore $\quad \tan \frac{1}{2}(A-B) = \dfrac{a-b}{a+b} \tan \frac{1}{2}(A+B)$.

OBLIQUE-ANGLED TRIANGLES.

122. To express the cosine of an Angle of a Triangle in terms of its sides.

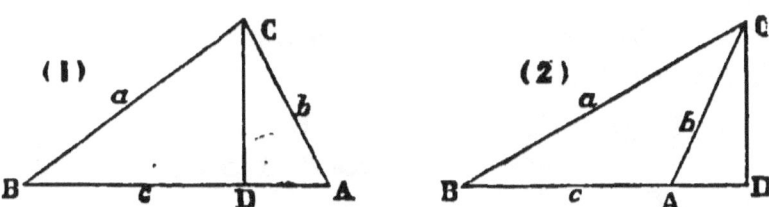

From the angle C of the triangle ABC, draw CD perpendicular to AB, produced if necessary.

From Fig. (1) we have by *Euc.* II., 13,

$$BC^2 = AC^2 + AB^2 - 2AB \cdot AD.$$

But $\qquad AD = AC \cos A,$
therefore $\qquad BC^2 = AC^2 + AB^2 - 2AB \cdot AC \cos A,$
that is, $\qquad a^2 = b^2 + c^2 - 2bc \cos A.$

Again, from Fig. (2) we have by *Euc.* II., 12,

$$BC^2 = AC^2 + AB^2 + 2AB \cdot AD.$$

But $\quad AD = AC \cos CAD = AC \cos(180° - A) = -AC \cos A,$
therefore $\qquad BC^2 = AC^2 + AB^2 - 2AB \cdot AC \cos A,$
that is, $\qquad a^2 = b^2 + c^2 - 2bc \cos A.$

Hence, in both cases, we have

$$a^2 = b^2 + c^2 - 2bc \cos A. \qquad (169)$$

In the same manner, we find

$$b^2 = a^2 + c^2 - 2ac \cos B. \qquad (170)$$

$$c^2 = a^2 + b^2 - 2ab \cos C. \qquad (171)$$

From the last three formulæ, we have

$$\cos A = \frac{b^2 + c^2 - a^2}{2bc}. \qquad (172)$$

$$\cos B = \frac{a^2 + c^2 - b^2}{2ac}. \qquad (173)$$

$$\cos C = \frac{a^2 + b^2 - c^2}{2ab}. \qquad (174)$$

123. The same results may be obtained independently of Euclid, as follows:

In Fig. (1) we have
$$c = BD + AD,$$
$$= a \cos B + b \cos A;$$

and from Fig. (2)
$$c = BD - AD,$$
$$= a \cos B - b \cos (180° - A),$$
$$= a \cos B + b \cos A;$$

and similarly for the other sides. Hence, in every triangle, we have

$$\left. \begin{array}{l} a = b \cos C + c \cos B. \\ b = c \cos A + a \cos C. \\ c = a \cos B + b \cos A. \end{array} \right\} \qquad (175)$$

Multiplying the first by a, the second by b, and the third by c, then subtracting the first result from the sum of the other two, we have

$$a^2 = b^2 + c^2 - 2bc \cos A,$$

whence $\cos A = \dfrac{b^2 + c^2 - a^2}{2bc}$, as before.

124. To express the Functions of half of an Angle of a Triangle in terms of its sides.

From (100) we have

$$2 \sin^2 \frac{A}{2} = 1 - \cos A,$$
$$= 1 - \frac{b^2 + c^2 - a^2}{2bc}, \text{ by (172)}$$
$$= \frac{2bc - b^2 - c^2 + a^2}{2bc},$$

$$= \frac{a^2 - (b-c)^2}{2bc},$$

$$= \frac{(a-b+c)(a+b-c)}{2bc}. \qquad (176)$$

This may be simplified by putting

$$a+b+c = 2s.$$

whence
$$-a+b+c = a+b+c - 2a = 2(s-a).$$
$$a-b+c = a+b+c - 2b = 2(s-b).$$
$$a+b-c = a+b+c - 2c = 2(s-c).$$

Hence (176) becomes, after dividing by 2 and extracting the square root,

$$\sin \frac{A}{2} = \sqrt{\frac{(s-b)(s-c)}{bc}}. \qquad (177)$$

In the same manner we find from (173) and (174)

$$\sin \frac{B}{2} = \sqrt{\frac{(s-a)(s-c)}{ac}}, \qquad (178)$$

$$\sin \frac{C}{2} = \sqrt{\frac{(s-a)(s-b)}{ab}}. \qquad (179)$$

From (99) we have

$$2 \cos^2 \frac{A}{2} = 1 + \cos A,$$

$$= 1 + \frac{b^2 + c^2 - a^2}{2bc}, \quad \text{by (172)}$$

$$= \frac{2bc + b^2 + c^2 - a^2}{2bc},$$

$$= \frac{(b+c)^2 - a^2}{2bc},$$

$$= \frac{(a+b+c)(-a+b+c)}{2bc},$$

$$= \frac{2s \cdot 2(s-a)}{2bc}.$$

Therefore, $$\cos \frac{A}{2} = \sqrt{\frac{s(s-a)}{bc}}. \tag{180}$$

Similarly, $$\cos \frac{B}{2} = \sqrt{\frac{s(s-b)}{ac}}. \tag{181}$$

$$\cos \frac{C}{2} = \sqrt{\frac{s(s-c)}{ab}}. \tag{182}$$

Dividing (177) by (180) we have

$$\tan \frac{A}{2} = \sqrt{\frac{(s-b)(s-c)}{s(s-a)}}, \tag{183}$$

and in a similar manner

$$\tan \frac{B}{2} = \sqrt{\frac{(s-a)(s-c)}{s(s-b)}}. \tag{184}$$

$$\tan \frac{C}{2} = \sqrt{\frac{(s-a)(s-b)}{s(s-c)}}. \tag{185}$$

125. To express the Sum and Difference of any two sides of a Triangle in terms of the other side and the opposite Angles.

From (165) we have

$$\frac{a+b}{c} = \frac{\sin A + \sin B}{\sin C},$$

$$= \frac{\sin \tfrac{1}{2}(A+B) \cos \tfrac{1}{2}(A-B)}{\sin \frac{C}{2} \cos \frac{C}{2}}, \text{ by (55) and (94)}$$

but $$A+B+C = 180°,$$

therefore $A+B = 180° - C$, and $\tfrac{1}{2}(A+B) = 90° - \frac{C}{2}$;

and $\sin \tfrac{1}{2}(A+B) = \cos \frac{C}{2}, \quad \cos \tfrac{1}{2}(A+B) = \sin \frac{C}{2}.$

Hence we have

$$\frac{a+b}{c} = \frac{\cos\frac{C}{2}\cos\frac{1}{2}(A-B)}{\cos\frac{1}{2}(A+B)\cos\frac{C}{2}},$$

$$= \frac{\cos\frac{1}{2}(A-B)}{\cos\frac{1}{2}(A+B)},$$

and $\qquad a+b = c \cdot \dfrac{\cos\frac{1}{2}(A-B)}{\cos\frac{1}{2}(A+B)}.$ (186)

In a similar manner we find

$$a-b = c \cdot \frac{\sin\frac{1}{2}(A-B)}{\sin\frac{1}{2}(A+B)}.$$ (187)

126. From (165) we have

$$\frac{a+b+c}{a} = \frac{\sin A + \sin B + \sin C}{\sin A},$$

$$= \frac{2\sin\frac{1}{2}(A+B)\cos\frac{1}{2}(A-B) + 2\sin\frac{1}{2}(A+B)\cos\frac{1}{2}(A+B)}{2\sin\frac{A}{2}\cos\frac{A}{2}},$$

$$= \frac{\sin\frac{1}{2}(A+B)\{\cos\frac{1}{2}(A-B) + \cos\frac{1}{2}(A+B)\}}{\sin\frac{A}{2}\cos\frac{A}{2}},$$

$$= \frac{2\cos\frac{C}{2}\cos\frac{A}{2}\cos\frac{B}{2}}{\sin\frac{A}{2}\cos\frac{A}{2}} = \frac{2\cos\frac{B}{2}\cos\frac{C}{2}}{\sin\frac{A}{2}}.$$ (188)

In a similar manner are obtained the following, which may be verified by the student:

$$\frac{-a+b+c}{a} = \frac{2\sin\frac{B}{2}\sin\frac{C}{2}}{\sin\frac{A}{2}}.$$ (189)

$$\frac{a-b+c}{a} = \frac{2\cos\frac{B}{2}\sin\frac{C}{2}}{\cos\frac{A}{2}}.$$ (190)

$$\frac{a+b-c}{a} = \frac{2 \sin \frac{B}{2} \cos \frac{C}{2}}{\cos \frac{A}{2}}. \tag{191}$$

If we write $2s$ for $(a+b+c)$, the last four formulæ become

$$\frac{s}{a} = \frac{\cos \frac{B}{2} \cos \frac{C}{2}}{\sin \frac{A}{2}}. \tag{192}$$

$$\frac{s-a}{a} = \frac{\sin \frac{B}{2} \sin \frac{C}{2}}{\sin \frac{A}{2}}. \tag{193}$$

$$\frac{s-b}{a} = \frac{\cos \frac{B}{2} \sin \frac{C}{2}}{\cos \frac{A}{2}}. \tag{194}$$

$$\frac{s-c}{a} = \frac{\sin \frac{B}{2} \cos \frac{C}{2}}{\cos \frac{A}{2}}. \tag{195}$$

From these we may at once deduce those for $\frac{s}{b}$, $\frac{s-c}{b}$, &c. Thus, writing b for a in (192) and (193) we have

$$\frac{s}{b} = \frac{\cos \frac{A}{2} \cos \frac{C}{2}}{\sin \frac{B}{2}}. \tag{196}$$

$$\frac{s-b}{b} = \frac{\sin \frac{A}{2} \sin \frac{C}{2}}{\sin \frac{B}{2}}. \tag{197}$$

&c.,   &c.

The product of (193) and (197) gives (179), &c.

AREA OF A TRIANGLE.

127. To express the sine of an Angle of a Triangle in terms of its sides.

From (94) we have

$$\sin A = 2 \sin \frac{A}{2} \cos \frac{A}{2},$$

$$= 2\sqrt{\frac{(s-b)(s-c)}{bc}} \cdot \sqrt{\frac{s(s-a)}{bc}}, \text{ by (177) and (180)}$$

$$= \frac{2}{bc} \sqrt{s(s-a)(s-b)(s-c)}. \qquad (198)$$

128. To express the Area of a Triangle in terms of its sides and angles.

Let Δ represent the area of a triangle; then, from the Figures of Art. 119, we have

$$\Delta = \tfrac{1}{2} AB \cdot CD, \text{ and } CD = AC \sin A.$$

Therefore $\qquad \Delta = \tfrac{1}{2} bc \sin A. \qquad (199)$

In a similar manner, we find

$$\Delta = \tfrac{1}{2} ac \sin B = \tfrac{1}{2} ab \sin C, \qquad (200)$$

that is, the area of a triangle is equal to half the product of any two sides into the sine of the included angle.

Again, from (199)

$$\Delta = \tfrac{1}{2} bc \sin A$$

$$= \frac{bc}{2} \cdot \frac{2}{bc} \sqrt{s(s-a)(s-b)(s-c)}, \text{ by (198)}$$

$$= \sqrt{s(s-a)(s-b)(s-c)}. \qquad (201)$$

From (165) we find $b = \dfrac{c \sin B}{\sin C}$, which substituted in (199) gives

$$\Delta = \frac{c^2 \sin A \sin B}{2 \sin C} = \frac{c^2 \sin A \sin B}{2 \sin (A+B)}. \qquad (202)$$

129. To express the Perpendicular of a Triangle in terms of the side on which it falls, and the Angles adjacent to that side.

Let p denote the perpendicular from C on the side c, then
$$pc = 2\Delta,$$
and
$$p = \frac{2\Delta}{c}, \qquad (203)$$

$$= \frac{c \sin A \sin B}{\sin (A+B)}, \quad \text{by (202)} \qquad (204)$$

If the triangle is right-angled at C, (204) becomes
$$p = c \sin A \sin B,$$
$$= \frac{c}{2} \sin 2A, \qquad (205)$$

since $\sin B = \cos A$.

Examples.

1. In any triangle, shew that $\dfrac{\sin^2 \frac{A}{2}}{a} + \dfrac{\sin^2 \frac{B}{2}}{b} = \dfrac{\cos^2 \frac{C}{2}}{s}$.

From Art. 124 we have
$$\frac{\sin^2 \frac{A}{2}}{a} = \frac{(s-b)(s-c)}{abc}$$

$$\frac{\sin^2 \frac{B}{2}}{b} = \frac{(s-a)(s-c)}{abc},$$

therefore $\dfrac{\sin^2 \frac{A}{2}}{a} + \dfrac{\sin^2 \frac{B}{2}}{b} = \dfrac{s-c}{abc}(s-a+s-b)$

$$= \frac{s-c}{ab}, \text{ since } 2s = a+b+c.$$

$$= \frac{s(s-c)}{ab} \cdot \frac{1}{s}$$

$$= \frac{\cos^2 \frac{C}{2}}{s}, \quad \text{by (182)}.$$

2. In any triangle, prove that
$$c^2 \cos(A-B) = 2ab + (a^2+b^2)\cos(A+B).$$

From (186) and (187) we have
$$c \cos \tfrac{1}{2}(A-B) = (a+b) \cos \tfrac{1}{2}(A+B)$$
$$c \sin \tfrac{1}{2}(A-B) = (a-b) \sin \tfrac{1}{2}(A+B).$$

Squaring and subtracting, we get
$$c^2 \{\cos^2 \tfrac{1}{2}(A-B) - \sin^2 \tfrac{1}{2}(A-B)\} = (a+b)^2 \cos^2 \tfrac{1}{2}(A+B)$$
$$- (a-b)^2 \sin^2 \tfrac{1}{2}(A+B)$$
$$c^2 \cos(A-B) = 2ab \{\cos^2 \tfrac{1}{2}(A+B) + \sin^2 \tfrac{1}{2}(A+B)\}$$
$$+ (a^2+b^2) \{\cos^2 \tfrac{1}{2}(A+B) - \sin^2 \tfrac{1}{2}(A+B)\}$$

or $\quad c^2 \cos(A-B) = 2ab + (a^2+b^2)\cos(A+B).$

3. In any triangle, prove that
$$c^2 = \frac{(a+b)^2 \sin^2 \frac{C}{2} - (a-b)^2 \cos^2 \frac{C}{2}}{\cos(A-B)}.$$

From the result of the last example we have
$$c^2 \cos(A-B) = 2ab - (a^2+b^2)\cos C,$$
since $(A+B)$ and C are supplementary angles.

Hence $\quad c^2 \cos(A-B) = 2ab\left(\cos^2 \frac{C}{2} + \sin^2 \frac{C}{2}\right)$
$$- (a^2+b^2)\left(\cos^2 \frac{C}{2} - \sin^2 \frac{C}{2}\right)$$
$$= (a+b)^2 \sin^2 \frac{C}{2} - (a-b)^2 \cos^2 \frac{C}{2},$$

therefore
$$c^2 = \frac{(a+b)^2 \sin^2 \frac{C}{2} - (a-b)^2 \cos^2 \frac{C}{2}}{\cos(A-B)}.$$

4. If the sides a, b, c of a triangle are in harmonical progression, prove that
$$\cos^2 \frac{B}{2} = \frac{\sin A \sin C}{\cos A + \cos C}.$$

Since a, b, c are in harmonical progression
$$b = \frac{2ac}{a+c}. \qquad (1)$$

As the sides of a triangle are proportional to the sines of the opposite angles (Art. 119), we evidently have from the above equation
$$\sin B = \frac{2 \sin A \sin C}{\sin A + \sin C}.$$

For (1) may be written
$$\frac{c}{b} = \frac{a+c}{2a} = \frac{1}{2}\left(\frac{a}{a} + \frac{c}{a}\right),$$

or
$$\frac{\sin C}{\sin B} = \frac{1}{2}\left(\frac{\sin A}{\sin A} + \frac{\sin C}{\sin A}\right), \quad \text{by (165)}$$

whence
$$\sin B = \frac{2 \sin A \sin C}{\sin A + \sin C}.$$

Hence
$$2 \sin \frac{B}{2} \cos \frac{B}{2} = \frac{2 \sin A \sin C}{2 \sin \frac{1}{2}(A+C) \cos \frac{1}{2}(A-C)},$$
$$= \frac{\sin A \sin C}{\cos \frac{B}{2} \cos \frac{1}{2}(A-C)},$$

therefore
$$\cos^2 \frac{B}{2} = \frac{\sin A \sin C}{2 \sin \frac{B}{2} \cos \frac{1}{2}(A-C)},$$

EXAMPLES. 143

$$= \frac{\sin A \sin C}{2 \cos \tfrac{1}{2}(A+C) \cos \tfrac{1}{2}(A-C)}$$

$$= \frac{\sin A \sin C}{\cos A + \cos C}, \quad \text{by (57)}$$

The student should notice the following transformations depending on (165), which will be found very useful in the solution of problems. Whenever the sides of a triangle and the sines of the opposite angles are homogeneously involved, and occur either in the numerator or denominator of a fraction, or on opposite sides of an equation, we may substitute the one for the other, as in the following examples:

5. In any triangle, prove that

$$(a^2 + b^2 + c^2) \sin A = a^2 \sin A + ab \sin B + ac \sin C.$$

Commencing with $\dfrac{a^2 + b^2 + c^2}{ab}$, we have

$$\frac{a^2 + b^2 + c^2}{ab} = \frac{a}{a} \cdot \frac{a}{b} + \frac{b}{a} \cdot \frac{b}{b} + \frac{c}{a} \cdot \frac{c}{b}$$

$$= \frac{\sin^2 A}{\sin A \sin B} + \frac{\sin^2 B}{\sin A \sin B} + \frac{\sin^2 C}{\sin A \sin B}$$

$$= \frac{1}{\sin A} \left(\frac{a \sin A}{b} + \frac{b \sin B}{b} + \frac{c \sin C}{b} \right),$$

whence $(a^2 + b^2 + c^2) \sin A = a^2 \sin A + ab \sin B + ac \sin C.$

6. Shew that the area of a triangle ABC

$$= \tfrac{1}{2}(b^2 - c^2) \frac{\sin A \sin B \sin C}{\sin^2 B - \sin^2 C}.$$

From (199) we have

$$\Delta = \tfrac{1}{2} bc \sin A$$

$$= \tfrac{1}{2}(b^2 - c^2) \frac{bc \sin A}{b^2 - c^2}$$

$$= \tfrac{1}{2}(b^2 - c^2) \frac{\sin A \sin B \sin C}{\sin^2 B - \sin^2 C}.$$

7. If the angles of a triangle ABC be bisected by the lines AD, BE, CF, meeting the opposite sides in the points D, E and F, shew that the area of the triangle ABC is to the area of the triangle DEF, as $(a+b)(a+c)(b+c) : 2abc$.

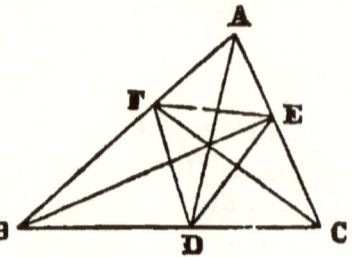

Here we have

$$1 = \frac{\sin BAD}{\sin DAC} = \frac{\sin BAD}{\sin ADB} \cdot \frac{\sin ADC}{\sin DAC}.$$

$$= \frac{BD}{AB} \cdot \frac{AC}{DC}, \quad \text{by (165)}$$

$$= \frac{BD}{DC} \cdot \frac{AC}{AB}$$

whence $\quad \dfrac{BD}{DC} = \dfrac{AB}{AC} = \dfrac{c}{b}$

and $\quad \dfrac{BD + DC}{BD} = \dfrac{b+c}{c}$

or $\quad \dfrac{a}{BD} = \dfrac{b+c}{c},$

therefore $\quad BD = \dfrac{ac}{b+c}, \quad \text{and} \quad DC = \dfrac{ab}{b+c},$

similarly $\quad EC = \dfrac{ab}{a+c}, \quad AE = \dfrac{bc}{a+c},$

$\quad AF = \dfrac{bc}{a+b}, \quad BF = \dfrac{ac}{a+b}.$

Let Δ denote the area of the triangle ABC, and δ that of the triangle DEF,

then $\quad \Delta = \tfrac{1}{2} bc \sin A, \quad$ whence $\quad \sin A = \dfrac{2\Delta}{bc}$

and $\quad \sin B = \dfrac{2\Delta}{ac}, \quad$ and $\quad \sin C = \dfrac{2\Delta}{ab}.$

Area of $AEF = \tfrac{1}{2} AE \cdot AF \sin A$,   by Art. 128

$$= \frac{\Delta bc}{(a+b)(a+c)}.$$

Area of $BFD = \tfrac{1}{2} BD \cdot BF \sin B$

$$= \frac{\Delta ac}{(a+b)(b+c)}.$$

Area of $CED = \tfrac{1}{2} CD \cdot CE \sin C$

$$= \frac{\Delta ab}{(a+c)(b+c)}.$$

Therefore the area of the three triangles, AEF, BDF, CED,

$$= \Delta \left(\frac{bc}{(a+b)(a+c)} + \frac{ac}{(a+b)(b+c)} + \frac{ab}{(a+c)(b+c)} \right),$$

and therefore

$$\delta = \Delta \left(1 - \frac{bc}{(a+b)(a+c)} - \frac{ac}{(a+b)(b+c)} - \frac{ab}{(a+c)(b+c)} \right)$$

$$= \frac{\Delta \cdot 2abc}{(a+b)(b+c)(a+c)},$$

or $\quad\quad \Delta : \delta :: (a+b)(b+c)(a+c) : 2abc.$

Examples.

In any triangle, right-angled at C, prove that

1. $\cos \dfrac{A}{2} = \sqrt{\dfrac{b+c}{2c}} \quad\quad \cos 2A = \dfrac{b^2 - a^2}{b^2 + a^2}.$

2. $\cos (A - B) = \dfrac{2ab}{c^2} \quad\quad \tan (A - B) = \dfrac{(a+b)(a-b)}{2ab}.$

3. $\tan 2A = \dfrac{2ab}{(b+a)(b-a)}. \quad\quad \tan 2A - \sec 2B = \dfrac{b+a}{b-a}.$

4. $\dfrac{a+b}{c} = \sqrt{2} \sin (A + 45°). \quad\quad c = \sqrt{2ab \operatorname{cosec} 2A}.$

5. $\Delta = \frac{1}{4}c^2 \sin 2A = \frac{b^2}{2} \tan A = \frac{1}{4}(a+b+c)(a+b-c)$.

In any triangle ABC, prove that

6. $\dfrac{\sin(A-B)}{\sin(A+B)} = \dfrac{a^2-b^2}{c^2}$. $\qquad \dfrac{\tan B}{\tan C} = \dfrac{a^2+b^2-c^2}{a^2-b^2+c^2}$.

7. $\dfrac{\text{versin } A}{\text{versin } B} = \dfrac{a(a-b+c)}{b(-a+b+c)}$. $\qquad \cos B + \sin B \cot C = \dfrac{a}{c}$.

8. $\tan \dfrac{A}{2} \tan \dfrac{B}{2} = \dfrac{s-c}{s}$. $\qquad \tan \dfrac{A}{2} \cot \dfrac{B}{2} = \dfrac{s-b}{s-a}$.

9. $\dfrac{\tan \dfrac{A}{2} + \tan \dfrac{B}{2}}{\tan \dfrac{A}{2} - \tan \dfrac{B}{2}} = \dfrac{c}{a-b}$. $\qquad \dfrac{\cot \dfrac{B}{2} + \cot \dfrac{C}{2}}{\cot \dfrac{A}{2}} = \dfrac{a}{s-a}$.

10. $\cos A + \cos B = 2\dfrac{a+b}{c} \sin^2 \dfrac{C}{2}$. $\qquad \tan \dfrac{A}{2} \tan \dfrac{B}{2} \tan \dfrac{C}{2} = \dfrac{\Delta}{s^2}$.

11. $\cot \dfrac{A}{2} + \cot \dfrac{B}{2} + \cot \dfrac{C}{2} = \dfrac{s^2}{\Delta} = \cot \dfrac{A}{2} \cot \dfrac{B}{2} \cot \dfrac{C}{2}$.

12. $\dfrac{\cot \dfrac{A}{2} + \cot \dfrac{C}{2}}{\cot \dfrac{B}{2} + \cot \dfrac{C}{2}} = \dfrac{b}{a}$. $\qquad b \cos B + c \cos C = a \cos(B-C)$.

13. $a \cos A + b \cos B + c \cos C = 2a \sin B \sin C$.

14. $(b+c) \cos A + (a+c) \cos B + (a+b) \cos C = a+b+c$.

15. $(s-a) \tan \dfrac{A}{2} = (s-b) \tan \dfrac{B}{2} = (s-c) \tan \dfrac{C}{2}$.

16. $\dfrac{b \cos C - c \cos B}{b-c} = \dfrac{b+c}{a}$.

17. $\cot A + \cot B = \dfrac{c}{a} \csc B$.

18. $c^2 = (a+b)^2 \sin^2 \dfrac{C}{2} + (a-b)^2 \cos^2 \dfrac{C}{2}$.

19. $\tfrac{1}{2}(a^2 + b^2 + c^2) = bc \cos A + ac \cos B + ab \cos C$.

20. $s(\cos A + \cos B + \cos C) = a \cos^2 \dfrac{A}{2} + b \cos^2 \dfrac{B}{2} + c \cos^2 \dfrac{C}{2}$.

21. $(b^2 - c^2) \cot A + (c^2 - a^2) \cot B + (a^2 - b^2) \cot C = 0$.

22. $\cos A + \cos B + \cos C = 1 - \dfrac{a}{s} \sin B \sin C$.

23. $(b-c) \cot \dfrac{A}{2} + (c-a) \cot \dfrac{B}{2} + (a-b) \cot \dfrac{C}{2} = 0$.

24. $\dfrac{\sin A + \sin C}{\cos A + \cos C} = \dfrac{\cos C - \cos A}{\sin A - \sin C} = \cot \dfrac{B}{2}$.

25. $\dfrac{\sin(A-B)}{\sin(B-C)} = \dfrac{a^2 - b^2}{b^2 - c^2} \cdot \dfrac{a}{c}$.

26. $a^2 \sin 2B + b^2 \sin 2A = 2ab \sin C$.

27. $\dfrac{a}{b} = \dfrac{\cos A \cos C + \cos B}{\cos B \cos C + \cos A}$.

28. $\dfrac{a^2 - b^2}{c^2} \sin C + \dfrac{b^2 - c^2}{a^2} \sin A + \dfrac{c^2 - a^2}{b^2} \sin B + 4 \sin \tfrac{1}{2}(A-B) \sin \tfrac{1}{2}(B-C) \sin \tfrac{1}{2}(C-A) = 0$.

29. $\dfrac{\cot \dfrac{A}{2} + \cot \dfrac{B}{2} + \cot \dfrac{C}{2}}{\sec \dfrac{A}{2} \sec \dfrac{B}{2} \sec \dfrac{C}{2}} = \dfrac{s^3}{abc}$.

30. $\dfrac{\cos^2 \dfrac{A}{2}}{a} + \dfrac{\cos^2 \dfrac{B}{2}}{b} + \dfrac{\cos^2 \dfrac{C}{2}}{c} = \dfrac{s^2}{abc}$.

31. $\cot B - \cot A = \dfrac{a^2 - b^2}{ab} \operatorname{cosec} C$.

32. $\sin\left(\dfrac{A}{2}+B\right)\operatorname{cosec}\dfrac{A}{2}=\dfrac{b+c}{a}$.

33. In any triangle, if $\cot\dfrac{A}{2}$, $\cot\dfrac{B}{2}$, $\cot\dfrac{C}{2}$ are in arithmetical progression, shew that $\cot\dfrac{A}{2}\cot\dfrac{C}{2}=3$; and if $\cot A$, $\cot B$, $\cot C$ are in arithmetical progression, shew that the squares of the sides are also in arithmetical progression.

34. If the line CE which bisects the angle C, of any triangle, meet the base in E, shew that

$$CE=\dfrac{2ab}{a+b}\cos\dfrac{C}{2},\quad \tan AEC=\dfrac{a+b}{a-b}\tan\dfrac{C}{2},$$

and $\quad\cos B-\cos A=2\cos\dfrac{C}{2}\cos AEC$.

35. In a triangle ABC, if $\sin A$, $\sin B$, $\sin C$ are in harmonical progression, so also are the versed sines.

36. In any triangle ABC, if $\tan\dfrac{A}{2}$, $\tan\dfrac{B}{2}$, $\tan\dfrac{C}{2}$ are in arithmetical progression, so also are $\cos A$, $\cos B$, $\cos C$.

37. If p, q, r denote the lines drawn from the angles of a triangle bisecting them and terminated by the opposite sides, shew that

$$\dfrac{\cos\dfrac{A}{2}}{p}+\dfrac{\cos\dfrac{B}{2}}{q}+\dfrac{\cos\dfrac{C}{2}}{r}=\dfrac{1}{a}+\dfrac{1}{b}+\dfrac{1}{c}.$$

38. In any triangle prove that

$$\Delta=s(s-a)\tan\dfrac{A}{2}.$$

39. $\Delta=\dfrac{abc}{s}\cos\dfrac{A}{2}\cos\dfrac{B}{2}\cos\dfrac{C}{2}$.

40. $\Delta = 2s^2 \dfrac{\sin A \sin B \sin C}{(\sin A + \sin B + \sin C)^2}$

41. $\Delta = \dfrac{s^2(\cos A + \cos B + \cos C - 1)}{\sin A + \sin B + \sin C}$.

42. $\Delta = \tfrac{1}{2}(a^2 - b^2) \dfrac{\sin A \sin B}{\sin (A - B)}$.

43. $\Delta = \dfrac{a^2 b^2 (\cot A + \cot B)}{2c^2 \operatorname{cosec}^2 C}$.

44. $\Delta = \dfrac{a^2 + b^2 - c^2}{4 \tan \tfrac{1}{2}(A + B - C)}$.

45. $\Delta = \dfrac{a^2 + b^2 + c^2}{4(\cot A + \cot B + \cot C)}$.

46. $\Delta = \dfrac{(a + b + c)^2}{4(\cot \tfrac{A}{2} + \cot \tfrac{B}{2} + \cot \tfrac{C}{2})}$.

47. $\Delta = \dfrac{ab + ac + bc}{2(\operatorname{cosec} A + \operatorname{cosec} B + \operatorname{cosec} C)}$.

48. $\Delta = \dfrac{c^2}{2(\cot A + \cot B)}$.

49. $\Delta = \tfrac{1}{4}(a^2 \sin 2B + b^2 \sin 2A)$.

50. $\Delta = s^2 \tan \dfrac{A}{2} \tan \dfrac{B}{2} \tan \dfrac{C}{2}$.

51. In any triangle, if p_1 be the perpendicular from the angle A upon the side a, shew that

$$p_1 = \dfrac{a + b + c}{\cot \dfrac{B}{2} + \cot \dfrac{C}{2}} = \dfrac{b^2 \sin C + c^2 \sin B}{b + c}.$$

52. If perpendiculars be drawn from the angles of a triangle to the opposite sides, shew that the sides of the triangle formed by joining the feet of these perpendiculars are $a \cos A$, $b \cos B$ and $c \cos C$ and that its area $= \frac{1}{2} ab \cos A \cos B \sin 2C$.

53. If the sides of a triangle are 3, 4 and 5, shew that the cotangents of the semi-angles of the triangles are 3, 2 and 1.

54. If perpendiculars be drawn from the angles of a triangle to the opposite sides, shew that the products of the alternate segments of the sides thus made $= abc \cos A \cos B \cos C$.

55. If p_1, p_2, p_3 be the perpendiculars drawn from the angles A, B, C of a triangle ABC to the opposite sides a, b, c respectively, shew that
$$a = \frac{p_1 \sin A + p_2 \sin B + p_3 \sin C}{\sin B \sin C}.$$

56. In any triangle shew that
$$(\sin A + \sin B + \sin C)^2 = \frac{2s^2}{abc}(a \cos A + b \cos B + c \cos C).$$
$$(a \sin A + b \sin B + c \sin C)^2 = (a^2 + b^2 + c^2)(\sin^2 A + \sin^2 B + \sin^2 C).$$

57. In any triangle, if $\dfrac{a}{c} = \dfrac{\sin(A - B)}{\sin(B - C)}$, shew that a^2, b^2, c^2 are in arithmetical progression.

58. In any triangle shew that the distance from the middle of the base (c) to the foot of the perpendicular drawn from the angle C to the opposite side, is $\dfrac{c}{2} \cdot \dfrac{\tan A - \tan B}{\tan A + \tan B}$.

59. In any triangle shew that $\dfrac{\sin(A - B)}{\sin C} = \dfrac{a^2 - b^2}{c^2}$.

60. In any triangle prove that
$$a \sin(B - C) + b \sin(C - A) + c \sin(A - B) = 0.$$

CHAPTER IX.

ON THE SOLUTION OF TRIANGLES.

RIGHT-ANGLED TRIANGLES.

130. The various cases of right-angled triangles are solved by (158)–(164).

Ex. 1.—Given $A = 65° \ 17' \ 20''$ and $c = 17.216$, to find the other parts. (See Fig. Art. 114.)

From the first of (158) we have

$$a = c \sin A,$$

or in logarithms
$$\log a = \log c + \text{Log} \sin A - 10.$$

$$\log c = 1.235932$$
$$\text{Log} \sin A = 9.958290$$

$$\log a = 1.194222$$

therefore $\quad a = 15.6395.$

Again, from the second of (158) we have

$$b = c \cos A,$$

or
$$\log b = \log c + \text{Log} \cos A - 10.$$

$$\text{Log } c = 1.235932$$
$$\text{Log} \cos A = 9.621221$$

$$\log b = 0.857153$$

therefore $\quad b = 7.197$

and $\quad B = 90° - A = 24° \ 42' \ 40''.$

Ex. 2.—Given $a = .1799$ and $b = .2465$, to find the other parts.

From the third of (158) we have

$$\tan A = \frac{a}{b},$$

or
$$\text{Log} \tan A = \log a - \log b + 10.$$

$$\text{Log } a = \overline{1}.255031$$
$$\log b = \overline{1}.391817$$

$$\text{Log} \tan A = 9.863214$$

therefore $\qquad A = 36°\ 7'\ 21''.5$
and $\qquad B = 53°\ 52'\ 38''.5.$

From the first of (158) we have

$$c = \frac{a}{\sin A},$$

or
$$\log c = \log a - \text{Log} \sin A + 10.$$

$$\log a = \overline{1}.255031$$
$$\text{Log} \sin A = 9.770495$$

$$\log c = \overline{1}.484536$$

therefore $\qquad c = .305166.$

Examples.

3. Given $c = 332.49$ and $a = 98.399$, to find the angles and the base. $\qquad$ *Ans.* $A = 17°\ 12'\ 51''$, $b = 317.6.$

4. Given $b = 374$ and $B = 52°\ 40'\ 18''$, to find the other parts.
$$\textit{Ans. } a = 285.2,\ c = 470.34.$$

5. Given $A = 25°\ 18'\ 48''$ and $a = 8.5623$, to find b.
$$\textit{Ans. } b = 18.1028.$$

6. Given $a = 48$ and $b = 36$, to find A.
$$\textit{Ans. } A = 53°\ 7'\ 48''.4.$$

RIGHT-ANGLED TRIANGLES. 153

7. Given $c = 197.01$ and $a = 196.64$, to find A.

By (159).
$$\log (c - a) = \overline{1}.568202$$
$$\log 2c = 2.595518$$
$$2) \overline{16.972684}$$
$$\text{Log sin} \left(45° - \frac{A}{2}\right) = 8.486342$$

therefore $\quad 45° - \dfrac{A}{2} = 1° \ 45' \ 21''.7$

and $\quad \dfrac{A}{2} = 43° \ 14' \ 38''.3$

$\quad A = 86° \ 29' \ 16''.6.$

8. Given $a = 984.1$ and $b = .3214$, to find A and c.

By (162).
$$\log (a - b) = 2.992919$$
$$\log (a + b) = 2.993203$$
$$\text{Log tan} (A - 45°) = 9.999716$$

therefore $\quad A - 45° = 44° \ 58' \ 52''.7$

and $\quad A = 89° \ 58' \ 52''.7.$

By (163).
$$\log a = 2.993039$$
$$\text{Log tan} \frac{A}{2} = 9.999858$$
$$\log (c - b) = 2.992897$$
$$c - b = 983.78$$

therefore $\quad c = 984.1014.$

9. Given $a = 4064$ and $A = 89° \ 58' \ 20''$, to find b and c.

By (163).
$$\log a = 3.608954$$
$$\text{Log tan} \frac{A}{2} = 9.999789$$
$$\log (c - b) = 3.608743$$
$$c - b = 4062.027$$

By (164).

$$\log a = 3.608954$$
$$\text{Log cot } \frac{A}{2} = 10.000211$$
$$\log (c+b) = 3.609165$$
$$c + b = 4065.98$$

therefore $b = 1.9765$ and $c = 4064.0035$.

10. Given $b = 100.56$ and $c = 100.64$, to find A.

Ans. $A = 2° \ 17' \ 5''.3$.

11. Given $c = 270$ and $A = 18' \ 40''$, to find b.

Ans. $b = 269.9960197$.

12. Given $A = 89° \ 20' \ 40''$ and $c = 10$, to find a.

Ans. $a = 9.9993674$.

OBLIQUE-ANGLED TRIANGLES.

131. The relation investigated in Art. 119, between the sides of any triangle and the sines of the opposite angles, furnishes us with two equations involving the three sides and the three angles, viz.:

$$\frac{\sin A}{a} = \frac{\sin B}{b} = \frac{\sin C}{c}.$$

We also have

$$A + B + C = 180°.$$

It is evident, then, that if any three of these quantities, except the three angles, be given, the other three may be found; for on substituting the three given parts we would have three equations, containing only three unknown quantities, which can therefore be found by solving the equations.

If the three angles only are given, we can find the ratios, but not the magnitudes of the sides, from the equations

$$\frac{a}{b} = \frac{\sin A}{\sin B}, \quad \frac{a}{c} = \frac{\sin A}{\sin C}.$$

OBLIQUE-ANGLED TRIANGLES. 155

Hence it is evident that *one* side, at least, or what is equivalent to one side, must be given, in order to solve any triangle.

The only cases then which can occur in oblique-angled triangles are the following:

(1) Given one side and two angles.
(2) Given two sides and an angle opposite to one of them.
(3) Given two sides and an included angle.
(4) Given the three sides.

We now proceed to the solution of the different cases.

CASE 1.—*Given one side and two angles, or a, B and C, to find the other parts.*

First Solution.

We have $\quad A = 180° - (B + C).$

From (165) we have

$$b = \frac{a \sin B}{\sin A} = a \sin B \operatorname{cosec} A,$$

or in logarithms

$$\log b = \log a + \operatorname{Log} \sin B + \operatorname{Log} \operatorname{cosec} A - 20.$$

and $\quad c = \dfrac{a \sin C}{\sin A} = a \sin C \operatorname{cosec} A.$

or $\quad \log c = \log a + \operatorname{Log} \sin C + \operatorname{Log} \operatorname{cosec} A - 20.$

Examples.

1. Given $a = 512.24$, $B = 54° \ 7' \ 35''$ and $C = 71° \ 1' \ 41''$, to find b, c and A.

$$A = 180° - (B + C)$$
$$= 180° - 125° \ 9' \ 16'' = 54° \ 50' \ 44''$$

$\log a =$ 2.709473	$\log a =$ 2.709473
$\operatorname{Log} \sin B =$ 9.908652	$\operatorname{Log} \sin C =$ 9.975744
$\operatorname{Log} \operatorname{cosec} A =$ 10.087457	$\operatorname{Log} \operatorname{cosec} A =$ 10.087457
$\log b =$ 2.705582	$\log c =$ 2.772674
$b =$ 507.67	$c =$ 592.58

2. Given $c = 71.984$, $B = 61°$, and $C = 58° \ 14'$, to find a and b. *Ans.* $a = 73.8838$, $b = 74.05$.

3. Given $A = 35° \ 42'$, $B = 76° \ 27'$ and $c = 142$, to find a and b. *Ans.* $a = 89.47$, $b = 149.05$.

132. Second Solution.

Given c, A and B, to find the other parts.

$$C = 180 - (A + B).$$

By (186) and (187)

$$a + b = c \frac{\cos \tfrac{1}{2}(A - B)}{\cos \tfrac{1}{2}(A + B)} = c \frac{\cos \tfrac{1}{2}(A - B)}{\sin \tfrac{C}{2}},$$

and

$$a - b = c \frac{\sin \tfrac{1}{2}(A - B)}{\sin \tfrac{1}{2}(A + B)} = c \frac{\sin \tfrac{1}{2}(A - B)}{\cos \tfrac{C}{2}},$$

then $a = \tfrac{1}{2}(a+b) + \tfrac{1}{2}(a-b)$ and $b = \tfrac{1}{2}(a+b) - \tfrac{1}{2}(a-b)$.

Examples.

1. Given $c = .4781$, $A = 70°$ and $B = 60° \ 40'$, to find a and b.

$$\tfrac{1}{2}(A - B) = 4° \ 40', \quad \frac{C}{2} = 24° \ 40'$$

$\log c =$ $\overline{1}.679518$	$\log c =$ $\overline{1}.679518$
Log cos $\tfrac{1}{2}(A - B) =$ 9.998558	Log sin $\tfrac{1}{2}(A - B) =$ 8.910404
Log cosec $\tfrac{C}{2} =$ 10.379512	Log sec $\tfrac{C}{2} =$ 10.041555
$\log (a+b) =$ 0.057588	$\log (a-b) =$ $\overline{2}.631477$
$a+b =$ 1.14179	$a-b =$ .04280

therefore $a = .59229$ and $b = .54949$.

2. Given $c = 100$, $A = 72°$ and $B = 30°$, to find a and b. *Ans.* $a = 97.23$, $b = 51.118$.

OBLIQUE-ANGLED TRIANGLES.

133. CASE II.—*Given two sides and an angle opposite to one of them, or a, b and A.*

From (165) we have

$$\sin B = \frac{b}{a} \sin A,$$

or in logarithms

$$\text{Log sin } B = \log b - \log a + \text{Log sin } A,$$

then $\qquad C = 180° - (A + B),$

and (165) $\qquad c = a \dfrac{\sin C}{\sin A} = a \sin C \operatorname{cosec} A,$

or $\qquad \log c = \log a + \text{Log sin } C + \text{Log cosec } A - 20.$

Since the sine of an angle is equal to the sine of its supplement, the equation

$$\sin B = \frac{b}{a} \sin A$$

cannot without other conditions, determine whether B is less or greater than 90°.

Hence this is called the *ambiguous* case, for it is evident that there may be two triangles which will fulfil the conditions of the problem. The ambiguity, however, does not always exist, as we now proceed to shew.

1st. When $A < 90°$.

Let CAE be the given angle A. Take AC equal to b, draw CD perpendicular to AE and denote it by p. Then from the right-angled triangle ACD we have

$$p = b \sin A.$$

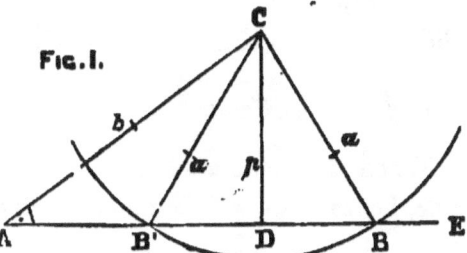

Fig. I.

With C as a centre and radius equal to a, describe a circle.

Now, if $a > p$ and $a < b$, the circle will cut AE in two points, B and B', on the *same* side of A, and there will be *two* solutions, since each of the triangles ABC, $AB'C$ fulfils the required conditions.

If $a > p$ and $a > b$, there will be but *one* solution; for, although there are two triangles ABC, $AB'C$, the latter is excluded by the condition that the given angle A is acute, while CAB' is obtuse, and there remains therefore but one triangle which satisfies the conditions of the problem.

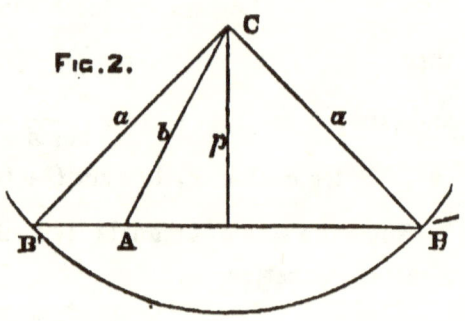
Fig. 2.

If $a = p$ there will be but one solution, and the triangle will be right-angled at B, as is evident from the figure.

If $a < p$ there will evidently be no solution, for the circle will neither intersect the opposite side nor be tangent to it.

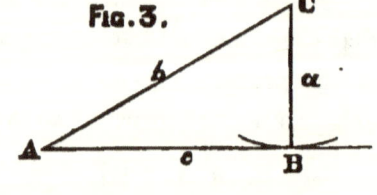
Fig. 3.

If $a = b$ there will be but one solution, and the triangle ABC will be isosceles.

2nd. When $A > 90°$.

When the given angle is obtuse the other angles must be *acute*, hence there will be no ambiguity, and a must be always greater than b. (*Euc.* I., 18.)

OBLIQUE-ANGLED TRIANGLES.

Examples.

1. Given $a = 925$, $b = 1256$ and $A = 30° \ 25'$, to find B, C and c.

or
$$p = b \sin A$$
$$\log p = \log b + \text{Log} \sin A - 10.$$

$$\log b = 3.098990$$
$$\text{Log} \sin A = 9.704395$$

$$\log p = 2.803385$$

therefore $\quad p = 635.9$

Since $a > p$ and $a < b$, there are two solutions.

To find B we have

$$\log b = 3.098990$$
$$\text{Log} \sin A = 9.704395$$
$$\text{ar. comp. } \log a = 7.033858 \quad \text{(Art. 106)}$$

$$\text{Log} \sin B = 9.837243$$

therefore $\quad B = 43° \ 25' \ 43''.5$
and $\quad B' = 136° \ 34' \ 16''.5$

Hence, (Fig. 1) $ACB = 180° - (A + B) = 106° \ 9' \ 16''.5$
and $\quad ACB' = 180° - (A + B') = 13° \ 0' \ 43''.5.$

To find c, we have

$\log a = 2.966142$	$\log a = 2.966142$
Log sin $ACB = 9.982504$	Log sin $ACB' = 9.352483$
Log cosec $A = 10.295605$	Log cosec $A = 10.295605$
$\log c = 3.244251$	$\log c = 2.614530$
$c = 1754.9$	$c = 411.65$

2. Given $a = .5412$, $b = .308$ and $A = 36° \ 50' \ 44''$, to find B and c. (Fig. 2.) $\quad Ans. \ B = 19° \ 57' \ 15''$, $c = .75519$.

3. Given $A = 107° \ 40'$, $a = 5.32$ and $b = 3.58$, to find B, C and c.   Ans. $B = 39° \ 52' \ 52''$, $C = 32° \ 27' \ 8''$, $c = 2.996$.

4. Given $A = 27° \ 44'$, $a = 17$ and $b = 40.25$. to find the other parts.   Ans. Impossible.

5. Given $A = 52° \ 19'$, $a = 325$ and $b = 333$, to find the other parts.

Ans. $B = 54° \ 10' \ 56''$ or $125° \ 49' \ 4''$,
$C = 73° \ 30' \ 4''$ or $1° \ 51' \ 56''$,
$c = 393.755$ or 13.3668.

134. CASE III.—*Given two sides and the included angle, or a, b and C.*

First Solution.

To find A and B, we have from (167)

$$\tan \tfrac{1}{2}(A - B) = \frac{a - b}{a + b} \cot \frac{C}{2}$$

or,   Log tan $\tfrac{1}{2}(A - B) = \log (a - b) - \log (a + b) + \text{Log cot} \frac{C}{2}$,

then $\quad A = \tfrac{1}{2}(A + B) + \tfrac{1}{2}(A - B)$
and $\quad B = \tfrac{1}{2}(A + B) - \tfrac{1}{2}(A - B)$.

To find c, we have from (165)

$$c = \frac{a \sin C}{\sin A}.$$

or $\quad \log c = \log a + \text{Log sin } C - \text{Log sin } A$.

Examples

1. Given $A = 176$, $b = 133$ and $C = 73°$, to find the remaining parts.

Here we have $a + b = 309$, $a - b = 43$ and $\dfrac{C}{2} = 36° \ 30'$

OBLIQUE-ANGLED TRIANGLES.

By (167).
log $(a-b)$ = 1.633468
ar. co. log $(a+b)$ = 7.510042
Log cot $\frac{C}{2}$ = 10.130791

Log tan $\frac{1}{2}(A-B)$ = 9.274301
therefore $\frac{1}{2}(A-B) = 10°\ 39'\ 2''.7$
and $\frac{1}{2}(A+B) = 53°\ 30'\ 0''$
hence $A = 64°\ 9'\ 2''.7$
$B = 42°\ 50'\ 57''.3$

By (165).
log a = 2.245513
Log sin C = 9.980596
———
12.226109
Log sin A = 9.954216
———
log c = 2.271893
therefore c = 187.022

2. Given $a = 16.86$, $b = 9.60$ and $C = 128°\ 4'$, to find the remaining parts.

$$Ans.\ A = 33°\ 34'\ 40'',$$
$$B = 18°\ 21'\ 20'',$$
$$C = 24.$$

When A and B have been found by the above method, c may be found by (186) or (187), from which we have

$$c = (a+b)\frac{\cos\frac{1}{2}(A+B)}{\cos\frac{1}{2}(A-B)}$$

and

$$c = (a-b)\frac{\sin\frac{1}{2}(A+B)}{\sin\frac{1}{2}(A-B)}.$$

135. Second Solution.—*Given a, b and C, to find the remaining parts.*

When the logarithms of a and b are given, which often occurs in the computation of a series of triangles, we may find A and B by (168) or (168 *bis*).

Example.

Given log $a = 2.245513$, log $b = 2.123852$ and $C = 73°$, to find the remaining parts. (Same as Ex. 1, Art. 134.)

By (168).

$\log b = 2.123852$
$\log a = 2.245513$
$\overline{}$
Log tan θ = 9.878339
$\theta = 37° \ 4' \ 39".7$

$45° - \theta = 7° \ 55' \ 20".3$

Log tan $(45° - \theta) = 9.143510$

Log cot $\dfrac{C}{2} = 10.130791$
$\overline{}$
Log tan $\tfrac{1}{2}(A - B) = 9.274301$
$\tfrac{1}{2}(A - B) = 10° \ 39' \ 2".7$

By (168 *bis*).

$\log b = 2.123852$
$\log a = 2.245513$
$\overline{}$
Log cos ϕ = 9.878339
$\phi = 40° \ 54' \ 54"$

$\dfrac{\phi}{2} = 20° \ 27' \ 27"$

2 Log tan $\dfrac{\phi}{2} = 9.143510$

Log cot $\dfrac{C}{2} = 10.130791$
$\overline{}$
Log tan $\tfrac{1}{2}(A - B) = 9.274301$
$\tfrac{1}{2}(A - B) = 10° \ 39' \ 2".7$

the same as by (167).

These methods are quite as short in practice as that of the preceding Article. The latter, however, should not be used when a is nearly equal to b. (Art. 115.)

136. Third Solution.—*Given a, b and C, to find c directly without finding the angles A and B.*

From (171) we have

$$c^2 = a^2 + b^2 - 2\,ab \cos C.$$

This, however, is not adapted to logarithmic computation. In the following forms it is easily computed by logarithms:

From (100) we have

$$\cos C = 1 - 2 \sin^2 \frac{C}{2},$$

then
$$c^2 = a^2 + b^2 - 2ab + 4ab \sin^2 \frac{C}{2}$$
$$= (a - b)^2 + 4ab \sin^2 \frac{C}{2}$$

$$= (a-b)^2 \left\{ 1 + \frac{4ab \sin^2 \frac{C}{2}}{(a-b)^2} \right\}.$$

Let
$$\tan^2 \theta = \frac{4ab \sin^2 \frac{C}{2}}{(a-b)^2}.$$

or
$$\tan \theta = \frac{2\sqrt{ab} \sin \frac{C}{2}}{a-b}, \qquad (206)$$

then we have
$$c^2 = (a-b)^2 (1 + \tan^2 \theta)$$
$$= (a-b)^2 \sec^2 \theta$$

and
$$c = (a-b) \sec \theta. \qquad (207)$$

When a is very nearly equal to b, the denominator $a-b$ will be very small and θ will be near $90°$; in that case the following form will be preferable:

$$c^2 = a^2 + b^2 - 2ab \cos C$$
$$= a^2 + b^2 - 2ab \left(2\cos^2 \frac{C}{2} - 1 \right) \quad \text{by (99)}$$
$$= (a+b)^2 - 4ab \cos^2 \frac{C}{2}$$
$$= (a+b)^2 \left\{ 1 - \frac{4ab \cos^2 \frac{C}{2}}{(a+b)^2} \right\}$$

Now, since the first member of this equation is positive, the second member must be so likewise, therefore $\dfrac{4ab \cos^2 \frac{C}{2}}{(a+b)^2}$ must be *less* than 1.

Hence we may assume
$$\sin^2 \theta = \frac{4ab \cos^2 \frac{C}{2}}{(a+b)^2}$$

164   PLANE TRIGONOMETRY.

or
$$\sin \theta = \frac{2\sqrt{ab}\cos\frac{C}{2}}{(a+b)}; \qquad (208)$$

then we have
$$c^2 = (a+b)^2(1-\sin^2\theta)$$
$$= (a+b)^2\cos^2\theta$$

and
$$c = (a+b)\cos\theta. \qquad (209)$$

Example.

Given $a = 176$, $b = 133$ and $C = 73°$, to find c. (Same as Ex. 1, Art. 134.)

By (208) and (209).

```
   log a = 2.245513              log (a+b) = 2.489958
   log b = 2.123852              Log cos θ = 9.781934
   ─────────────                 ─────────────────────
   2)4.369365                    log c = 2.271892
                                 c = 187.022
   log √ab = 2.184682
   log 2 = 0.301030
   Log cos C/2 = 9.905179        The remaining angles may
                                 now be found by (165).
   2.390891
   log (a+b) = 2.489958
   ─────────────
   Log sin θ = 9.900933
   θ = 52° 45′ 12″.
```

137. CASE IV.—*Given the three sides, or a, b and c, to find A, B, C.*

First Solution.

By (198) we have

$$\sin A = \frac{2}{bc}\sqrt{s(s-a)(s-b)(s-c)}, \qquad (210)$$

which may be used when A is not near 90°. Similar formulæ of course exist for $\sin B$ and $\sin C$.

Second Solution.

From Art. 124 we have

$$\left.\begin{aligned} \sin\frac{A}{2} &= \sqrt{\frac{(s-b)(s-c)}{bc}}. \\ \sin\frac{B}{2} &= \sqrt{\frac{(s-a)(s-c)}{ac}}. \\ \sin\frac{C}{2} &= \sqrt{\frac{(s-a)(s-b)}{ab}}. \end{aligned}\right\} \quad (211)$$

Third Solution.

$$\left.\begin{aligned} \cos\frac{A}{2} &= \sqrt{\frac{s(s-a)}{bc}}. \\ \cos\frac{B}{2} &= \sqrt{\frac{s(s-b)}{ac}}. \\ \cos\frac{C}{2} &= \sqrt{\frac{s(s-c)}{ab}}. \end{aligned}\right\} \quad (212)$$

Fourth Solution.

$$\left.\begin{aligned} \tan\frac{A}{2} &= \sqrt{\frac{(s-b)(s-c)}{s(s-a)}}. \\ \tan\frac{B}{2} &= \sqrt{\frac{(s-a)(s-c)}{s(s-b)}}. \\ \tan\frac{C}{2} &= \sqrt{\frac{(s-a)(s-b)}{s(s-c)}}. \end{aligned}\right\} \quad (213)$$

in which $s = \frac{1}{2}(a+b+c)$.

When all the angles are required, the last group will be the most convenient, since only four different logarithms are required from the tables. The first group is to be preferred when half the angle is *less* than 45°, and the second when it is *greater* than 45°. The third group is accurate for all angles.

Examples.

1. The sides of a triangle are 13, 14 and 15; find the angles.

Let $a = 13$, $b = 14$ and $c = 15$; then

$$s = 21, \ s - a = 8, \ s - b = 7, \ s - c = 6.$$

To find A.

From the first of the last group, we have

$$\begin{aligned}
\log (s - b) &= 0.845098 \\
\log (s - c) &= 0.778151 \\
\text{ar. co. } \log s &= 8.677781 \quad \text{(Art. 106)} \\
\text{ar. co. } \log (s - a) &= 9.096910 \\
\hline
2)&19.397940 \\
\log \tan \tfrac{A}{2} &= 9.698970
\end{aligned}$$

therefore $\quad \tfrac{A}{2} = 26° \ 33' \ 54''$

and $\quad A = 53° \ 7' \ 48''.$

To find B.

$$\begin{aligned}
\log (s - a) &= 0.903090 \\
\log (s - c) &= 0.778151 \\
\text{ar. co. } \log s &= 8.677781 \\
\text{ar. co. } \log (s - b) &= 9.154902 \\
\hline
2)&19.513924 \\
\log \tan \tfrac{B}{2} &= 9.756962
\end{aligned}$$

therefore $\quad \tfrac{B}{2} = 29° \ 44' \ 41''.6$

and $\quad B = 59° \ 29' \ 23''.2.$

To find C.

$$\log(s-a) = 0.903090$$
$$\log(s-b) = 0.845098$$
$$\text{ar. co. } \log s = 8.677781$$
$$\text{ar. co. } \log(s-c) = 9.221849$$
$$\overline{\quad 2)19.647818 \quad}$$
$$\text{Log tan} \frac{C}{2} = 9.823909$$

therefore $\quad \dfrac{C}{2} = 33°\ 41'\ 24''.4$

and $\quad C = 67°\ 22'\ 48''.8.$

Verification $\quad A + B + C = 180°.$

The following transformation of (213) will facilitate the computation when all the angles are required.

Multiply the numerator and denominator of the second member of the first of (213) by $s-a$, and we have

$$\tan\frac{A}{2} = \frac{1}{s-a}\sqrt{\frac{(s-a)(s-b)(s-c)}{s}}.$$

Put $\quad \sqrt{\dfrac{(s-a)(s-b)(s-c)}{s}} = r.$

then
also
$$\left.\begin{array}{l} \tan\dfrac{A}{2} = \dfrac{r}{s-a}, \\[4pt] \tan\dfrac{B}{2} = \dfrac{r}{s-b}. \\[4pt] \tan\dfrac{C}{2} = \dfrac{r}{s-c}. \end{array}\right\} \quad (214)$$

2. Given $a = 1468$, $b = 1359$ and $c = 1263$, to find the angles.

Ans. $A = 67°\ 58'\ 51''$, $B = 59°\ 7'\ 4''$, $C = 52°\ 54'\ 5''$.

3. Given $a = \sqrt{56}$, $b = 1$ and $c = 7$, to find A.

Ans. $A = 115° \ 22' \ 37''$.

4. Given $a = 5$, $b = 4.037$ and $c = 3.9575$, to find A.

Ans. $A = 77° \ 25' \ 12''$.

5. Given $a = 25$, $b = 30$ and $c = 20$, to find A, B and C.

Ans. $A = 55° \ 46' \ 18''$, $B = 82° \ 49' \ 8''$, $C = 41° \ 24' \ 34''$.

6. Given $a = 39$, $b = 35$ and $c = 27$, to find A, B and C.

Ans. $A = 76° \ 45' \ 21''$, $B = 60° \ 52' \ 33''$, $C = 42° \ 22' \ 6''$.

138. *Given two sides and the included angle of a triangle, to find its area.*

From Art. 128 we have

$$\Delta = \tfrac{1}{2} bc \sin A = \tfrac{1}{2} ac \sin B = \tfrac{1}{2} ab \sin C.$$

Ex. 1.—Given $a = 30$, $b = 40$ and $C = 28° \ 57'$, to find the area.

$$\log a = 1.477121$$
$$\log b = 1.602060$$
$$\text{Log sin } C = 9.684887$$
$$\text{ar. co. log } 2 = 9.698970$$
$$\log \Delta = 2.463038$$
$$\Delta = 290.427$$

Ex. 2.—The sides of a triangle are 103.5 and 90, and the included angle is 100°, find the area.

Ans. 4586.75.

139. *Given two angles and the included side of a triangle, to find the area.*

From (202) we have

$$\Delta = \frac{c^2}{2} \cdot \frac{\sin A \ \sin B}{\sin (A + B)}.$$

EXAMPLES.

Ex. 1.—Given $A = 80°$, $B = 60°$ and $c = 32$ feet, to find the area.

$$\begin{aligned}
\text{Log sin } A &= 9.993351 \\
\text{Log sin } B &= 9.937531 \\
\text{Log cosec } (A + B) &= 10.191933 \\
2 \log c &= 3.010300 \\
\text{ar. co. } \log 2 &= 9.698970 \\
\hline
\log \Delta &= 2.832085 \\
\Delta &= 679.33 \text{ square feet.}
\end{aligned}$$

Examples.

1. In the ambiguous case of triangles, given a, b and A, shew that if c_1, c_2 be the two values of the third side,

$$c_1 + c_2 = 2b \cos A, \quad c_1 c_2 = b^2 - a^2,$$

and that the area of both triangles $= \frac{1}{2} b^2 \sin 2A$.

From (169) we have

$$a^2 = b^2 + c^2 - 2bc \cos A$$

or $c^2 - 2b \cos A \cdot c + b^2 - a^2 = 0$ (1)

If c_1 and c_2 be the two values of c in this equation, we have by the theory of quadratics,

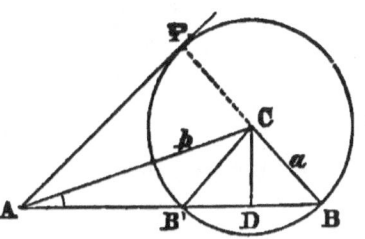

$$c_1 + c_2 = 2b \cos A \text{ and } c_1 c_2 = b^2 - a^2.$$

Now, $AD = \dfrac{c_1 + c_2}{2} = b \cos A$,

therefore $c_1 + c_2 = 2b \cos A$,

Again, $c_1 c_2 = AB \times AB' = AP^2$. (*Euc.* III. 36.)
$$= AC^2 - PC^2$$
$$= b^2 - a^2.$$

By solving (1) for c we have

$$c = b \cos A \pm \sqrt{a^2 - b^2 \sin^2 A}. \qquad (2)$$

Hence

$c_1 = b \cos A + \sqrt{a^2 - b^2 \sin^2 A}$ and $c_2 = b \cos A - \sqrt{a^2 - b^2 \sin^2 A}$.

Area $ABC = \tfrac{1}{2}bc_1 \sin A = \tfrac{1}{2}b^2 \sin A \cos A + \tfrac{1}{2}b \sin A \sqrt{a^2 - b^2 \sin^2 A}$.

Area $AB'C = \tfrac{1}{2}bc_2 \sin A = \tfrac{1}{2}b^2 \sin A \cos A - \tfrac{1}{2}b \sin A \sqrt{a^2 - b^2 \sin^2 A}$.

therefore the area of both triangles $= b^2 \sin A \cos A$
$$= \tfrac{1}{2}b^2 \sin 2A.$$

From (2) it is seen that when $a = b \sin A$, $c = b \cos A$, which gives only one value for c and the triangle is right-angled at B.

2. Given the perpendiculars from the angles of a triangle upon the opposite sides, to find the sides and angles.

Let p_a, p_b, p_c denote the perpendiculars on the sides a, b, c respectively; then by (203)

$$p_a = \frac{2\Delta}{a}, \quad p_b = \frac{2\Delta}{b}, \quad p_c = \frac{2\Delta}{c},$$

or

$$a = \frac{2\Delta}{p_a}, \quad b = \frac{2\Delta}{p_b}, \quad c = \frac{2\Delta}{p_c},$$

Substituting these values of a, b, c in (172) we have

$$\cos A = \frac{\dfrac{4\Delta^2}{p_b^2} + \dfrac{4\Delta^2}{p_c^2} \cdot \dfrac{4\Delta^2}{p_a^2}}{2 \cdot \dfrac{4\Delta^2}{p_b p_c}}$$

$$= \frac{\dfrac{1}{p_b^2} + \dfrac{1}{p_c^2} - \dfrac{1}{p_a^2}}{\dfrac{2}{p_b p_c}} = \frac{p_a^2 p_c^2 + p_a^2 p_b^2 - p_b^2 p_c^2}{2 p_a^2 p_b p_c},$$

which determines A.

Then
$$b = \frac{p_c}{\sin A}, \quad c = \frac{p_b}{\sin A}, \quad \&c.$$

3. Given the sum of the sides a and $b = k$, the side c and the angle C, of a triangle, shew that

$$a = k \cos^2 \frac{\phi}{2}, \quad b = k \sin^2 \frac{\phi}{2}$$

where
$$\sin \phi = \pm \frac{\sqrt{k^2 - c^2}}{k} \sec \frac{C}{2}.$$

From (171)
$$c^2 = a^2 + b^2 - 2ab \cos C$$
$$= a^2 + 2ab + b^2 - 2ab(1 + \cos C)$$
$$= k^2 - 4ab \cos^2 \frac{C}{2},$$

therefore
$$4ab = (k^2 - c^2) \sec^2 \frac{C}{2}$$

and
$$\frac{4ab}{(a+b)^2} = \frac{(k^2 - c^2) \sec^2 \frac{C}{2}}{k^2} = \sin^2 \phi,$$

which is possible since $(a+b)^2 > 4ab$,

then
$$1 - \frac{4ab}{(a+b)^2} = 1 - \sin^2 \phi$$

or
$$\frac{a-b}{a+b} = \cos \phi$$

$$\frac{2a}{a+b} = 1 + \cos \phi,$$

or
$$\frac{2a}{k} = 2 \cos^2 \frac{\phi}{2}$$

and
$$a = k \cos^2 \frac{\phi}{2}.$$

Also
$$\frac{2b}{a+b} = 1 - \cos \phi$$

or
$$\frac{2b}{k} = 2 \sin^2 \frac{\phi}{2}$$

and
$$b = k \sin^2 \frac{\phi}{2}.$$

172   PLANE TRIGONOMETRY.

This problem may also be easily solved by (186) and (187). Thus, from (186) we have

$$\cos \tfrac{1}{2}(A - B) = \frac{k}{c} \cos \tfrac{1}{2}(A + B)$$

$$= \frac{k}{c} \sin \frac{C}{2}.$$

Hence the angles are known.

From (187) we have

$$a - b = c \, \frac{\sin \tfrac{1}{2}(A - B)}{\cos \dfrac{C}{2}}.$$

Hence the sides a and b are known.

4. The sides of a triangle are 13, 14 and 15, find the area.
Ans. 84.

5. The sides of a triangle are 13, 20 and 21, find the perpendicular on the longest side and the area.
Ans. Perpendicular = 12, area = 126.

6. The sides of a triangle are 25, 51 and 52, find the area and the angle opposite the shortest side.
Ans. Area = 624, angle = 28° 4′ 21″.

7. The sides of a triangle are 137, 111 and 124, find the area. *Ans.* 6510.

8. The angles of a triangle are 70°, 60° and 50°, and the perimeter is 150, find the sides. (See formula 192.)
Ans. 54.81, 50.51, 44.68.

9. The sides of a triangle are 85 and 75 and the included angle is 75°, find the other angles.
Ans. 57° 10′ and 47° 50′.

EXAMPLES.

10. The ratio of two sides of a triangle is $7:3$, and the angle they contain is $6°\ 37'\ 24''$, find the other angles.

Ans. $168°\ 27'\ 25''.4$ and $4°\ 55'\ 10''.6$.

11. The sides a, b, c of a triangle are as $8:10:16$, find the angle B. *Ans.* $B = 55°\ 46'\ 16''$.

12. Given two sides of a triangle and the difference of their opposite angles, to solve the triangle. (See formula 166.)

13. In any triangle prove that $\tan\dfrac{B}{2} = \dfrac{(b+d)(b-d)}{2bp}$, where d is the difference between the sides a and c, and p is the perpendicular upon the side b, and that $\tan B = \dfrac{b \sin A}{c - b \cos A}$.

14. If a, β, γ be the perpendiculars from the angles of a triangle upon the opposite sides, prove that

$$\frac{a^2}{\beta\gamma} = \frac{bc}{a^2}.$$

15. In the ambiguous case, if Δ and δ be the areas of the two triangles, prove that

$$\Delta^2 + \delta^2 - 2\Delta\delta \cos 2A = \frac{a^2}{b^2}(\Delta + \delta)^2$$

a, b and A being given.

16. In the ambiguous case, if c_1, c_2 be the values found for the side c, when a, b and A are given, prove that

$$c_1^2 + c_2^2 - 2c_1 c_2 \cos 2A = 4a^2 \cos^2 A$$

and $\qquad c_1^2 - c_2^2 = 4b \cos A \sqrt{a^2 - b^2 \sin^2 A}$.

17. In any triangle, given a, B and the sum of the other two sides (equal to m), to solve the triangle.

Ans. $\tan\dfrac{C}{2} = \dfrac{m-a}{m+a}\cot\dfrac{B}{2}$.

18. If a, β, γ be the distances from the angles A, B, C of a

triangle to a point P within it, from which the sides subtend equal angles, find the sides and the area.

$$\text{Ans. The sides are } \sqrt{(\beta^2 + \gamma^2 + \beta\gamma)}$$
$$\sqrt{(\alpha^2 + \gamma^2 + \alpha\gamma)}$$
$$\sqrt{\alpha^2 + \beta^2 + \alpha\beta}$$

and the area is $\dfrac{\sqrt{3}}{4}(\alpha\beta + \beta\gamma + \alpha\gamma)$.

19. In a triangle, given C, c and $ab = m^2$, to find the sides a and b.

$$\text{Ans. } a + b = c \sec\theta,$$
$$a - b = c \cos\phi,$$

where $\tan\theta = \dfrac{2m}{c}\cos\dfrac{C}{2}$ and $\sin\phi = \dfrac{2m}{c}\sin\dfrac{C}{2}$.

20. In a triangle, given C, the perpendicular from $C = p$ and $a + b = m$, to find c.

$$\text{Ans. } c = m\tan\dfrac{\theta}{2}, \text{ where } \tan\theta = \dfrac{m}{p}\tan\dfrac{C}{2}.$$

21. The perimeter of a triangle is 100 rods, $A = 102°\ 51'\ 30''$, $B = 25°\ 42'\ 45''$ and $C = 51°\ 25'\ 45''$; find the sides.

$$\text{Ans. } a = 44.51,\ b = 19.8,\ c = 35.69.$$

22. The perimeter of a right-angled triangle is 24 rods, and one of the angles $= 30°$; find the sides.

$$\text{Ans. } 5.072,\ 8.784,\ 10.144.$$

23. Given the base a, the vertical angle A, and the difference of the other two sides $= d$, to find the sides. (See formula 187.)

24. Two adjacent sides of a parallelogram are a and b, and the included angle is θ; shew that the diagonal drawn from this angle is

$$\sqrt{(a^2 + b^2 + 2ab\cos\theta)},$$

and that the other diagonal is

$$\sqrt{(a^2 + b^2 - 2ab\cos\theta)}.$$

25. The sides of a triangle are 17, 25 and 28; find the area of the triangle formed by joining the feet of the perpendiculars drawn from the angles to the opposite sides. (See problem 52, Chapter VIII.) *Ans.* $24\frac{264}{1445}$.

26. In a right-angled triangle, the sum of the hypothenuse and base is 2986, and the angle at the base is 52°, find the perpendicular. *Ans.* 1456.37.

27. Two sides of a triangle are 356 and 294, and the angle opposite the latter side is 51° 27′, find the other side.
Ans. 316.309 or 127.4079.

28. The perimeter of a triangle is 128, the angle $C = 28° 4′ 21″$ and the perpendicular from C on $c = 49.92$, find the side c.
Ans. 25.

29. Given $b = 39$, $c = 51$ and $B = 115°$, to solve the triangle.
Ans. Impossible. Why?

30. In a triangle given C, c and $a^2 - b^2 = m^2$, solve the triangle.
Ans. $\sin(A - B) = \dfrac{m^2}{c^2} \sin C$.

When will there be two solutions?

CHAPTER X.

APPLICATION OF TRIGONOMETRY TO SURVEYING, NAVIGATION AND ASTRONOMY.

140. In this chapter we will give some examples shewing how the formulæ of Trigonometry are applied in the solution of some very interesting and useful problems in Surveying, Navigation and Astronomy. Horizontal and vertical angles are most conveniently measured with a theodolite. This instrument is composed essentially of two graduated circles having their planes perpendicular to each other. When the instrument is in use, one of the circles is placed in a horizontal position by means of spirit levels. On this circle horizontal angles are measured; on the other are measured vertical angles whether of elevation or depression, that is, whether the object is above or below the horizontal line passing through the centre of the axis of the vertical circle. A telescope is attached to the axis of the vertical circle, by means of which a clearer view of the object may be had. The sextant and reflecting circle are employed to measure angles in any plane whatever. They are the only instruments which can be conveniently used at sea for measuring the altitudes of the heavenly bodies above the horizon. For the descriptions, adjustments and modes of using these instruments, the student is referred to treatises on surveying, such as Gillespie's Land Surveying, or Sims's Treatise on Mathematical Instruments.

Their construction and adjustments will, however, be best learned by a careful study of the instruments themselves. We

HEIGHTS AND DISTANCES. 177

shall therefore suppose that the manner of adjusting and applying them to practice is known, and proceed at once to give a collection of problems which they enable us to solve.

141. *To find the distance of an inaccessible object upon a horizontal plane.*

Let it be required to find the distance from A to an object C situated on the opposite bank of a river. Measure a base AB, whose length we will denote by c. Measure also the horizontal angles CAB, CBA; then (Art. 131)

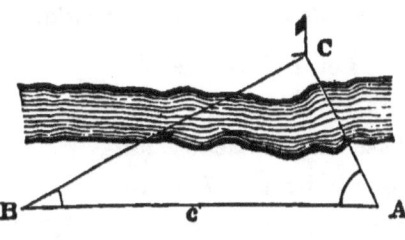

$$\frac{AC}{AB} = \frac{\sin B}{\sin C},$$

whence

$$AC = c \frac{\sin B}{\sin (A + B)},$$

or in logarithms

$$\log AC = \log c + \text{Log sin } B - \text{Log sin } (A + B).$$

This problem may also be solved without the use of any instruments for measuring angles, as follows:

Having measured the base AB as before, measure any length Ad along AC and an equal one Ag along AB; then measure dg. Similarly measure Bm and Bn, equal to each other, and then measure mn.

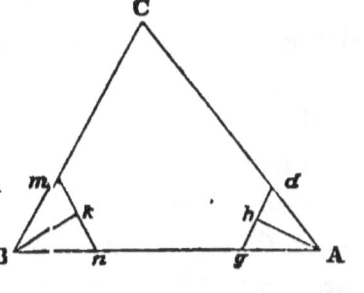

Bisect dg in h and join Ah, then since Adg is an isosceles triangle Ah bisects the angle A, and we have

$$\sin\frac{A}{2} = \frac{hd}{Ad} = \frac{gd}{2\ Ad},$$

similarly
$$\sin\frac{B}{2} = \frac{mk}{Bm} = \frac{mn}{2\ Bm}.$$

Hence the angles A and B are known by the aid of the tables, and the solution may now be completed as before.

142. *To find the height and the distance of an inaccessible object standing on a horizontal plane.*

Let DC be the object. Measure any base AB, and denote its length by c. Measure the horizontal angles CAB (α) ABC (β) and the angle of elevation CAD (γ). Then from the triangle ABC we have

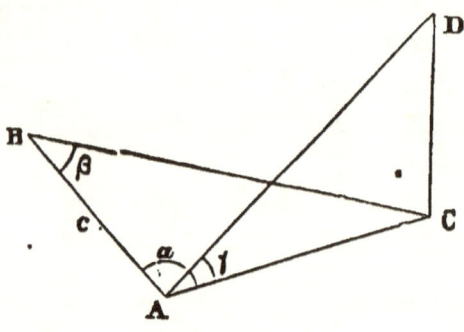

$$AC = c\frac{\sin B}{\sin C} = c\frac{\sin \beta}{\sin(\alpha+\beta)}.$$

In the right-angled triangle ACD we have AC, just found, and the angle CAD (γ) to find DC.

Hence
$$DC = AC \tan CAD$$
$$= c\frac{\sin \beta \tan \gamma}{\sin(\alpha+\beta)}.$$

Or thus:

At any convenient point A measure the angle of elevation CAD (α); then measure a base AB (c) directly from the object, and at B measure the vertical angle

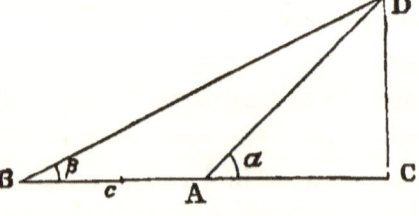

ABD (β). Then the angle $ADB = \alpha - \beta$, and from the triangle ABD we have by (165)

$$AD = \frac{c \sin \beta}{\sin (\alpha - \beta)},$$

then in the right-angled triangle ADC

$$DC = AD \sin CAD$$
$$= \frac{c \sin \alpha \sin \beta}{\sin (\alpha - \beta)},$$

and
$$AC = AD \cos CAD$$
$$= \frac{c \cos \alpha \sin \beta}{\sin (\alpha - \beta)}.$$

143. *To find the distance between two inaccessible objects on a horizontal plane.*

Let C and D be the two objects. Measure the base AB equal to a, and the angles CAB, DAB, ABC, ABD, which put equal to α, β, γ, δ respectively. Then from the triangle ABC we have by (165)

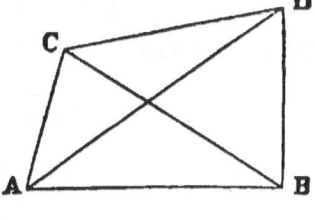

$$AC = AB \frac{\sin ABC}{\sin ACB} = a \frac{\sin \gamma}{\sin (\alpha + \gamma)};$$

similarly,
$$AD = AB \frac{\sin ABD}{\sin ADB} = a \frac{\sin \delta}{\sin (\beta + \delta)}.$$

Then in the triangle ACD, the sides AC, AD and the included angle $CAD = (\alpha - \beta)$, are known, therefore CD can be found by the methods of Arts. 134–136.

144. *Three inaccessible objects A, B, C (not in the same straight line), situated on a horizontal plane and at known distances from each other, are visible from another point P on the plane, and the angles APC, BPC are given; to find the distance from P to each of the objects A, B, C.*

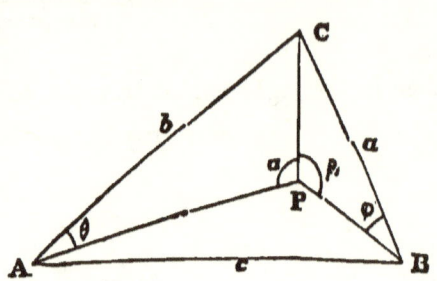

Let A, B, C be the three inaccessible objects, and P the point of observation. Let the observed angles APC, BPC be denoted by α, β respectively PAC by θ, and PBC by ϕ.

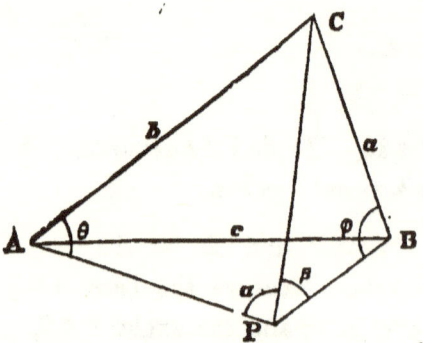

Since the sides of the triangle ABC are known, the angle C can be found by (179) or (182)

$$\theta + \phi + \alpha + \beta + C = 360°,$$

therefore $\quad \tfrac{1}{2}(\theta + \phi) = 180° - \tfrac{1}{2}(\alpha + \beta + C).\quad(1)$

From the triangles APC, BPC, we have by (165)

$$PC = \frac{b \sin \theta}{\sin \alpha}, \quad \text{and} \quad PC = \frac{a \sin \phi}{\sin \beta},$$

therefore $\quad \dfrac{\sin \theta}{\sin \phi} = \dfrac{a}{b} \cdot \dfrac{\sin \alpha}{\sin \beta}$

$$= \tan \psi, \text{ suppose};$$

then $\quad \dfrac{\sin \phi - \sin \theta}{\sin \phi + \sin \theta} = \dfrac{1 - \tan \psi}{1 + \tan \psi},$

whence $\dfrac{\tan \frac{1}{2}(\phi - \theta)}{\tan \frac{1}{2}(\phi + \theta)} = \tan (45° - \psi)$, by (59) and (81)

therefore $\tan \frac{1}{2}(\phi - \theta) = \tan (45° - \psi) \tan \frac{1}{2}(\phi + \theta)$. (2)

From (1) and (2) θ and ϕ become known, and therefore by (165) AP, BP and CP can be found.

When θ and ϕ are supplementary, $\psi = 45°$, and the solution fails when the point P is without the triangle, in which case the quadrilateral $ACBP$ is inscriptible in a circle.

145. *To determine the height and distance of an inaccessible object, by having given its angles of elevation observed at three points at given distances from each other, and in the same straight line.*

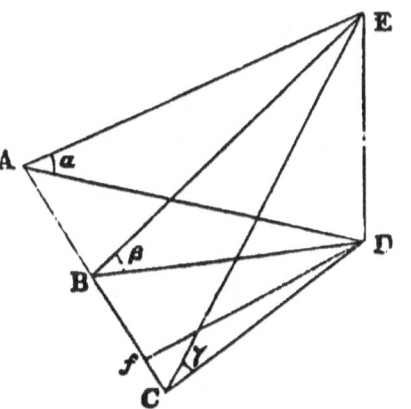

Let DE be the object, A, B, C, the three stations of observation: let $AB = a$, $BC = b$, and $DE = x$; and let the observed angles of elevation at A, B, C, be α, β, γ, respectively:

Then $AD = x \cot \alpha$, $BD = x \cot \beta$, $CD = x \cot \gamma$.

Draw Df perpendicular to AC, then in the triangles ABD, BCD, we have by *Euc.* II., 12, 13,

$$x^2 \cot^2 \alpha = a^2 + x^2 \cot^2 \beta + 2a \cdot Bf$$
$$x^2 \cot^2 \gamma = b^2 + x^2 \cot^2 \beta - 2b \cdot Bf.$$

Eliminating Bf, and solving for x, we have

$$x = \sqrt{\dfrac{ab(a+b)}{a \cot^2 \gamma - (a+b) \cot^2 \beta + b \cot^2 \alpha}}.$$

And thus AD, BD, and CD become known.

146. *A flagstaff of known length, standing upon the top of a tower of known height, subtends at the eye of a spectator on the same horizontal plane as the tower, a given angle; to find the distance of the tower.*

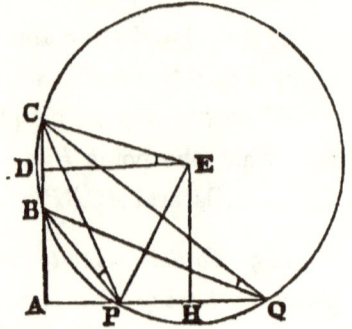

Let $AB = a$, the height of tower; $BC = h$, the length of the staff; P, the position of the observer, at which the staff subtends an angle $BPC = \beta$.

Let $APB = \theta$, then $APC = \theta + \beta$.

$$AP = a \cot \theta \text{ and } AP = (a+h) \cot (\theta + \beta),$$

whence
$$a \tan (\theta + \beta) = (a+h) \tan \theta \quad \times$$

and
$$\tan \theta = \frac{h \cot \beta \pm \sqrt{h^2 \cot^2 \beta - 4a(a+h)}}{2(a+h)},$$

therefore $AP = \tfrac{1}{2} \{h \cot \beta \mp \sqrt{h^2 \cot^2 \beta - 4a(a+h)}\}$,

from which we see that there are two values of AP, unless $h^2 \cot^2 \beta = 4a(a+h)$.

If a segment of a circle be described on BC containing an angle equal to β, it will in general cut the horizontal line in two points, P and Q, corresponding to the two values of AP.

When $h \cot \beta = 2\sqrt{a(a+h)}$, the circle will touch the horizontal line, the points P and Q will coincide, and β will then have its maximum value, so that $\tan \theta = \sqrt{\dfrac{a}{a+h}}$,

and then $\quad AP = \sqrt{a(a+h)}.\quad$ (*Euc.* III., 36)

Or we may proceed thus:

From the centre E draw ED, EH perpendicular to CB and

✳ *See page 335.*

AQ; join CE and EP; the angle $CED = CPB = CQB = \beta$. In the right-angled triangle, CDE, we have

$$DE \tan \beta = \frac{h}{2}, \text{ or } DE = \frac{h}{2} \cot \beta = AH,$$

and $CE \sin \beta = \frac{h}{2}$, or $CE = \frac{h}{2} \operatorname{cosec} \beta = EP$.

$$\begin{aligned} PH &= \sqrt{EP^2 - EH^2} \\ &= \sqrt{EP^2 - AD^2} \\ &= \sqrt{\frac{h^2}{4} \operatorname{cosec}^2 \beta - \left(a + \frac{h}{2}\right)^2} \\ &= \tfrac{1}{2}\sqrt{h^2 \cot^2 \beta - 4a(a+h)} \end{aligned}$$

therefore $AP = AH - PH = \tfrac{1}{2}\{h \cot \beta - \sqrt{h^2 \cot^2 \beta - 4a(a+h)}\}$

and $AQ = AH + PH = \tfrac{1}{2}\{h \cot \beta + \sqrt{h^2 \cot^2 \beta - 4a(a+h)}\}$

the same as before.

147. In Surveying, the course or bearing of a line, is the angle which it makes with the meridian passing through one extremity, and is reckoned from the north or south point of the horizon, toward the east or west.

Thus, if NS represent the meridian and the angle NPQ be 30°, then the bearing of PQ from P is 30° to the east of north, or as usually read "north thirty degrees east" and written N. 30° E.

The reverse bearing of a line is the bearing taken from the other extremity. Thus the bearing of PQ from Q is S. 30° W.

In Navigation, the course of a ship is generally referred to the points of the Mariner's Compass which consists of a circular piece of card board attached to a magnetic needle, and is so balanced that it can move freely in any direction.

The circumference is divided into thirty-two equal parts called *points*, and each *point* into four equal parts called quarter points.

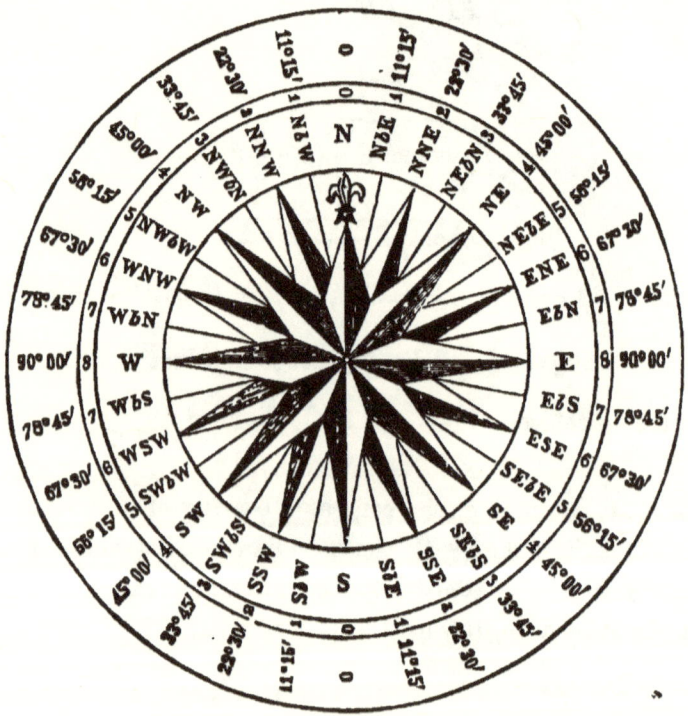

The points are read as follows, beginning at the north and going east: "north, north by east, north-north-east, north east by north, north east," and so on as seen in the figure.

The angle between two adjacent points is $\frac{360}{32}°$ or $11°\ 15'$.

A quarter point is therefore $2°\ 48'\ 45''$, a half point $5°\ 37'\ 30''$, and a three-quarter point $8°\ 26'\ 15''$.

LENGTH OF A DEGREE OF LONGITUDE.

148. *To find the length of a degree of longitude on any parallel of latitude.*

Let P be the pole of the earth, C the centre, EQ an arc of the equator containing one degree, and AB an arc of one degree on any parallel of latitude. Join CE, CQ and draw AD and BD perpendicular to the radius PC; then it is evident that the angle ADB is equal to the angle ECQ, and therefore

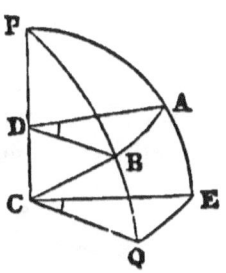

$$CQ : DB :: EQ : AB.$$

But regarding the radius CQ as unity, DB is the sine of the angle BCP or the cosine of the angle BCQ, that is, the cosine of the latitude; hence

$$1 : \text{cosine of the latitude} :: EQ : AB$$

and therefore $\qquad AB = EQ \cos. \text{lat.}$

From which it follows that similar portions of different parallels of latitude are to each other as the cosines of the latitudes.

Ex.—Find the length of a degree of longitude at Toronto, latitude 43° 39′ 24″.

The equatorial radius of the earth is 3962.8 miles, but in consequence of the flattening in the direction of the polar diameter it is only 3956.514 miles at Toronto: and considering the earth a sphere with a radius of 3956.514 miles, the length of a degree of longitude on the equator would be 69.054 miles $= EQ$.

Hence
$$\log EQ = 1.839190$$
$$\text{Log cos lat.} = 9.859432$$
$$\overline{\log AB = 1.698622}$$

therefore $\quad AB = 49.96$ or 50 miles very nearly.

As a degree in longitude makes a difference of four minutes in time, it follows that fifty miles east or west on the parallel of Toronto is equivalent to four minutes difference in time.

149. *To find the Moon's distance from the Earth, and her diameter.*

Let PP' be the earth's axis, EQ the equator, and A and B the positions of two observatories on the same meridian and

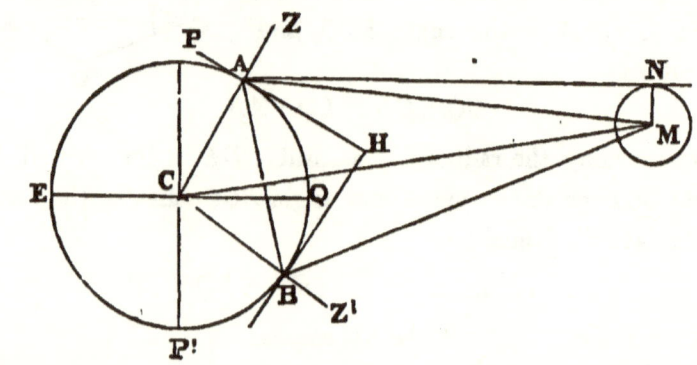

whose latitudes ACQ, BCQ are known. Draw the radii, CA, CB, and produce them to meet the celestial sphere in Z and Z'; through A and B draw AH, BH at right angles to CZ and CZ' respectively, then Z and Z' are the zeniths, and AH and BH are the horizons of A and B respectively; let M be the moon, and when she is on the meridian of the observatories, let her altitudes MAH, MBH be measured with a sextant, theodolite or any other suitable instrument. Join AB, then regarding the earth a sphere in order to render the problem as simple as possible, we have in the isosceles triangle ACB, the sides AC, CB and the angle ACB, the sum of the latitudes given, hence AB and the angles CAB, CBA can be found. The angle $HAB = HBA =$ half of ACB, and since MAH and MBH are determined by observation, the angles MAB, MBA are known; hence AM and BM can be found. Finally, in the triangle ACM, the sides AM, AC and the included angle MAC are now known, therefore MC can be found, which is the distance sought.

From A draw AN tangent to the moon and join MN, then ANM is a right angle and the angle MAN the moon's apparent semi-diameter can be determined by observation,

therefore $$MN = AM \sin MAN,$$

whence her radius is known and therefore her diameter.

Parallax.

150. The parallax of a celestial body is, in general, the apparent angular displacement which is produced by viewing the body from two different points. The term is used in astronomy to express the difference of altitude or zenith distance of a celestial body when seen from the surface and the centre of the earth respectively.

Let C be the centre of the earth, P the place of an observer on its surface, M a celestial object seen in the horizon, M' the same seen at the zenith distance $M''PM'$, and M'' the same object seen in the zenith. Now it is evident that when the object is at M'' it will appear in the same direction whether viewed from P or C; there is then no displacement and therefore no parallax.

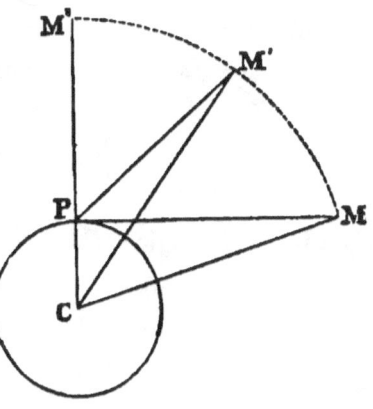

If the object be at M', its apparent direction is PM' while its *true* direction is CM', and the angular displacement $PM'C$ is the parallax due to the zenith distance $M''PM'$ or to the altitude $M'PM$.

It is evident that the angle $PM'C$ increases as the body

approaches the horizon, and when it is in the horizon as at M the parallax PMC has its maximum value, and is then called the horizontal parallax which is, in fact, the largest angle which the earth's radius subtends at the object.

Parallax increases the zenith distance and consequently diminishes the altitude. The angle $M''PM'$ is the apparent and $M''CM'$ the true zenith distance of the object when at M'.

Hence, to obtain the *true* zenith distance from the apparent, the parallax must be *subtracted;* and to obtain the *true* altitude from the apparent, the parallax must be added.

It is evident from the figure that the effect of parallax is wholly in the vertical circle passing through the observer and the body, the azimuth or angular distance from the meridian of the observer is therefore not affected.

151. *To find the parallax at any altitude when the horizontal parallax is given.*

Let $P = CMP$, the horizontal parallax,
$p = CM'P$, the parallax in altitude,
$Z' = M'PM''$, the apparent zenith distance,
$R = CM = CM'$, the distance of the object,
$r = CP$, the earth's radius,

then in the triangle CPM' we have

$$\frac{\sin CM'P}{\sin CPM'} = \frac{CP}{CM'}$$

or

$$\frac{\sin p}{\sin Z'} = \frac{r}{R} = \sin P,$$

whence
$$\sin p = \sin P \sin Z'. \qquad (215)$$

In the case of the sun and planets, the horizontal parallax is so small that we may, without sensible error, use the more convenient formula

$$p'' = P'' \sin Z'.$$

PARALLAX IN ALTITUDE.

If the true zenith distance $M'CM'' = Z$ be given, instead of the apparent, we have

$$\frac{\sin CM'P}{\sin CPM'} = \frac{CP}{CM'},$$

or

$$\frac{\sin p}{\sin (Z+p)} = \frac{r}{R} = \sin P; \qquad (a)$$

hence

$$\frac{\sin (Z+p) + \sin p}{\sin (Z+p) - \sin p} = \frac{1 + \sin P}{1 - \sin P},$$

or

$$\frac{\tan(\frac{Z}{2} + p)}{\tan \frac{Z}{2}} = \tan^2 (45° + \frac{P}{2}),$$

whence

$$\tan (\frac{Z}{2} + p) = \tan^2 (45° + \frac{P}{2}) \tan \frac{Z}{2}. \quad (216)$$

This equation makes known $\frac{1}{2}Z + p$, from which we obtain p by subtracting $\frac{Z}{2}$.

Another solution of equation (a) will be given in a subsequent chapter.

152. *To find the augmentation of the Moon's semi-diameter on account of her altitude above the horizon.*

The *apparent* diameter of the moon is the *angle* which her disk subtends, and is not the same for all points on the earth, on account of their different distances from her.

Supposing the moon's distance from the centre of the earth to remain constant, her distance from the observer must diminish as her altitude above the horizon increases, and therefore her *apparent* diameter must increase. This increase of the apparent diameter is called the augmentation of the diameter due to altitude.

In the figure of Article 151, let δ denote the moon's apparent semi-diameter at M, and δ' the augmented semi-diameter when at M', then we shall evidently have

$$\delta : \delta' :: PM' : CM' \text{ or } CM.$$

$$\frac{\delta}{\delta'} = \frac{PM'}{CM'} = \frac{\sin PCM'}{\sin CPM'} = \frac{\sin Z'}{\sin Z},$$

whence
$$\delta' = \delta \frac{\sin Z}{\sin Z'}. \qquad (217)$$

Examples.

1. To determine the distance of a ship at anchor at C, a base line AB of 100 rods is measured along the shore, and the angle ABC was observed to be 83° 18' and the angle BAC 32° 10'; find the ship's distance from B. *Ans.* 58.968 rods.

2. From the decks of two ships A and B, 880 yards apart, the angle of elevation of a mountain C due north of both ships was observed at each; at A the angle was 35° and at B 64°; required the height of the mountain above the surface of the sea, the deck of each ship being 21 feet above it.
Ans. 942.75 yds.

3. A tower subtends an angle of 39° at a distance of 200 feet from its base, what angle will it subtend at a distance of 350 feet from its base? *Ans.* 24° 49' 53".5.

4. There are two monuments whose heights are 100 feet and 50 feet respectively, and the line joining their tops, when produced, makes with the horizontal plane on which they stand an angle of 37°; find their distance apart.
Ans. 66.352 feet.

5. To determine the distance between two inaccessible rocks C and D, in the sea, a base line AB of 670 yards was measured

on the shore and the following angles were measured at the extremities of the base: $BAD = 40° \ 16'$, $BAC = 97° \ 56'$; $ABC = 42° \ 22'$, $ABD = 113° \ 29'$; find CD.

Ans. 1174.26 yards.

6. The hypothenuse of a right-angled triangle rests on a horizontal plane, and is 100 feet long; one of the angles is 36° 40' and the inclination of the triangle to the horizontal plane is 60°; find the height of the right angle above the plane.

Ans. 41.48 feet.

7. An object stands on the top of an inaccessible hill, and from a certain point on the horizontal plane the angle of elevation of the top of the hill is 40°, and that of the top of the object 51°. Going back 100 yards in a direct line from the object, the angle of elevation of the top of the object was then 33° 45'; find the height of the object. *Ans.* 46.663 yards.

8. The angle of elevation of a tower 100 feet high, due north of an observer, was 50°; what will be its angle of elevation after walking due west 300 feet? *Ans.* 17° 47' 50".5.

9. Wishing to find the distance between a battery at B and a fort at F, which cannot be seen from the battery in consequence of the ground between the battery and fort being covered with a forest, distances BA and AC to points A and C where both the fort and battery were visible, were measured, the former being 2000 yards and the latter 3000, and the angle BAF was found to be 34° 10', FAC 74° 42', and FCA 80° 10'; find the distance between B and F. *Ans.* 5422.3 yards.

10. The elevation of a balloon was observed to be 20° bearing N.E., and by another observer 4000 yards due south of the former its bearing was found to be N.b.E.; required its height. *Ans.* 511.24 yards.

11. Coming in from sea, at a certain point P I observed

two headlands A and B, and inland at C a tower which appeared between the headlands: the distance between the headlands was 5.35 miles, and the distance from A to the tower was 2.8 miles, and from B to the tower 3.47 miles. I observed the angles APC, BPC, the former being 12° 15' and the latter 15° 30'; required my distance from each of the three objects.

Ans. $AP = 11.257$; $CP = 12.7523$; $BP = 11.034$ miles.

12. Required the distance of the three objects A, B, C from the point P situated within the triangle ABC, from the following data: $AB = 267$ rods, $AC = 346$ rods, $BC = 209$ rods, angle $APC = 128°\ 40'$, angle $APB = 91°\ 20'$.

Ans. $AP = 248.854$; $BP = 91.134$; $CP = 130.81$ rods.

13. While sailing along a coast a headland C was observed to bear N.E.b.N.; having run E.b.N. 15 miles to B, the headland bore W.N.W.; find the distance from the headland at each observation. *Ans.* 8.499, 10.81 miles.

14. A and B are two points lying N. and S., and 50 rods apart; what must be the distance of a third point C from each, that it may bear N.b.W. from B and W.b.S. from A?

Ans. 9.754 and 49.04 rods.

15. The courses of two ships are N. and E., and their rates of sailing are equal, the bearing of the former from the latter was E.N.E.; but after each had sailed ten miles the bearing of the latter from the former was S.S.E.; find the distance between the ships at the first observation. *Ans.* 7.653 miles.

16. The elevations of two mountains, in the same line with the observer, are 9° 30' and 20°; on approaching two miles they both have an elevation of 38° 15'; find their heights and the distance between them.

Ans. The nearer, 747.77 yds.; the more remote, 2380 yds.
The distance between them, 2070.48 yds.

EXAMPLES.

17. From a ship sailing N.W., two islands appeared in sight, one bearing W.N.W., the other N, and after sailing eight miles farther the first bore W.b.S., and the other N.E.; required their bearing and distance from each other.

Ans. S. 58° 40′ 50″.4 W.; distance, 12.95 miles.

18. Two ships sail from the same port at the same time, the one due N. at the rate of six miles per hour, the other on a course N. 60° E. at the rate of ten miles per hour for two hours, she then tacks to cut the other off or to overtake her; how far must she sail to do it, and on what course?

Ans. 23.75 miles, on a course N. 46° 49′ 35″ W.

19. From two stations on the deck of a ship 100 feet apart, the bearings of an object on shore were N.E. and N.N.E., and the ship's head was N.b.W.; find the distances of the object from each station. *Ans.* 145.177 and 217.27 feet.

20. A cape C bears from a headland H, W. $\frac{1}{4}$ S. 4.23 miles; how must the cape bear from a ship which runs in towards the headland on a course N.b.W.$\frac{1}{2}$ W., until the headland is 2.3 miles distant from the ship? *Ans.* W.N.W.

21. A ship was 2640 yards due south of a lighthouse, and after sailing N.W.b.N. 800 yards, its angle of elevation was 5° 25′; required its height. *Ans.* 191.94 yards.

22. A tower, 65 feet high, subtends an angle of 60° at the eye of an observer standing on the same horizontal plane as the tower; find his distance from it, his eye being five feet above the plane. *Ans.* 44.3 feet.

23. A lighthouse standing on a rock, is observed from two points in a line with it, and one quarter of a mile apart; from the nearer point the elevations of the top and bottom are 52° 14′ and 48° 38′ respectively, and from the more remote point the elevation of the top is 16° 28'; find its height and elevation above the sea.

Ans. Height, 60.82 feet; height above sea, 445.2 feet.

24. From the top and foot of a lighthouse 58 feet high, standing on a cliff by the sea coast, the angles of depression of a ship's hull measured from the *visible* horizon are 5° 47' and 5° 8'; find the ship's distance, the diameter of the earth being taken at 7926 miles. *Ans.* 1688.7 yards.

25. At noon, a column in the direction E.S.E. from an observer, cast a shadow the extremity of which lay in the direction N.E. from him; the elevation of the column was found to be 45°, and the length of the shadow 80 feet; find the height of the column. *Ans.* 61.23 feet.

26. From the top of a hill, two telegraph posts standing on the horizontal plane below, were seen in a line with the observer; their angles of depression were 30° and 45° respectively and their distance from each other 176 yards; find the height of the hill. *Ans.* 240.416 yards.

27. From a ship a lighthouse bore N.N.E.; after sailing E.b.S. seven miles it bore N.W.b.N.; find its distance from the ship at each station. *Ans.* 5.953 miles; 8.257 miles.

28. A flagstaff 24 feet long, standing on a cliff by the sea shore, subtends at a ship an angle of 38'; the elevation of the cliff is 14°; find the height of the cliff and the ship's distance from it. *Ans.* Height, 508 feet; distance, 2038 feet.

29. From the top of a hill, a vertical pillar 220 feet high, standing on the horizontal plane below, subtends an angle of 1° 12', and the depression of its top is 12° 20'; what is its distance, and the height of the hill?

Ans. Distance, 9977 feet; height, 2401 feet.

30. If R denote the earth's radius, h the height of a mountain and θ the dip of the horizon from its summit, show that

$$h = 2R \sin^2 \frac{\theta}{2} \sec \theta = R \tan \theta \tan \frac{\theta}{2}.$$

31. A tower standing on a horizontal plane is surmounted by a flagstaff; from a certain point on the plane, the tower subtends an angle β, and the flagstaff an angle a; from another point c feet nearer to the base of the tower, the flagstaff subtends the same angle a; shew that the height of the tower is

$$\frac{c \tan \beta}{1 - \tan \beta \tan (a + \beta)} \text{ feet.}$$

32. From the bottom of a tower 100 feet high, the angular elevation of the summit of a hill was $32°$, and on retiring 180 feet from the foot of the tower its top is seen to be in a straight line with the top of the hill; find its height.

Ans. 901.5 feet.

33. A person walking along a straight road observes the *greatest* angular elevation of a tower to be a, and from another straight road he observes the *greatest* angular elevation of the tower to be β. The distances of the points of observation from the intersection of the two roads, are a and b; shew that the height of the tower is

$$\sqrt{\frac{a^2 - b^2}{\cot^2 a - \cot^2 \beta}}.$$

34. Three observers A, B, C situated in the same straight line, A and C being each at a distance of 1000 yards from B, find at the same instant the angular elevations of a balloon to be at A $48° 10'$, at B $54°$ and at C $50° 30'$; find the height of the balloon. *Ans.* 2169.05 yards.

35. A person standing on a horizontal plane observes that

the top of a telegraph post standing on the same plane, is in a line with the top of a building which stands on a hill at a greater distance than the post; the distance of his station from the foot of the post is b and the angle subtended by the height of the building is β. He then moves to a station farther off from the foot of the post by a distance a, and in the same line with the former station and the foot of the post, and he there observes that the angle subtended by the building is the same as before, and that the top of the post is in a line with the foot of the building. Shew that the height of the post is $\sqrt{ab + b^2}$, and the height of the building

$$\frac{a(a+2b)\tan\beta}{a - 2\sqrt{ab + b^2}\tan\beta}.$$

36. The angular altitude and breadth of a cylindrical tower are observed to be α and β respectively, and at a point c feet nearer to the tower they are γ and δ; shew that its height is

$$\frac{c}{\cot\alpha - \cot\gamma},$$ and its breadth

$$2c \frac{\sin\frac{\beta}{2}\sin\frac{\delta}{2}}{\sin\frac{\delta}{2} - \sin\frac{\beta}{2}}.$$

37. A tower, 51 feet high, has a mark at the height of 25 feet from the ground; find at what distance the two parts subtend equal angles to an eye at the height of 5 feet from the ground. *Ans.* 160 feet.

38. The angular elevation of a tower at a place A due south of it is α, and at a place B due west of A, and at a distance c from it, the elevation is β; shew that the height of the tower is

$$\frac{c \sin\alpha \sin\beta}{\sqrt{\{\sin(\alpha+\beta)\sin(\alpha-\beta)\}}}.$$

EXAMPLES. 197

39. A ship sailing on a S.S.W. course bore due south, and the angle subtended by the ship was 20' 15", and her length was known to be 160 feet; find her distance.

Ans. 1.94 miles.

40. From the top of a mountain three miles high the true depression of the horizon was 2° 13' 27"; required the diameter of the earth, supposing it to be a sphere. *Ans.* 7952 miles.

41 At noon, on the shortest day of the year, the shadow of a perpendicular post was seven times as long as its shadow at noon on the longest day; find the latitude, the sun's declination being 23° 28'. *Ans.* 38° 27' 47".5.

42. When the sun's declination was 15° 7' 12".5 N. the shadow of a perpendicular post was to the height of the post as 5 to 3; find the latitude.

Ans. 74° 9' 23" N., or 43° 54' 58" S., according as the shadow fell N. or S. of the post.

43 The length of the shadow of a perpendicular object was 4 feet, and its longest shadow when sloping was 5 feet; find the sun's altitude. *Ans.* 36° 52' 11".5.

44 If a ship sail from a certain place 174 miles eastward on a parallel of latitude, then due south 5°, and then westward on a parallel of latitude 194 miles, and reach the same longitude; required the latitude of the place arrived at, supposing the earth a sphere 7912 miles in diameter.

Ans. 48° 43' 22" N.

45. A ship sails westward on a parallel of latitude 125 miles which are found to be equal to 2° 30' of longitude; find her latitude. *Ans.* 43° 36' nearly.

46. An object 12 feet high, standing on the top of a tower, subtends an angle of 1° 54' 10" at a point 250 feet from the base; find the height of the tower. *Ans.* 160.85 feet.

47. A statue 10 feet high, standing on a column 100 feet high, subtends at the eye of an observer in the horizontal plane on which the column stands, the same angle as a man 6 feet high standing at the foot of the column; find the distance of the observer from the column, his height being 6 feet.

Ans. 121.095 feet.

48. A flagstaff 8 feet high, standing on the top of a tower, subtends an angle of 57' 17".75 at 100 yards from the foot of the tower; find its height. *Ans.* 232 feet.

49. A tree leans towards the north, and at two points due south at distances a and b respectively from the base, the angular elevations of the tree are α and β. If θ be the inclination of the tree, and h the perpendicular height, shew that

$$\tan \theta = \frac{b-a}{b \cot \alpha - a \cot \beta}, \quad h = \frac{b-a}{\cot \beta - \cot \alpha}.$$

50. Four inaccessible objects A, B, C, D are situated in the same straight line, and visible from only one point E. The distance between A and B is 20 chains, and between C and D 12 chains; the angle $AEB = 20°$, $AEC = 38°$, and $AED = 50°$; find the distance between B and C. *Ans.* 16.48 chains.

51. If the three segments AB, BC, CD of the straight line in the last problem, be represented by a, x, b respectively, and subtend at E angles α, β, γ respectively, shew that

$$x^2 + (a+b)x = ab \frac{\sin \beta \sin (\alpha + \beta + \gamma)}{\sin \alpha \sin \gamma},$$

and $\quad \dfrac{\operatorname{cosec}^2 \beta}{ab} = \left(\dfrac{\cot \alpha + \cot \beta}{a+x} \right) \left(\dfrac{\cot \beta + \cot \gamma}{b+x} \right).$

52. A vertical pillar standing on a horizontal plane appears of the same breadth all the way up, to a spectator standing at a given point on the plane; shew that at any point of the pillar

EXAMPLES. 199

whose angular elevation is θ, the radius is $a \sec \theta$, where a is the radius of the base.

53. From the top of a mountain the angles of depression of two stations on the plane at its foot are observed to be α and β, and the difference of their bearings is γ. If a be the distance between the stations, shew that the height of the mountain is

$$\frac{a \sin \alpha \sin \beta}{\sin (\alpha + \beta) \cos \theta}, \text{ where } \sin^2 \theta = \frac{\sin 2\alpha \sin 2\beta}{\sin^2 (\alpha + \beta)} \cos^2 \frac{\gamma}{2}.$$

54. The boundaries of a tract of land are described in a deed as follows: "Commencing at the intersection of two roads, thence on a course N. 52° E. 21.28 chains; thence S. 29° 45' E. 8.18 chains; thence S. 31° 45' W. 15.36 chains; thence to the point of beginning." Find the bearing of the last course and the area of the tract of land.

Ans. Bearing, N. 61° W.; area = 19 ac. 2 r. 36 p.

55. Calculate the bearings of the last two courses and the area from the following field notes: Commencing at a post, thence N. 45° W. 20 chains; thence N. 18° E. 12.25 chains; thence E. 12.80 chains; thence N. 32° E. 6.50 chains; thence S. 45° 30' E. 13.20 chains; thence 14.75 chains; and thence to the point of beginning 16.30 chains.

Ans. S. and S. 65° 25' W.; area = 59 acres.

56. The boundaries of a tract of land are: AB, W. 25 chains; BC, N. 32° 15' W. 16.09 chains; CD, N. 20° E. 15.50 chains; DE, E. 25 chains; EF, S. 30° E.; and FA, S. 25° W. to the point of beginning. A line is run from A cutting off 70 acres 1 rood 33 perches from the west side; find the second point in which this line cuts the boundary.

Ans. The side DE, 18 chains east of D.

57. A ship sailed from latitude 51° 24' N. as follows: S.E. 40 miles; N.E. 28 miles; S.W.b.W. 52 miles; N.W.b.W. 30

miles; S.S.E. 36 miles; S.E.b.E. 58 miles; find her bearing and distance from the point of departure.

Ans. Bearing S. 25° 59′ E.; distance = 95.87 miles.

58. A ship sailed 320 miles on a parallel of latitude from longitude 81° 36′ W. to longitude 90° W.; in what latitude was she? *Ans.* 56° 30′ 47″.3.

59. Twilight ceases when the sun is 18° below the horizon, find the latitude of the places at which twilight lasts all night at the time of the summer solstice, the sun's declination being 23° 27′ N. *Ans.* 48° 33′ N

60. The meridian altitude of a star is 64° 10′, and its depression below the horizon at midnight is 28° 30′; find the latitude of the place and the declination of the star.

Ans. Latitude, 43° 40′ N; declination, 17° 50′ N.

61. At two places on the same meridian, one in latitude 59° 31′ 30″ N., the other in latitude 33° 56′ S, the moon's meridian altitude was observed to be, at the northern station 43° 47′ 40″, and at the southern station 41° 21′ 40″; find the moon's distance from the northern station, the earth being regarded a sphere whose radius is 3956 miles.

Ans. 238020 miles.

62. When the moon's horizontal parallax is 57′ 32″, and the apparent altitude 50° 40′, what is the parallax in altitude?

Ans. 36′ 28″.

63. Find the augmented semi-diameter of the moon when her true semi-diameter is 15′ 42″, her horizontal parallax and apparent altitude being the same as in the last question.

Ans. 12″.41.

64. When the moon's horizontal parallax is 58′ 10″ and her *true* altitude 50°, find her parallax in altitude.

Ans. 37′ 52″.6.

CHAPTER XI.

CIRCLES INSCRIBED IN AND CIRCUMSCRIBED ABOUT A TRIANGLE, POLYGONS, AREA OF A CIRCLE, ETC.

152. To find the radius of the Circle inscribed in a Triangle.

The centre O is in the intersection of the three lines bisecting the angles of the triangle. Let the inscribed circle touch the sides of the triangle in the points D, E and F. Join OD, OE, OF; the angles at D, E and F are right-angles. (*Euc.* III., 18.)

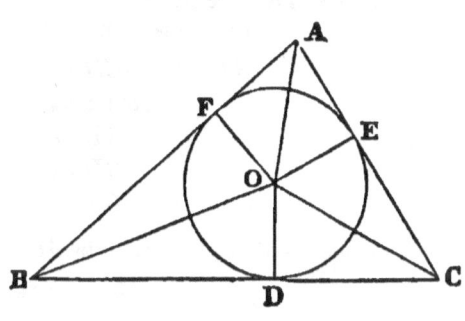

Let $OD = OE = OF = r$, the radius of the circle;

then area of triangle $BOC = \tfrac{1}{2}BC.OD = \dfrac{ar}{2}$,

area of triangle $COA = \tfrac{1}{2}AC.OE = \dfrac{br}{2}$,

area of triangle $AOB = \tfrac{1}{2}AB.OF = \dfrac{cr}{2}$;

therefore, by addition,

$$(a+b+c)\frac{r}{2} = \text{area of triangle } ABC = \Delta.$$

Put, as before, $\quad a+b+c = 2s$,

then $\quad r = \dfrac{\Delta}{s}.$ \hfill (218)

That is, the radius of the inscribed circle is equal to the area of the triangle divided by half the sum of the sides. Different forms can therefore be obtained for the radius by employing the various expressions already given for the area of the triangle.

Again, from the triangle BOC we have by (204)
$$OD = \frac{BC \sin OBD \sin OCD}{\sin (OBD + OCD)},$$

or
$$r = \frac{a \sin \frac{B}{2} \sin \frac{C}{2}}{\sin \frac{1}{2}(B+C)} = \frac{a \sin \frac{B}{2} \sin \frac{C}{2}}{\cos \frac{A}{2}}. \qquad (219)$$

From the figure we have
$$AE = AF, \quad BD = BF, \quad CD = CE;$$
hence
$$AE + BD + CD = s,$$
or
$$AE + a = s,$$
therefore
$$AE = s - a = AF.$$
Similarly
$$\left. \begin{array}{c} BD = s - b = BF. \\ CE = s - c = CD. \end{array} \right\} \qquad (220)$$

From the right-angled triangle AOE, we have
$$OE = AE \tan OAE$$

or
$$\left. r = (s-a) \tan \frac{A}{2}. \right.$$

Similarly
$$\left. \begin{array}{c} r = (s-b) \tan \frac{B}{2}. \\ r = (s-c) \tan \frac{C}{2}. \end{array} \right\} \qquad (221)$$

Multiplying the last three equations together, we have
$$r^3 = (s-a)(s-b)(s-c) \tan \frac{A}{2} \tan \frac{B}{2} \tan \frac{C}{2}$$
$$= \frac{\Delta^2}{s} \tan \frac{A}{2} \tan \frac{B}{2} \tan \frac{C}{2}$$
$$= r^2 s \tan \frac{A}{2} \tan \frac{B}{2} \tan \frac{C}{2},$$

or
$$r = s \tan \frac{A}{2} \tan \frac{B}{2} \tan \frac{C}{2}. \qquad (222)$$

From the figure we have

$$AO = AE \sec OAE$$
$$= (s-a) \sec \frac{A}{2} = \frac{s-a}{\cos \frac{A}{2}}, \qquad (223)$$

Similarly
$$\left. \begin{array}{l} = \dfrac{bc}{s} \cos \dfrac{A}{2} = \sqrt{\dfrac{s-a}{s} \cdot bc}. \\[2mm] BO = \dfrac{ac}{s} \cos \dfrac{B}{2} = \sqrt{\dfrac{s-b}{s} \cdot ac}. \\[2mm] CO = \dfrac{ab}{s} \cos \dfrac{C}{2} = \sqrt{\dfrac{s-c}{s} \cdot ab}. \end{array} \right\} \qquad (224)$$

153. To find the radii of the Escribed Circles.

The escribed circles are the three circles which touch one side of a triangle and the other two produced, and are exterior to the triangle.

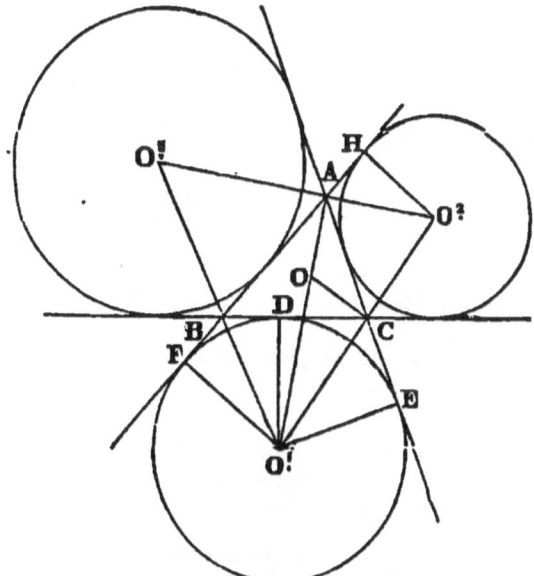

Let ABC be a triangle, bisect the exterior angles at B and C by BO_1, CO_1, then O_1, will be the centre of the escribed circle which touches BC and the other two sides AB, AC, produced. Join AO_1 which will also bisect the angle A, and draw O_1D, O_1E, O_1F, to the points of contact D, E, F; then the angles at D, E and F are right angles.

Let $O_1D = O_1E = O_1F = r_1$, the radius of that circle which touches the side a, and let r_2, r_3 denote the radii of those that touch the sides b and c respectively.

The quadrilateral ABO_1C = the triangles AO_1C, AO_1B,
= the triangles ABC, BO_1C,
therefore, the triangles AO_1C, AO_1B = the triangles ABC, BO_1C,

or
$$\frac{br_1}{2} + \frac{cr_1}{2} = \Delta + \frac{ar_1}{2},$$

$$(b+c-a)r_1 = 2\Delta,$$

therefore $\quad r_1 = \dfrac{\Delta}{s-a}.$

In like manner we find $\quad r_2 = \dfrac{\Delta}{s-b},\qquad$ (225)

and $\quad r_3 = \dfrac{\Delta}{s-c}.$

Bisect the angle ACB by CO, then O is the centre of the inscribed circle, and OCO_1 is a right angle; hence the angle $CBO_1 = 90° - \dfrac{B}{2}$ and $BCO_1 = 90° - \dfrac{C}{2}$.

From the triangle BCO_1 we have by (204)

$$O_1D = a \frac{\sin CBO_1 \, \sin BCO_1}{\sin (CBO_1 + BCO_1)},$$

or $\quad r_1 = \dfrac{a \sin\left(90° - \dfrac{B}{2}\right) \sin\left(90° - \dfrac{C}{2}\right)}{\sin\{180° - \tfrac{1}{2}(B+C)\}}$

that is
$$r_1 = \frac{a \cos\frac{B}{2} \cos\frac{C}{2}}{\cos\frac{A}{2}}.$$

Similarly
$$r_2 = \frac{b \cos\frac{A}{2} \cos\frac{C}{2}}{\cos\frac{B}{2}}.$$
(226)

$$r_3 = \frac{c \cos\frac{A}{2} \cos\frac{B}{2}}{\cos\frac{C}{2}}.$$

From the figure we have

$$AE = AC + CE = AC + CD,$$
$$AF = AB + BF = AB + BD.$$

But $\quad AE = AF,$

therefore $\quad 2AE = 2AF = AB + AC + BC$
$$= a + b + c,$$

and $\quad AE = AF = \tfrac{1}{2}(a+b+c) = s.$

Also $\quad CD = CE = AE - b = s - b,$ (227)

and $\quad BD = FB = AF - c = s - c.$

From the right-angled triangle AO_1E we have

$$O_1E = AE \tan EAO_1$$

or $\quad r_1 = s \tan\dfrac{A}{2}.$

Similarly $\quad r_2 = s \tan\dfrac{B}{2}.$ (228)

$\quad r_3 = s \tan\dfrac{C}{2}.$

From the right-angled triangles AEO_1, CDO_1, BDO_1 we easily find

$$\left.\begin{array}{l} AO_1 = AE \sec EAO_1 = s \sec \dfrac{A}{2}, \\[2mm] BO_1 = BD \sec DBO_1 = (s-c) \csc \dfrac{B}{2}, \\[2mm] CO_1 = DC \sec DCO_1 = (s-b) \csc \dfrac{C}{2}. \end{array}\right\} \quad (229)$$

Similar expressions may be easily deduced for AO_2, BO_2, &c.

From (228) we find

$$\dfrac{r_1}{r_2} = \dfrac{\tan \dfrac{A}{2}}{\tan \dfrac{B}{2}}, \quad \text{or} \quad r_1 \tan \dfrac{B}{2} = r_2 \tan \dfrac{A}{2},$$

that is, $\quad BF = AH$, and therefore $AF = BH = s$.

154. **To find the distance between the centre of the Inscribed Circle and that of one of the Escribed Circles.**

From the figure we have

$$OO_1 = AO_1 - AO$$

$$= s \sec \dfrac{A}{2} - (s-a) \sec \dfrac{A}{2}, \quad \text{by (229) and (223)}$$

$$\left.\begin{array}{l} = a \sec \dfrac{A}{2}. \\[2mm] \text{Similarly} \quad OO_2 = b \sec \dfrac{B}{2}. \\[2mm] OO_3 = c \sec \dfrac{C}{2}. \end{array}\right\} \quad (230)$$

CIRCUMSCRIBED CIRCLE. 207

155. To find the radius of the Circle circumscribed about a Triangle.

Let ABC be a triangle and O the centre of the circumscribing circle which is found geometrically by *Euc.* IV., 5.

Draw the diameter BOA' and join $A'C$, then the angle $BA'C =$ angle BAC, because they are in the same segment, and BCA' being in a semi-circle, is a right angle. Denote the radius BO by R,

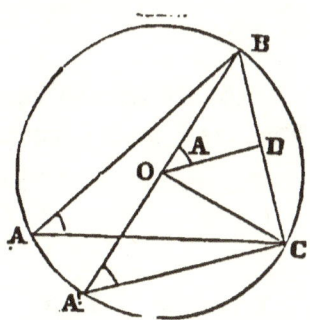

then $\qquad A'B \sin BA'C = BC,$

or $\qquad 2R \sin A = a,$

whence $\qquad R = \dfrac{a}{2 \sin A}$.

$\qquad\qquad\qquad = \dfrac{b}{2 \sin B}$. $\qquad$ (231)

$\qquad\qquad\qquad = \dfrac{c}{2 \sin C}$,

by (165).

Or we may proceed as follows:

Draw OD perpendicular to BC,

then $\qquad OB \sin BOD = BD,$

or $\qquad R \sin A = \dfrac{a}{2}$,

whence $\qquad R = \dfrac{a}{2 \sin A}$, as before.

To express R in terms of the sides and area we have

$$R = \dfrac{a}{2 \sin A}$$

$$= \dfrac{abc}{2bc \sin A}$$

$$= \dfrac{abc}{4\Delta} \quad \text{by (199).} \qquad (232)$$

From (180), (181) and (182) we have

$$\cos \frac{A}{2} \cos \frac{B}{2} \cos \frac{C}{2} = \frac{\Delta s}{abc},$$ which combined with (232) gives

$$R = \frac{s}{4} \sec \frac{A}{2} \sec \frac{B}{2} \sec \frac{C}{2}. \qquad (233)$$

156. Relations between the radii of the Inscribed, Escribed and Circumscribed Circles.

The product of (218) and (225) is

$$rr_1r_2r_3 = \frac{\Delta^4}{s(s-a)(s-b)(s-c)} = \Delta^2. \qquad (234)$$

From (225) we have

$$\frac{1}{r_1} + \frac{1}{r_2} + \frac{1}{r_3} = \frac{s-a+s-b+s-c}{\Delta}$$

$$= \frac{3s-(a+b+c)}{\Delta} = \frac{s}{\Delta} = \frac{1}{r}. \qquad (235)$$

From (222) and (228) we have

$$r_1+r_2+r_3-r = s(\tan\frac{A}{2}+\tan\frac{B}{2}+\tan\frac{C}{2}-\tan\frac{A}{2}\tan\frac{B}{2}\tan\frac{C}{2}).$$

Dividing by (233) we obtain

$$\frac{r_1+r_2+r_3-r}{R} = 4(\sin\frac{A}{2}\cos\frac{B}{2}\cos\frac{C}{2}+\cos\frac{A}{2}\sin\frac{B}{2}\cos\frac{C}{2}$$

$$+\cos\frac{A}{2}\cos\frac{B}{2}\sin\frac{C}{2}-\sin\frac{A}{2}\sin\frac{B}{2}\sin\frac{C}{2})$$

$$= 4(\sin\frac{A}{2}\cos\frac{B+C}{2}+\cos\frac{A}{2}\sin\frac{B+C}{2})$$

$$= 4(\sin^2\frac{A}{2}+\cos^2\frac{A}{2}) = 4,$$

therefore
$$r_1+r_2+r_3-r = 4R. \qquad (236)$$

From (231) we have

$$R = \frac{a}{2\sin A} = \frac{a}{4\sin\frac{A}{2}\cos\frac{A}{2}},$$

INSCRIBED AND CIRCUMSCRIBED CIRCLES. 209

whence $a = 4R \sin\dfrac{A}{2} \cos\dfrac{A}{2}$, which substituted in (219) gives

$$\left. \begin{aligned} r &= 4R \sin\dfrac{A}{2} \sin\dfrac{B}{2} \sin\dfrac{C}{2} \\ &= R(\cos A + \cos B + \cos C - 1) \quad \text{by (112 bis).} \end{aligned} \right\} \quad (237)$$

In a similar manner we obtain from (226)

$$\left. \begin{aligned} r_1 &= 4R \sin\dfrac{A}{2} \cos\dfrac{B}{2} \cos\dfrac{C}{2}, \\ r_2 &= 4R \cos\dfrac{A}{2} \sin\dfrac{B}{2} \cos\dfrac{C}{2}, \\ r_3 &= 4R \cos\dfrac{A}{2} \cos\dfrac{B}{2} \sin\dfrac{C}{2}. \end{aligned} \right\} \quad (238)$$

157. To find the distance between the centres of the Inscribed and Circumscribed Circles.

Let O be the centre of the inscribed, and Q that of the circumscribed circle. Draw OE, and QF perpendicular to AB. Let $OQ = D$. The angle $AQF = C$, therefore $FAQ = 90° - C$, and $EAO = \dfrac{A}{2}$;

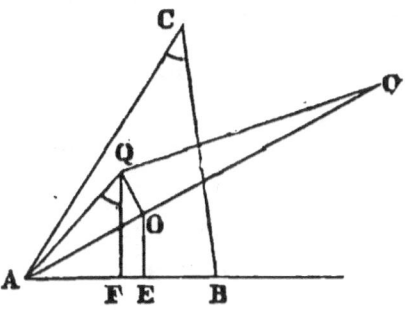

hence $OAQ = 90° - C - \dfrac{A}{2} = \tfrac{1}{2}(B - C)$, and $AO = \dfrac{r}{\sin\dfrac{A}{2}}$.

By (169) we have

$$OQ^2 = AQ^2 + AO^2 - 2AQ \cdot AO \cos OAQ$$

or $$D^2 = R^2 + \dfrac{r^2}{\sin^2\dfrac{A}{2}} - \dfrac{2Rr \cos \tfrac{1}{2}(B-C)}{\sin\dfrac{A}{2}}; \qquad (a)$$

but from (237) we have $$\dfrac{r^2}{\sin^2\dfrac{A}{2}} = \dfrac{4Rr \sin\dfrac{B}{2} \sin\dfrac{C}{2}}{\sin\dfrac{A}{2}},$$

15

therefore $\quad D^2 = R^2 + \dfrac{4Rr \sin \dfrac{B}{2} \sin \dfrac{C}{2}}{\sin \dfrac{A}{2}} - \dfrac{2Rr \cos \tfrac{1}{2}(B-C)}{\sin \dfrac{A}{2}}$

$$= R^2 - 2Rr \dfrac{\cos \tfrac{1}{2}(B+C)}{\sin \dfrac{A}{2}} = R^2 - 2Rr. \qquad (239)$$

Otherwise as follows. From (a) we have

$$(D^2 - R^2) \sin^2 \dfrac{A}{2} = r^2 - 2Rr \sin \dfrac{A}{2} \cos \tfrac{1}{2}(B - C)$$

or $\quad \tfrac{1}{2}(D^2 - R^2)(1 - \cos A) = r^2 - 2Rr \cos \tfrac{1}{2}(B+C) \cos \tfrac{1}{2}(B - C)$

$$= r^2 - Rr(\cos B + \cos C).$$

In a similar manner we find

$$\tfrac{1}{2}(D^2 - R^2)(1 - \cos B) = r^2 - Rr(\cos A + \cos C),$$

the difference of which is

$$\tfrac{1}{2}(D^2 - R^2)(\cos A - \cos B) = -Rr(\cos A - \cos B),$$

whence $\quad D^2 = R^2 - 2Rr.$

158. To find the distance between the centres of the Escribed and Circumscribed Circles.

Let O_1 be the centre of the escribed circle touching the side a (Fig. of last Art.), and $O_1Q = D_1$, then since $AO_1 = \dfrac{r_1}{\sin \dfrac{A}{2}}$, we have from the triangle O_1AQ

$$D_1^2 = R^2 + \dfrac{r_1^2}{\sin^2 \dfrac{A}{2}} - \dfrac{2Rr_1 \cos \tfrac{1}{2}(B-C)}{\sin \dfrac{A}{2}};$$

but from (238) we have $\quad \dfrac{r_1^2}{\sin^2 \dfrac{A}{2}} = \dfrac{4Rr_1 \cos \dfrac{B}{2} \cos \dfrac{C}{2}}{\sin \dfrac{A}{2}},$

INSCRIBED AND CIRCUMSCRIBED CIRCLES. 211

therefore
$$D_1^2 = R^2 + \frac{4Rr_1 \cos\frac{B}{2} \cos\frac{C}{2}}{\sin\frac{A}{2}} - \frac{2Rr_1 \cos\frac{1}{2}(B-C)}{\sin\frac{A}{2}}.$$

Similarly
$$\left.\begin{array}{l} = R^2 + 2Rr_1 \,. \\ D_2^2 = R^2 + 2Rr_2 \,. \\ D_3^2 = R^2 + 2Rr_3 \,. \end{array}\right\} \qquad (240)$$

The sum of (239) and (240) gives

$$D^2 + D_1^2 + D_2^2 + D_3^2 = 4R^2 + 2R(r_1 + r_2 + r_3 - r)$$
$$= 12R^2, \quad \text{by (236)} \qquad (241)$$

159. To shew that the distances between the centres of the Escribed Circles and the centre of the Inscribed Circle are bisected by the circumference of the Circumscribed Circle.

Let Q be the centre of the circumscribed circle of the triangle ABC. Bisect CB in D and draw DE perpendicular to CB, and join AE; then the angle CAB is bisected by AE, and the centres of the inscribed and escribed circles are in AE; let them be at O and O_1 respectively. Join EB and OB.

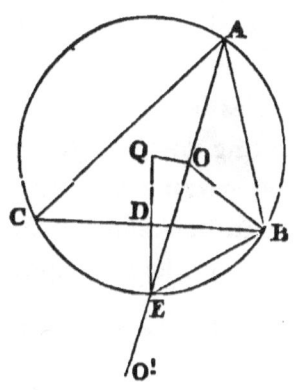

Since the angle CBE is equal to the angle CAE, we have

and
$$EBO = CBE + CBO = \tfrac{1}{2}(A+B),$$
$$EOB = OAB + OBA = \tfrac{1}{2}(A+B), \quad (Euc. \text{ I.}, 32)$$
therefore
$$EO = EB = DB \sec DBE$$

or
$$OE = \frac{a}{2} \sec \frac{A}{2}$$

but by (230)
$$OO_1 = a \sec \frac{A}{2},$$

therefore
$$OO_1 = 2OE.$$

212 PLANE TRIGONOMETRY.

We may here notice, for the sake of problems, that if OQ be joined and produced both ways to meet the circumference, we have by *Euc.* III., 35,

$$(R+OQ)(R-OQ) = EO \cdot AO$$

$$= \frac{a}{2} \sec \frac{A}{2} \cdot \frac{r}{\sin \frac{A}{2}}$$

$$= \frac{ar}{2 \sin \frac{A}{2} \cos \frac{A}{2}} = \frac{2ar}{2 \sin A} = 2Rr,$$

whence $\quad OQ^2 = R^2 - 2Rr.$

160. To find the Perimeter and Area of a Regular Polygon of any number of sides, which is inscribed in or described about a Circle of given radius.

Let AEB be an arc of a circle whose centre is O; AB a side of the inscribed regular polygon of n sides; OE at right angles to AB, and therefore bisecting it; CD a tangent at E, and meeting OA and OB produced in C and D; then CD is a side of the circumscribed regular polygon of n sides.

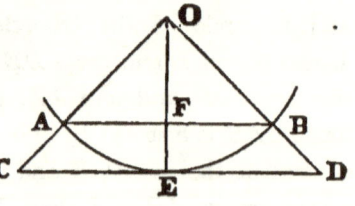

Let $AO = r$. The angle $AOB = \frac{2\pi}{n}$, and therefore $AOF = \frac{\pi}{n}$,

$$AB = 2AF = 2r \sin AOF = 2r \sin \frac{\pi}{n};$$

therefore perimeter of inscribed polygon $= 2nr \sin \frac{\pi}{n}.$ \quad (242)

Area of the inscribed polygon $= n \times$ triangle AOB

$$= n \frac{AO \cdot BO}{2} \sin AOB$$

$$= \tfrac{1}{2} n r^2 \sin \frac{2\pi}{n}. \quad (243)$$

AREA OF A CIRCLE.

Again, $CD = 2CE = 2r \tan COE = 2r \tan \dfrac{\pi}{n}$,

therefore

perimeter of circumscribed polygon $= 2nr \tan \dfrac{\pi}{n}$. (244)

Area of the circumscribed polygon $= n \times$ triangle COD

$$= n \cdot OE \cdot CE = nr^2 \tan \dfrac{\pi}{n}. \quad (245)$$

From (243) and (245) we have

$$\dfrac{\text{area inscribed polygon}}{\text{area circumscribed polygon}} = \dfrac{\text{area triangle } AOF}{\text{area triangle } COE} = \dfrac{OF^2}{OE^2}$$

$$= \dfrac{OF^2}{OA^2} = \cos^2 \dfrac{\pi}{n}. \quad (246)$$

161. Circumference and Area of a Circle.

The circumference of the circle is evidently intermediate in length to the perimeters of the inscribed and circumscribed polygons, hence the circumference of the circle lies between

$$2nr \sin \dfrac{\pi}{n} \text{ and } 2nr \tan \dfrac{\pi}{n},$$

that is, between

$$2\pi r \dfrac{\sin \dfrac{\pi}{n}}{\dfrac{\pi}{n}} \text{ and } 2\pi r \dfrac{\tan \dfrac{\pi}{n}}{\dfrac{\pi}{n}}.$$

Now, let the number of sides be indefinitely increased, then $\dfrac{\pi}{n}$ is very small, and when $n = \infty$, $\dfrac{\pi}{n} = 0$, therefore (Art. 74)

$$\dfrac{\sin \dfrac{\pi}{n}}{\dfrac{\pi}{n}} = \dfrac{\tan \dfrac{\pi}{n}}{\dfrac{\pi}{n}} = 1, \text{ when } \dfrac{\pi}{n} = 0.$$

Hence when n is infinite, the perimeters of both polygons between which the circumference of the circle lies, become $2\pi r$, therefore the circumference of the circle $= 2\pi r$.

Again, the area of the circle lies between

$$\tfrac{1}{2}nr^2 \sin \frac{2\pi}{n} \text{ and } nr^2 \tan \frac{\pi}{n},$$

or between
$$\pi r^2 \frac{\sin \dfrac{2\pi}{n}}{\dfrac{2\pi}{n}} \text{ and } \pi r^2 \frac{\tan \dfrac{\pi}{n}}{\dfrac{\pi}{n}},$$

each of which becomes πr^2 when $n = \infty$.

Therefore the area of the circle which is always intermediate in magnitude to the areas of the two polygons

$$= \pi r^2. \qquad (246 \; bis)$$

Hence, if θ be the circular measure of the angle of a sector such as AOE in the last Figure, then the area of the sector $AOE = \tfrac{1}{2}\theta r^2$.

162. To find the Angles and the Area of a Quadrilateral inscribed in a Circle.

Let $AB=a$, $BC=b$, $CD=c$, $DA=d$; join AC: then

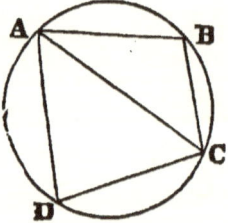

Area $ABCD$ = area $ABC + ADC$
$= \tfrac{1}{2}ab \sin B + \tfrac{1}{2}cd \sin D$
$= \tfrac{1}{2}(ab+cd) \sin B.$ \qquad (1)

We have by (169)
$$AC^2 = a^2 + b^2 - 2ab \cos B = c^2 + d^2 + 2cd \cos B$$

since $\cos D = -\cos B$, (Art. 39)

therefore
$$\cos B = \frac{a^2 + b^2 - c^2 - d^2}{2(ab+cd)},$$

hence
$$2 \sin^2 \frac{B}{2} = 1 - \cos B = \frac{(c+d)^2 - (a-b)^2}{2(ab+cd)} \qquad (2)$$

and
$$2 \cos^2 \frac{B}{2} = 1 + \cos B = \frac{(a+b)^2 - (c-d)^2}{2(ab+cd)}, \qquad (3)$$

therefore
$$\tan^2 \frac{B}{2} = \frac{(c+d)^2 - (a-b)^2}{(a+b)^2 - (c-d)^2}$$

$$= \frac{(-a+b+c+d)(a-b+c+d)}{(a+b-c+d)(a+b+c-d)}$$

$$= \frac{(s-a)(s-b)}{(s-c)(s-d)}, \qquad (247)$$

where $2s = a+b+c+d$.

Multiplying (2) and (3) together we easily find

$$\sin B = \frac{2}{ab+cd} \sqrt{(s-a)(s-b)(s-c)(s-d)}$$

which substituted in (1) gives

$$\text{Area } ABCD = \sqrt{(s-a)(s-b)(s-c)(s-d)}. \qquad (248)$$

If we substitute the value of $\cos B$, in the expression for AC^2, we readily find

$$AC^2 = \frac{(ac+bd)(ad+bc)}{ab+cd}.$$

If R_1 denote the radius of the circle, we have

$$R_1 = \frac{AC}{2 \sin B} = \frac{1}{4} \sqrt{\frac{(ab+cd)(ac+bd)(ad+bc)}{(s-a)(s-b)(s-c)(s-d)}}. \qquad (249)$$

Examples.

1. In any triangle prove that

(1) $a \cot A + b \cot B + c \cot C = 2(R+r)$,

(2) $a \cos A + b \cos B + c \cos C = 4R \sin A \sin B \sin C$.

From (1) we have

$$a \cot A + b \cot B + c \cot C = \frac{a \cos A}{\sin A} + \frac{b \cos B}{\sin B} + \frac{c \cos C}{\sin C}$$

$$= \frac{a}{\sin A} (\cos A + \cos B + \cos C), \text{ by (165)}$$

$$= 2R(1 + \frac{r}{R}), \text{ by (231) and (237)}$$

$$= 2(R+r).$$

From (2) we have
$$a \cos A + b \cos B + c \cos C = \frac{a \sin A \cos A}{\sin A} + \frac{b \sin B \cos B}{\sin B} + \frac{c \sin C \cos C}{\sin C}$$
$$= \frac{a}{2 \sin A} (\sin 2A + \sin 2B + \sin 2C)$$
$$= 4R \sin A \sin B \sin C. \quad \text{by (112).}$$

2. If a, β, γ be the distances of the angles of a triangle from the centre of the inscribed circle, and a, b, c the sides opposite to them, prove that
$$a^2 a + \beta^2 b + \gamma^2 c = abc.$$

In the Figure of Art. 152, let $AO = a$, $BO = \beta$ and $CO = \gamma$;

then
$$a = \frac{r}{\sin \frac{A}{2}}, \quad \beta = \frac{r}{\sin \frac{B}{2}}, \quad \gamma = \frac{r}{\sin \frac{C}{2}},$$

hence
$$a^2 a + \beta^2 b + \gamma^2 c = r^2 \left(\frac{a}{\sin^2 \frac{A}{2}} + \frac{b}{\sin^2 \frac{B}{2}} + \frac{c}{\sin^2 \frac{C}{2}} \right)$$
$$= 2r^2 \left(\frac{a}{\sin A} \cot \frac{A}{2} + \frac{b}{\sin B} \cot \frac{B}{2} + \frac{c}{\sin C} \cot \frac{C}{2} \right)$$
$$= \frac{2ar^2}{\sin A} \left(\cot \frac{A}{2} + \cot \frac{B}{2} + \cot \frac{C}{2} \right).$$

But $r \cot \frac{A}{2} = AE$, $r \cot \frac{B}{2} = BF$, $r \cot \frac{C}{2} = DC$,

therefore $r (\cot \frac{A}{2} + \cot \frac{B}{2} + \cot \frac{C}{2}) = AE + BF + DC = s$,

hence
$$a^2 a + \beta^2 b + \gamma^2 c = \frac{2a}{\sin A} rs = \frac{2a}{\sin A} \Delta,$$
$$= \frac{2a}{\sin A} \cdot \frac{bc}{2} \sin A = abc.$$

3. Prove that the radius of the circle which circumscribes the triangle $O_1 O_2 O_3$, formed by joining the centres of the escribed circles of a triangle ABC, is $\frac{abc}{2\Delta}$, or equal to the diameter of the

EXAMPLES. 217

circle which circumscribes the triangle ABC; and that the area of the triangle $O_1O_2O_3$ is $\dfrac{abc}{2r}$, where r is the radius of the inscribed circle. (See Fig. of Art. 153.)

In the Figure we find the angle $BCO_1 = \tfrac{1}{2}(A+B) = ACO_2$, $CAO_2 = \tfrac{1}{2}(B+C) = BAO_3$, $BO_1C = \tfrac{1}{2}(B+C)$, $AO_2C = \tfrac{1}{2}(A+C)$.

From the triangles AO_2C, BO_1C we find by (165)

$$CO_2 = \frac{b \cos \dfrac{A}{2}}{\cos \dfrac{B}{2}} \quad \text{and} \quad CO_1 = \frac{a \cos \dfrac{B}{2}}{\cos \dfrac{A}{2}},$$

therefore $O_1O_2 = a \dfrac{\cos \dfrac{B}{2}}{\cos \dfrac{A}{2}} + b \dfrac{\cos \dfrac{A}{2}}{\cos \dfrac{B}{2}} = \dfrac{a \cos^2 \dfrac{B}{2} + b \cos^2 \dfrac{A}{2}}{\cos \dfrac{A}{2} \cos \dfrac{B}{2}}$

$$= \frac{\dfrac{as(s-b)}{ac} + \dfrac{bs(s-a)}{bc}}{\sqrt{\dfrac{s(s-a) \cdot s(s-b)}{bc \cdot ac}}} = \frac{2s - a - b}{\sqrt{\dfrac{(s-a)(s-b)}{ab}}} = \frac{c}{\sin \dfrac{C}{2}}.$$

Similarly $O_1O_3 = \dfrac{b}{\sin \dfrac{B}{2}}$ and $O_2O_3 = \dfrac{a}{\sin \dfrac{A}{2}}$.

Let R_1 denote the radius of the circumscribing circle, then

$$R_1 = \frac{O_2O_3}{2 \sin BO_1C} = \frac{a \operatorname{cosec} \dfrac{A}{2}}{2 \cos \dfrac{A}{2}} = \frac{a}{2 \sin \dfrac{A}{2} \cos \dfrac{A}{2}}$$

$$= \frac{a}{\sin A} = \frac{abc}{2\Delta} = 2R.$$

Area of the triangle $O_1O_2O_3 = \tfrac{1}{2} O_1O_3 \cdot O_1O_2 \sin O_2O_1O_3$

$$= \tfrac{1}{2} \frac{b}{\sin \dfrac{B}{2}} \cdot \frac{c}{\sin \dfrac{C}{2}} \cos \dfrac{A}{2}$$

$$= \frac{bc}{2} \cdot \frac{\cos\frac{A}{2}}{\sin\frac{B}{2}\sin\frac{C}{2}}$$

$$= \frac{abc}{2r}, \quad \text{by (219)}$$

$$= \frac{abc}{4\Delta}(a+b+c).$$

4. The sides of a triangle are 13, 14 and 15, find the radii of the inscribed and circumscribed circles. *Ans.* 4 and $8\frac{1}{8}$.

5. Find the radii of the inscribed, circumscribed, and of an escribed circle, when the triangle is equilateral.

Ans. $\frac{a}{6}\sqrt{3},\ \frac{a}{3}\sqrt{3},\ \frac{a}{2}\sqrt{3}$.

6. In the Fig. of Art. 152 prove that $AO = c\sin\frac{B}{2}\sec\frac{C}{2}$.

7. In the Fig. of Art. 153 prove that $AO_1 = c\cos\frac{B}{2}\csc\frac{C}{2}$.

8. In any triangle prove that $r = \dfrac{a-b}{\cot\frac{B}{2}-\cot\frac{A}{2}}$,

and $\quad r_1\cot\frac{A}{2} = r_2\cot\frac{B}{2} = r_3\cot\frac{C}{2} = r\left(\cot\frac{A}{2}+\cot\frac{B}{2}+\cot\frac{C}{2}\right)$.

9. In the *ambiguous* case of triangles, when a, b and A are given, shew that the circles circumscribing both triangles are equal, and that the distance between their centres is

$$\sqrt{(a^2\csc^2 A - b^2)}.$$

10. Prove that the perpendicular from an angle of a triangle on the opposite side, is a harmonic mean between the radii of the adjacent escribed circles.

11. In the Fig. of Art. 153 prove that $OA \cdot OO_1 = 4Rr$.

12. In any triangle prove that $(r_1-r)(r_2-r)(r_3-r) = 4Rr^2$.

13. A circle is described passing through the vertex A of a triangle and touching the base BC at its middle point; prove that its radius is $\dfrac{2(b^2+c^2)-a^2}{8b\sin C}$.

EXAMPLES. 219

14. The sides of a triangle are 13, 20 and 21; find the radii of the inscribed, circumscribed and escribed circles, and the distance between the centres of the inscribed and circumscribed circles.

Ans. $r=4\frac{2}{3}$; $R=10\frac{5}{8}$, $r_1=9$, $r_2=18$, $r_3=21$, and $D=\frac{1}{8}\sqrt{65}$.

15. An inaccessible tower standing on a horizontal plane subtends an angle of 36° at each of three points on the plane, whose distances from each other are 29, 35 and 48 yards; find its height.

Ans. $24\frac{1}{6}\sqrt{5-2\sqrt{5}}$ yards.

16. In any triangle prove that

(1) $\Delta = Rr(\sin A + \sin B + \sin C)$;

(2) $= r^2 \cot\dfrac{A}{2} \cot\dfrac{B}{2} \cot\dfrac{C}{2}$;

(3) $= 2R^2 \sin A \sin B \sin C$;

(4) $= \frac{1}{2}R^2(\sin 2A + \sin 2B + \sin 2C)$.

17. In any triangle prove that

(1) $R = \frac{1}{2}\sqrt[3]{\dfrac{abc}{\sin A \sin B \sin C}}$;

(2) $= \dfrac{a+b+c}{2(\sin A + \sin B + \sin C)}$;

(3) $= \dfrac{abc}{(a+b+c)^2}\left(\cot\dfrac{A}{2} + \cot\dfrac{B}{2} + \cot\dfrac{C}{2}\right)$;

(4) $\dfrac{R}{r} = \dfrac{\sin A + \sin B + \sin C}{2\sin A \sin B \sin C}$;

(5) $2Rr = \dfrac{abc}{a+b+c}$.

18. In any triangle prove that

(1) $r = \dfrac{c \sin A \sin B}{\sin A + \sin B + \sin C}$;

(2) $= \dfrac{abc}{(a+b+c)^2}(\sin A + \sin B + \sin C)$.

19. If the radii of the escribed circles be in arithmetical progression, the tangents of the semi-angles of the triangle are in arithmetical progression; if in harmonical progression, the cotangents of the semi-angles are in arithmetical progression.

20. If p, q, r denote the lines drawn from A, B, C, bisecting the angles of a triangle ABC, and terminated by the circumference of the circumscribed circle, prove that

$$p\cos\frac{A}{2}+q\cos\frac{B}{2}+r\cos\frac{C}{2}=a+b+c.$$

21. A circle is described about a triangle ABC, and another triangle is formed by joining the points of bisection of the arcs subtended by the sides of ABC; prove that the sides of the new triangle are

$$\frac{a}{2}\operatorname{cosec}\frac{A}{2},\ \frac{b}{2}\operatorname{cosec}\frac{B}{2},\ \frac{c}{2}\operatorname{cosec}\frac{C}{2},$$

and that its area $=\dfrac{\Delta}{8}\operatorname{cosec}\dfrac{A}{2}\operatorname{cosec}\dfrac{B}{2}\operatorname{cosec}\dfrac{C}{2},$

where Δ is the area of the triangle ABC.

22. In the Fig. of Art. 155, if BO meet AC in E, prove that
$$R = EO\cos(A-C)\sec B.$$

23. Perpendiculars are drawn from the angles A, B, C of an acute-angled triangle to the opposite sides, and produced to meet the circumscribing circle; if the produced parts are α, β, γ respectively, prove that

$$\frac{a}{\alpha}+\frac{b}{\beta}+\frac{c}{\gamma}=2(\tan A+\tan B+\tan C).$$

24. Perpendiculars AD, BE, CF are drawn from the angles of a triangle ABC to the opposite sides, meeting one another in G; if R and R' be the radii of the circles which circumscribe the triangles ABC and DEF respectively, and r' the radius of the circle inscribed in the triangle DEF, prove that

(1) $AG = 2R\cos A$;   (4) $DG = 2R\cos B\cos C$;

(2) $EF = R\sin 2A$;   (5) Area $ABC = \dfrac{R}{2}(DE+EF+FD)$;

(3) $R = 2R'$;   (6) $r' = 2R\cos A\cos B\cos C$;

(7) $AG+BG+CG = 2(R+r).$

Shew also that the circle which circumscribes the triangle DEF, bisects the sides of the triangle ABC and the lines AG, BG, CG.

25. Prove that

(1) $\tan^2 \dfrac{A}{2} = \dfrac{r r_1}{r_2 r_3}$; (2) $r_1 = \dfrac{b \tan \dfrac{A}{2} \cot \dfrac{B}{2}}{\tan \dfrac{A}{2} + \tan \dfrac{C}{2}} = \dfrac{c \tan \dfrac{A}{2} \cot \dfrac{C}{2}}{\tan \dfrac{A}{2} + \tan \dfrac{B}{2}}$;

(3) $\dfrac{a}{r_1} + \dfrac{b}{r_2} + \dfrac{c}{r_3} = 2\left(\tan \dfrac{A}{2} + \tan \dfrac{B}{2} + \tan \dfrac{C}{2}\right)$.

26. In the Fig. of Art. 153, if r and r' denote the radii of the circles inscribed in the triangles ABC and $O_1 O_2 O_3$ respectively, prove that

$$\dfrac{r}{r'} = \left(\cos \dfrac{A}{2} + \cos \dfrac{B}{2} + \cos \dfrac{C}{2}\right) \tan \dfrac{A}{2} \tan \dfrac{B}{2} \tan \dfrac{C}{2}.$$

27. If the middle points of the sides of a triangle be joined with the opposite angles, and R_1, R_2, R_3, &c., be the radii of the circles described about the six triangles so formed, and r_1, r_2, r_3, &c., the radii of the circles inscribed in the same, prove that

$$R_1 R_3 R_5 = R_2 R_4 R_6, \text{ and } \dfrac{1}{r_1} + \dfrac{1}{r_3} + \dfrac{1}{r_5} = \dfrac{1}{r_2} + \dfrac{1}{r_4} + \dfrac{1}{r_6}.$$

28. If a and a' are homologous sides of two similar triangles described, one about and the other within a circle, prove that

$$a' = 4a \sin \dfrac{A}{2} \sin \dfrac{B}{2} \sin \dfrac{C}{2}.$$

29. If α, β, γ denote the distances from the angles of a triangle to the points of contact of the inscribed circle, prove that

$$r^2 = \dfrac{\alpha \beta \gamma}{\alpha + \beta + \gamma}.$$

30. If a', b', c' denote the distances between the centres of the escribed centres of a triangle, prove that

$$r_1 r_2 r_3 = \dfrac{a' b' c'}{8} \sin A \sin B \sin C.$$

31. With the same notation as in Example 2, prove that

(1) $r = \tfrac{1}{2} \dfrac{\alpha \beta \gamma}{abc}(a+b+c)$; (2) $\dfrac{1}{a \alpha^2} + \dfrac{1}{b \beta^2} + \dfrac{1}{c \gamma^2} = \dfrac{1}{abc}\left(1 + \dfrac{4R}{r}\right)$;

(3) $\alpha^2 \left(\dfrac{1}{b} - \dfrac{1}{c}\right) + \beta^2 \left(\dfrac{1}{c} - \dfrac{1}{a}\right) + \gamma^2 \left(\dfrac{1}{a} - \dfrac{1}{b}\right) = 0.$

32. In any triangle prove that $r = \dfrac{r_1 r_2 r_3}{r_1 r_2 + r_1 r_3 + r_2 r_3}$.

33. The radii of the escribed circles of a triangle are 16, 48 and 52; find the sides. *Ans.* 25, 51 and 52.

34. In any triangle prove that $r_1 r_2 r_3 = r^3 \cot^2 \dfrac{A}{2} \cot^2 \dfrac{B}{2} \cot^2 \dfrac{C}{2}$.

35. If r be the radius of the inscribed circle, and r_a the radius of the circle inscribed between this circle and the sides containing the angle A, shew that

$$r_a = r \dfrac{1 - \sin \dfrac{A}{2}}{1 + \sin \dfrac{A}{2}} = r \tan^2 \left(45° - \dfrac{A}{4}\right).$$

36. The sides of a triangle are in arithmetical progression, and the distance between the centres of the inscribed and circumscribed circles is a geometric mean between the greatest and least; shew that the sides are as

$$\sqrt{5} - 1 : \sqrt{5} : \sqrt{5} + 1.$$

37. In any triangle prove that

(1) $\quad \dfrac{r}{R} = 2 - 2\left(\sin^2 \dfrac{A}{2} + \sin^2 \dfrac{B}{2} + \sin^2 \dfrac{C}{2}\right)$;

(2) $\quad ab + ac + bc = rr_1 + rr_2 + rr_3 + r_1 r_2 + r_1 r_3 + r_2 r_3$;

(3) $\quad 4R = \dfrac{(r_1 + r_2)(r_1 + r_3)(r_2 + r_3)}{r_1 r_2 + r_1 r_3 + r_2 r_3}$.

38. If p_1, p_2, p_3 be the perpendiculars from the angles of a triangle on the sides a, b, c respectively, prove that

(1) $\quad \dfrac{1}{p_1} + \dfrac{1}{p_2} + \dfrac{1}{p_3} = \dfrac{1}{r_1} + \dfrac{1}{r_2} + \dfrac{1}{r_3}$;

(2) $\quad R = \dfrac{2 r r_2 r_3}{p_1 p_2 p_3}$;

(3) $\quad p_1 = \dfrac{2 r_2 r_3}{r_2 + r_3}$;

(4) $\quad \dfrac{1}{p_1} + \dfrac{1}{p_2} - \dfrac{1}{p_3} = \dfrac{1}{r_3}$;

(5) $\quad \dfrac{p_1^2}{p_2 p_3} + \dfrac{p_2^2}{p_1 p_3} + \dfrac{p_3^2}{p_1 p_2} = \dfrac{bc}{a^2} + \dfrac{ac}{b^2} + \dfrac{ab}{c^2}$.

EXAMPLES. 223

39. If r_1, r_2, r_3 be the radii of three circles which touch one another externally, shew that the area of the triangle formed by joining their centres is $\sqrt{(r_1+r_2+r_3)\,r_1r_2r_3}$.

40. If O and Q be the centres of the inscribed and circumscribed circles of a triangle, and if r_a, r_b, r_c be the radii of the circles which circumscribe the triangles BOC, AOC, AOB respectively, and R_a, R_b, R_c the radii of the circles which circumscribe the triangles BQC, AQC, AQB respectively, prove that

$$\frac{a}{R_a}+\frac{b}{R_b}+\frac{c}{R_c}=\frac{abc}{R^3}, \text{ and } \frac{r_a r_b r_c}{abc}=\frac{R}{a+b+c},$$

where R is the radius of the circle which circumscribes the triangle whose sides are a, b and c.

41. Shew that there is only one point within a triangle from which, if perpendiculars be drawn to the sides, circles can be inscribed in each of the three resulting quadrilaterals; and if r_1, r_2, r_3 be the radii of these circles, and r that of the inscribed circle of the triangle, then

$$\left(\frac{1}{r_1}-\frac{1}{r}\right)\left(\frac{1}{r_2}-\frac{1}{r}\right)+\left(\frac{1}{r_2}-\frac{1}{r}\right)\left(\frac{1}{r_3}-\frac{1}{r}\right)+\left(\frac{1}{r_3}-\frac{1}{r}\right)\left(\frac{1}{r_1}-\frac{1}{r}\right)=\frac{1}{r^2}.$$

42. Shew that the area of the triangle $O_1O_2O_3$ (Fig. of Art. 153)

$$=\Delta\left(1+\frac{a}{-a+b+c}+\frac{b}{a-b+c}+\frac{c}{a+b-c}\right),$$

where Δ represents the area of the triangle ABC.

43. $ABCD$ is a quadrilateral inscribed in a circle, shew that
$$AC\sin A = BD\sin B.$$

44. Shew that the perimeters of an equilateral triangle, a square and a hexagon, each containing the same area, are as
$$\sqrt{27} : \sqrt{16} : \sqrt{12}.$$

45. In the Fig. of Art. 152, prove that
$$AO^2+BO^2+CO^2 = ab+ac+bc-\frac{6abc}{a+b+c}.$$

46. Prove that the area of a regular hexagon inscribed in a circle is geometric mean between the areas of the inscribed and circumscribed equilateral triangles.

47. In a regular polygon of n sides, of which a side is $2a$, prove that $R+r=a \cot \dfrac{\pi}{2n}$, where R and r are the radii of the circumscribed and inscribed circles respectively.

48. If A, A_1, A_2, A_3 denote the areas of the four circles which touch the sides of a triangle, A being that of the inscribed circle, prove that
$$\dfrac{1}{\sqrt{A}}=\dfrac{1}{\sqrt{A_1}}+\dfrac{1}{\sqrt{A_2}}+\dfrac{1}{\sqrt{A_3}}.$$

49. The external bisectors of the angles of a triangle are produced to meet the circumference of the circumscribing circle; shew that the area of the triangle formed by joining the three points thus obtained is $\dfrac{R}{4}(a+b+c)$.

50. If r_a, r_b, r_c denote the radii of the circles inscribed in the triangles BOC, AOC, AOB, in the Fig. of Art. 152, prove that
$$\dfrac{a}{r_a}+\dfrac{b}{r_b}+\dfrac{c}{r_c}=2(\cot \dfrac{A}{4}+\cot \dfrac{B}{4}+\cot \dfrac{C}{4}).$$

51. The square of the side of a pentagon inscribed in a circle is equal to the sum of the squares of the sides of a regular hexagon and decagon inscribed in the same circle.

52. Shew that the areas of the circles in *Euc.* IV., 10, are as $5+\sqrt{5} : 2$.

53. If through any point O within a triangle three straight lines be drawn from the angles A, B, C, meeting the opposite sides in D, E, F respectively, prove that
$$\dfrac{OD}{AD}+\dfrac{OE}{BE}+\dfrac{OF}{CF}=1.$$

54. If α, β, γ be the angles which the sides of a triangle subtend at the centre of the inscribed circle, prove that
$$4\sin \alpha \sin \beta \sin \gamma = \sin A + \sin B + \sin C.$$

CHAPTER XII.

INVERSE TRIGONOMETRICAL FUNCTIONS.

163. If $\sin \theta = a$, then 'θ is an angle whose sine is a'; but instead of writing this description of θ at full length, the following notation is employed to express this relation:

$$\theta = \sin^{-1} a.$$

Similarly, if $\cos \phi = b$ and $\tan \psi = c$, then $\phi = \cos^{-1} b$ and $\psi = \tan^{-1} c$.

The student must be careful to remember that the index -1 does not here imply an *algebraic operation*, but merely expresses the relation between the angle and its functions as above enunciated.

The functions $\sin^{-1} a$, $\cos^{-1} b$, &c., are called *inverse* trigonometrical functions, from the nature of the notation which is analagous to that employed in Algebra, where x^{-1} is the inverse of x.

Since there are several angles which have the same sine, several which have the same cosine, &c., it follows that the above equations are more correctly written as follows:

$$\sin^{-1} a = n\pi + (-1)^n \theta,$$
$$\cos^{-1} b = 2n\pi \pm \phi,$$
$$\tan^{-1} c = n\pi + \psi,$$

where n is any integer, either positive or negative.

Any relation between the trigonometrical functions may be expressed by the inverse notation. Thus, we know that

$$\tan \frac{\theta}{2} = \frac{\sin \theta}{1 + \cos \theta}$$

which may be written

$$\frac{\theta}{2} = \tan^{-1} \left(\frac{\sin \theta}{1 + \cos \theta} \right),$$

and therefore

$$\theta = 2 \tan^{-1} \left(\frac{\sin \theta}{1 + \cos \theta} \right),$$

which is read thus,

θ is equal to twice the angle whose tangent is $\dfrac{\sin \theta}{1 + \cos \theta}$.

PLANE TRIGONOMETRY.

164. *Given $\sin^{-1} a$ and $\sin^{-1} b$, to find $\sin^{-1} a \pm \sin^{-1} b$ and $\cos^{-1} a \pm \cos^{-1} b$.*

Let $\qquad\qquad \theta = \sin^{-1} a \quad$ and $\quad \phi = \sin^{-1} b$,
then $\qquad\qquad \sin\theta = a \quad$ and $\sin\phi = b$,
hence $\qquad\qquad \cos\theta = \pm\sqrt{(1-\sin^2\theta)} = \pm\sqrt{(1-a^2)}$,
and $\qquad\qquad \cos\phi = \pm\sqrt{(1-\sin^2\phi)} = \pm\sqrt{(1-b^2)}$,
$$\sin(\theta+\phi) = \sin\theta\cos\phi + \cos\theta\sin\phi$$
$$= \pm a\sqrt{(1-b^2)} \pm b\sqrt{(1-a^2)},$$
therefore $\qquad \theta+\phi = \sin^{-1}\{\pm a\sqrt{(1-b^2)} \pm b\sqrt{(1-a^2)}\}$,
or $\quad \sin^{-1} a + \sin^{-1} b = \sin^{-1}\{\pm a\sqrt{(1-b^2)} \pm b\sqrt{(1-a^2)}\}.\qquad(250)$

In a similar manner we find

$$\sin^{-1} a - \sin^{-1} b = \sin^{-1}\{\pm a\sqrt{(1-b^2)} \mp b\sqrt{(1-a^2)}\}. \qquad(251)$$
$$\cos^{-1} a \pm \cos^{-1} b = \cos^{-1}\{ab \mp \sqrt{(1-a^2)(1-b^2)}\}. \qquad(252)$$

165. *Given $\tan^{-1} a$ and $\tan^{-1} b$, to find $\tan^{-1} a \pm \tan^{-1} b$.*

Let $\qquad\qquad \theta = \tan^{-1} a \quad$ and $\quad \phi = \tan^{-1} b$,
then $\qquad\qquad \tan\theta = a \quad$ and $\tan\phi = b$,
$$\tan(\theta\pm\phi) = \frac{\tan\theta \pm \tan\phi}{1 \mp \tan\theta\tan\phi}$$
$$= \frac{a \pm b}{1 \mp ab},$$
therefore $\qquad\qquad \theta\pm\phi = \tan^{-1}\left(\frac{a\pm b}{1\mp ab}\right)$,
$$\tan^{-1} a \pm \tan^{-1} b = \tan^{-1}\left(\frac{a\pm b}{1\mp ab}\right). \qquad(253)$$

In a similar manner we find

$$\cot^{-1} a \pm \cot^{-1} b = \cot^{-1}\left(\frac{ab\mp 1}{b\pm a}\right). \qquad(254)$$

166. Here we may remark in regard to (253), that since there are several angles whose tangent is a, several whose tangent is b, and several whose tangent is $\dfrac{a\pm b}{1\mp ab}$, it follows that the sum or difference of any *two* of the angles whose tangents are a and b respectively, is equal to some *one* of the angles whose tangent is $\dfrac{a+b}{1-ab}$ or $\dfrac{a-b}{1+ab}$ respectively.

Similar remarks apply, of course, to the other formulæ of this chapter, and in fact to all formulæ containing the inverse functions.

Examples.

1. Prove that

$$\tan^{-1}\{(\sqrt{2}+1)\tan\theta\} - \tan^{-1}\{(\sqrt{2}-1)\tan\theta\} = \tan^{-1}\sin 2\theta.$$

By (253) we have

$$\tan^{-1}\{(\sqrt{2}+1)\tan\theta\} - \tan^{-1}\{(\sqrt{2}-1)\tan\theta\}$$
$$= \tan^{-1}\left\{\frac{(\sqrt{2}+1)\tan\theta - (\sqrt{2}-1)\tan\theta}{1+(\sqrt{2}+1)(\sqrt{2}-1)\tan^2\theta}\right\}$$
$$= \tan^{-1}\left(\frac{2\tan\theta}{1+\tan^2\theta}\right) = \tan^{-1}\sin 2\theta.$$

2. Given $\operatorname{versin}^{-1}\dfrac{x}{a} - \operatorname{versin}^{-1}(1-b) = \operatorname{versin}^{-1}\dfrac{bx}{a}$, find x.

Since the versin $=1-\cos$, the above becomes

$$\cos^{-1}\left(1-\frac{x}{a}\right) - \cos^{-1} b = \cos^{-1}\left(1-\frac{bx}{a}\right),$$

or $\cos^{-1}\left\{\left(1-\dfrac{x}{a}\right)b + \sqrt{(1-b^2)\left(1-\left(1-\dfrac{x}{a}\right)^2\right)}\right\} = \cos^{-1}\left(1-\dfrac{bx}{a}\right)$, by (252)

Cancelling $\cos^{-1}$ and solving the equation for x, we have

$$x = a\left(1 \pm \sqrt{\frac{2b}{1+b}}\right).$$

3. Prove that $\tan^{-1}\dfrac{a\cos\phi}{1-a\sin\phi} - \tan^{-1}\dfrac{a-\sin\phi}{\cos\phi} = \phi.$

Let $\tan^{-1}\dfrac{a\cos\phi}{1-a\sin\phi} = x$ and $\tan^{-1}\dfrac{a-\sin\phi}{\cos\phi} = y$,

then $\dfrac{\sin x}{\cos x} = \dfrac{a\cos\phi}{1-a\sin\phi}$ and $\dfrac{\sin y}{\cos y} = \dfrac{a-\sin\phi}{\cos\phi}$,

whence $\sin x = a\cos(\phi-x)$ and $a\cos y = \sin(\phi+y)$,

twice the product of which is
$$2 \sin x \cos y = 2 \sin (\phi+y) \cos (\phi - x)$$
or
$$\sin (x+y)+\sin (x-y)=\sin (2\phi-x+y)+\sin (x+y),$$
whence
$$\sin (x-y)=\sin (2\phi-x+y)$$
or
$$x-y = 2\phi - x+y,$$
that is
$$x-y = \phi$$
or $\tan^{-1} \dfrac{a \cos \phi}{1-a \sin \phi} - \tan^{-1} \dfrac{a-\sin \phi}{\cos \phi} = \phi$.

4. Prove that $\tan^{-1} \dfrac{x}{y} = \tan^{-1} \dfrac{a_1 x - y}{a_1 y + x} + \tan^{-1} \dfrac{a_2 - a_1}{a_2 a_1 + 1}$

$+ \tan^{-1} \dfrac{a_3 - a_2}{a_3 a_2 + 1} + \ldots \tan^{-1} \dfrac{a_n - a_{n-1}}{a_n a_{n-1} + 1} + \tan^{-1} \dfrac{1}{a_n}$.

where $a_1, a_2, \ldots a_n$ are any quantities whatever.

By (253) we have
$$\tan^{-1} \frac{x}{y} - \tan^{-1} \frac{1}{a_1} = \tan^{-1} \frac{a_1 x - y}{a_1 y + x},$$
$$\tan^{-1} \frac{1}{a_1} - \tan^{-1} \frac{1}{a_2} = \tan^{-1} \frac{a_2 - a_1}{a_2 a_1 + 1},$$
$$\tan^{-1} \frac{1}{a_2} - \tan^{-1} \frac{1}{a_3} = \tan^{-1} \frac{a_3 - a_2}{a_3 a_2 + 1},$$
$$\ldots \ldots \ldots \ldots \ldots \ldots$$
$$\tan^{-1} \frac{1}{a_{n-1}} - \tan^{-1} \frac{1}{a_n} = \tan^{-1} \frac{a_n - a_{n-1}}{a_n a_{n-1} + 1},$$

by addition we have

$\tan^{-1} \dfrac{x}{y} - \tan^{-1} \dfrac{1}{a_n} = \tan^{-1} \dfrac{a_1 x - y}{a_1 y + x} + \tan^{-1} \dfrac{a_2 - a_1}{a_2 a_1 + 1} + \ldots + \tan^{-1} \dfrac{a_n - a_{n-1}}{a_n a_{n-1} + 1}$,

therefore by transposition

$\tan^{-1} \dfrac{x}{y} = \tan^{-1} \dfrac{a_1 x - y}{a_1 y + x} + \tan^{-1} \dfrac{a_2 - a_1}{a_2 a_1 + 1} + \ldots + \tan^{-1} \dfrac{a_n - a_{n-1}}{a_n a_{n-1} + 1} + \tan^{-1} \dfrac{1}{a_n}$.

5. Prove that $\cot (\theta + \tan^{-1} \tan^3 \theta) = 2 \cot 2\theta$.

6. Prove that $\cos \sin^{-1} \cos \sin^{-1} \theta = \pm \theta$.

7. Prove that $\cos^{-1} x = 2 \sin^{-1} \sqrt{\dfrac{1-x}{2}} = 2 \cos^{-1} \sqrt{\dfrac{1+x}{2}}$.

EXAMPLES. — 229

8. Find the general values of $\cos^{-1}\sin\theta$ and $\tan^{-1}\cot\theta$.

$$\text{Ans. } 2n\pi \pm (\frac{\pi}{2} - \theta),\ n\pi + (\frac{\pi}{2} - \theta).$$

9. Prove that $\tan^{-1}\frac{3}{4} = 2\tan^{-1}\frac{1}{3}$.

10. Prove that

(1) $\sin^{-1}\frac{3}{5} + \sin^{-1}\frac{4}{5} = \frac{\pi}{2}$.

(2) $\sec^{-1}3 + \tan^{-1}2\sqrt{2} = \tan^{-1}(-\frac{4}{7}\sqrt{2})$.

(3) $\tan^{-1}\frac{2}{3} = \frac{1}{2}\tan^{-1}\frac{12}{5}$.

(4) $2\tan^{-1}\frac{1}{3} + \tan^{-1}\frac{1}{7} = \frac{\pi}{4}$.

(5) $\sin^{-1}(\frac{1}{2}\sqrt{2+\sqrt{2}}) + \sin^{-1}(\frac{1}{2}\sqrt{2-\sqrt{2}}) = \frac{\pi}{2}$.

(6) $2\tan^{-1}\frac{1}{8} + 2\tan^{-1}\frac{1}{5} + \tan^{-1}\frac{1}{7} = \frac{\pi}{4}$.

11. If $\sin(\pi\cos\theta) = \cos(\pi\sin\theta)$, shew that $\theta = -\frac{1}{2}\sin^{-1}\frac{3}{4}$.

12. If $2\sin\frac{\theta}{2} = \cos\theta$, shew that $\theta = 2\cos^{-1}\sqrt{\cos\frac{\pi}{6}}$.

13. Find the values of

$\tan(\tan^{-1}\theta + \cot^{-1}\theta)$ and $\sin(\sin^{-1}\frac{1}{2} + \cos^{-1}\frac{1}{2})$.

$$\text{Ans. } \infty,\ \text{and } 1,\ \text{or } -\frac{1}{2}.$$

14. If $\tan(n\cot x) = \cot(n\tan x)$, shew that

$$x = \frac{m\pi}{2} + (-1)^m \frac{1}{2}\sin^{-1}\frac{4n}{(2r+1)\pi},$$

m and r being any integers

15. If $2\tan^{-1}x = \sin^{-1}2y$, shew that $y = \frac{x}{1+x^2}$.

16. Find the value of x in the following equations:

(1) $\tan^{-1} 2x + \tan^{-1} 3x = \dfrac{\pi}{4}$.   Ans. $x = -1$ or $\dfrac{1}{6}$.

(2) $\tan^{-1} \dfrac{1}{x-1} - \tan^{-1} \dfrac{1}{x+1} = \dfrac{\pi}{12}$.   Ans. $x = \pm(1+\sqrt{3})$.

(3) $\sin^{-1} 2x - \sin^{-1} \sqrt{3}\, x = \sin^{-1} x$.   Ans. $x = 0$ or $\pm\dfrac{1}{2}$.

(4) $\sec^{-1} \dfrac{x}{a} - \sec^{-1} \dfrac{x}{b} + \sec^{-1} a - \sec^{-1} b = 0$.   Ans. $x = \pm ab$.

(5) $\sin 2\cos^{-1} \cot 2\tan^{-1} x = 0$.   Ans. $x = \pm 1$, or $\pm(1\pm\sqrt{2})$.

(6) $\operatorname{versin}^{-1}(1+x) - \operatorname{versin}^{-1}(1-x) = \tan^{-1} 2\sqrt{1-x^2}$.

             Ans. $x = \dfrac{1}{2}$ or -1.

(7) $\sin^{-1} x + \tan^{-1} x = \dfrac{\pi}{2}$.   Ans. $x = \sqrt{\dfrac{\sqrt{5}-1}{2}}$.

(8) $\sin(\tan^{-1} x) + \tan(\sin^{-1} x) = mx$.

        Ans. $x = \dfrac{1}{m}\sqrt{m^4 - 2m^2 \mp 2\sqrt{1+2m^2} - 2}$.

17. If $\sec\theta - \operatorname{cosec}\theta = 2\sqrt{2}$, shew that $\theta = \dfrac{1}{2}\sin^{-1}\dfrac{1}{2}$.

18. Shew that $4\tan^{-1}\dfrac{1}{5} - \tan^{-1}\dfrac{1}{239} = \dfrac{\pi}{4}$.

19. Shew that $\tan^{-1}\dfrac{1}{3} + \tan^{-1}\dfrac{1}{5} + \tan^{-1}\dfrac{1}{7} + \tan^{-1}\dfrac{1}{8} = \dfrac{\pi}{4}$.

20. Shew that $\tan(2\tan^{-1} a) = 2\tan(\tan^{-1} a + \tan^{-1} a^3)$.

21. Shew that $\tan^{-1}\left(\dfrac{x}{a}\right)^{\frac{1}{2}} = \sin^{-1}\left(\dfrac{x}{a+x}\right)^{\frac{1}{2}}$.

22. Shew that $\tan^{-1}\dfrac{\sqrt{3}+\sqrt{2}}{\sqrt{3}-\sqrt{2}} + \tan^{-1}\dfrac{\sqrt{3}}{\sqrt{2}} = \dfrac{3\pi}{4}$.

23. Shew that $\cot^{-1} 3 + \operatorname{cosec}^{-1}\sqrt{5} = \dfrac{\pi}{4}$.

CHAPTER XIII.

DIVISION OF ANGLES — SOLUTION OF EQUATIONS — AUXILIARY ANGLES — ELIMINATION OF TRIGONOMETRICAL FUNCTIONS.

Division of Angles.

167. We shall here determine, *a priori*, how many values any assigned trigonometrical function can have when determined from any other function of the angle or of a submultiple of the angle.

168. *Given $\sin A$, to find how many values $\sin \dfrac{A}{2}$ can have when expressed in terms of it.*

Let a be the *circular measure* of the least angle whose sine is equal to $\sin A$; then all the angles whose sines are equal to $\sin A$ are included in the general expression

$$n\pi + (-1)^n a. \quad \text{(Art. 42.)}$$

Hence all the values which $\sin \dfrac{A}{2}$ can have when expressed in terms of $\sin A$, are included in

$$\sin\left(\frac{n\pi + (-1)^n a}{2}\right).$$

Now n must be of one of the forms 2λ or $2\lambda+1$, since every number is either divisible by 2 or divisible by 2 with a remainder 1.

Let $n = 2\lambda$, then

$$\sin\left(\frac{n\pi + (-1)^n a}{2}\right) = \sin\left(\lambda\pi + \frac{a}{2}\right) = \pm \sin \frac{a}{2},$$

according as λ is even or odd.

Let $n = 2\lambda + 1$, then
$$\sin\left(\frac{n\pi + (-1)^n a}{2}\right) = \sin\left(\lambda\pi + \frac{\pi - a}{2}\right)$$
$$= \pm \sin\left(\frac{\pi}{2} - \frac{a}{2}\right) = \pm \cos\frac{a}{2},$$
according as λ is even or odd.

Therefore $\sin\frac{A}{2}$, when expressed in terms of $\sin A$, has four different values, viz., $\pm \sin\frac{a}{2}$ and $\pm \cos\frac{a}{2}$.

169. *Given $\cos A$, to find how many values $\cos\frac{A}{3}$ can have when expressed in terms of it.*

Let a be the *circular measure* of the least angle whose cosine is equal to $\cos A$; then all the angles whose cosines are equal to $\cos A$ are included in the expression $2n\pi \pm a$, and therefore all the different values which $\cos\frac{A}{3}$ can have when expressed in terms of $\cos A$ are included in
$$\cos\frac{2n\pi \pm a}{3}.$$

Now n must be of one of the forms $3p$, $3p+1$, $3p+2$, since every number is either exactly divisible by 3, or divisible by 3 with a remainder 1 or 2.

Taking $n = 3p$, we have
$$\cos\frac{2n\pi \pm a}{3} = \cos\left(2p\pi \pm \frac{a}{3}\right) = \cos\left(\pm\frac{a}{3}\right) = \cos\frac{a}{3}.$$

Taking $n = 3p+1$, we have
$$\cos\frac{2n\pi \pm a}{3} = \cos\left(2p\pi + \frac{2\pi \pm a}{3}\right) = \cos\frac{2\pi \pm a}{3}.$$

Taking $n = 3p+2$, we have
$$\cos\frac{2n\pi \pm a}{3} = \cos\left(2p\pi + \frac{4\pi \pm a}{3}\right) = \cos\frac{4\pi \pm a}{3}$$
$$= \cos\left(2\pi - \frac{2\pi \mp a}{3}\right) = \cos\frac{2\pi \mp a}{3}.$$

DIVISION OF ANGLES. 233

Therefore $\cos \dfrac{A}{3}$, when expressed in terms of $\cos A$, has three different values, viz.:

$$\cos \dfrac{a}{3}, \ \cos \dfrac{2\pi+a}{3} \ \text{and} \ \cos \dfrac{2\pi-a}{3}.$$

170. *Given $\sin A$, to determine how many values $\sin \frac{3}{4}A$ can have when expressed in terms of it.*

Let a be the *circular measure* of the least angle whose sine is equal to $\sin A$; then all the values which $\sin \frac{3}{4}A$ can have are included in

$$\sin \tfrac{3}{4}\{n\pi + (-1)^n a\},$$

where n is of one of the forms $4p$, $4p+1$, $4p+2$, $4p+3$, since every number must be exactly divisible by 4, or divisible by 4 with a remainder 1, 2 or 3.

If $n = 4p$, which is even,

$$\sin \tfrac{3}{4}\{n\pi + (-1)^n a\} = \sin(3p\pi + \tfrac{3}{4}a) = \pm \sin \tfrac{3}{4}a,$$

according as p is even or odd.

If $n = 4p+1$, which is odd,

$$\sin \tfrac{3}{4}\{n\pi + (-1)^n a\} = \sin\{3p\pi + \tfrac{3}{4}(\pi - a)\}$$
$$= \pm \sin \tfrac{3}{4}(\pi - a),$$

according as p is even or odd; and so on.

Hence we find that $\sin \frac{3}{4}A$, when expressed in terms of $\sin A$, has eight different values, viz.,

$$\pm \sin \tfrac{3}{4}a, \ \pm \sin \tfrac{3}{4}(\pi - a), \ \pm \cos \tfrac{3}{4}a, \ \pm \sin \tfrac{1}{4}(\pi - 3a).$$

171. *To find the number of values which $\cos \dfrac{2r\pi + \theta}{n}$ has when successive integral values are assigned to n.*

Here r, being an integer, must be of the form $mn + p$ where m is 0 or any integer, and p is 0 or any integer less than n; that is, r must be exactly divisible by n, or divisible by n with a remainder which is 1, 2, 3 ... or $n-1$.

Hence giving r, the values, $0, 1, 2, 3 \ldots n-1, n, n+1, \ldots$ &c., in succession, we have

when $\quad r=0, \qquad \cos\dfrac{2r\pi+\theta}{n}=\cos\dfrac{\theta}{n},$

$\qquad\qquad r=1, \qquad \cos\dfrac{2r\pi+\theta}{n}=\cos\dfrac{2\pi+\theta}{n},$

$\qquad\qquad r=2, \qquad \cos\dfrac{2r\pi+\theta}{n}=\cos\dfrac{4\pi+\theta}{n},$

$\qquad\qquad r=3, \qquad \cos\dfrac{2r\pi+\theta}{n}=\cos\dfrac{6\pi+\theta}{n},$

$\qquad\qquad$ &c., $\qquad\quad$ &c., $\qquad\quad$ &c.

$\qquad\qquad r=n-3, \quad \cos\dfrac{2r\pi+\theta}{n}=\cos\dfrac{6\pi-\theta}{n},$

$\qquad\qquad r=n-2, \quad \cos\dfrac{2r\pi+\theta}{n}=\cos\dfrac{4\pi-\theta}{n},$

$\qquad\qquad r=n-1, \quad \cos\dfrac{2r\pi+\theta}{n}=\cos\dfrac{2\pi-\theta}{n},$

$\qquad\qquad r=n, \qquad \cos\dfrac{2r\pi+\theta}{n}=\cos\dfrac{\theta}{n},$

$\qquad\qquad r=n+1, \quad \cos\dfrac{2r\pi+\theta}{n}=\cos\dfrac{2\pi+\theta}{n},$

$\qquad\qquad$ &c., $\qquad\quad$ &c., $\qquad\quad$ &c.

Therefore there are n and *only* n *different* values of $\cos\dfrac{2r\pi+\theta}{n}$, corresponding to the values $0, 1, 2, \ldots n-1$, of r; for the same values of the function recur in the same order when r is successively made equal to n, $n+1$, &c.

In a similar manner we may shew that $\sin\dfrac{2r\pi+\theta}{n}$ has also n different values.

The preceding examples are quite sufficient to shew the mode of proceeding in any assigned case.

Examples.

1. Shew that $\sin A$, when determined from $\tan A$, has two values.

2. Prove, *a priori*, that $\sin mA$, when expressed in terms of $\sin A$, will have one or two values, according as m is odd or even; and that $\cos mA$, in terms of $\cos A$, will have only one value, m being in each case a positive integer.

3. Prove that $\tan \dfrac{A}{4}$, when expressed in terms of $\sin A$, will have four different values.

4. If $\tan A = \sin 2A$, find A. *Ans.* $n\pi$ or $n\pi + (-1)^n \dfrac{\pi}{4}$.

5. If $\cos\theta + \cos 2\theta + \cos 3\theta = 0$. then will, r being any integer,
$$\theta = \{11 + (-1)^r 5\} \left\{\left(7 + (-1)^{r-1} 1\right)\dfrac{n}{2} \pm 1\right\} \dfrac{\pi}{24}.$$

Solution of Equations.

172. An equation in which the unknown quantity is a trigonometrical function of an angle, is, in general, readily solved by the aid of the ordinary trigonometrical transformations. We shall here illustrate the mode of solving a few easy equations, such as are most frequently met with in Spherical Astronomy.

173. *Given* $\sin\theta = \cos\beta \sin(\theta + a)$, *to find* θ.

Developing the second member by (45) we have
$$\sin\theta = \cos a \cos\beta \sin\theta + \sin a \cos\beta \cos\theta,$$
or $\qquad \tan\theta = \cos a \cos\beta \tan\theta + \sin a \cos\beta,$

whence $\qquad \tan\theta = \dfrac{\sin a \cos\beta}{1 - \cos a \cos\beta}.$

Now it is evident that if θ is not limited to any particular quadrant by the nature of the problem under consideration, there will be an indefinite number of solutions; for all the angles θ, $\theta + 180°$, $\theta + 360°$, $\theta + 540°$, &c., that is, all the angles included by the expression $\theta + n\pi$, have the same tangent. (Art. 44.) In practice, however, only the first two values of θ, viz., θ and $\theta + 180°$, or those less than $360°$, are considered; and the conditions of the problem are generally such as enable us to determine which of these is to be taken.

It is evident then, that when an angle is determined by a single trigonometrical function, there will be two values less than 360°; but if the values of *two* functions of the required angle, which have not the same *sign*—such as the sine and tangent, or the cosine and cotangent—can be found from the problem, the solution is determinate under 360°.

Suppose, for example, that the required angle is found by its sine and tangent; if the sine is positive and the tangent negative, the angle will evidently be in the second quadrant or between 90° and 180°; if both functions are negative, then the angle will lie between 270° and 360°, and so on.

The solution of the last equation cannot be effected by logarithms; a formula adapted to logarithms is easily deduced as follows:

Put the given equation in the form

$$\frac{\sin \theta}{\sin (\theta+a)} = \cos \beta,$$

then

$$\frac{\sin (\theta+a)+\sin \theta}{\sin (\theta+a)-\sin \theta} = \frac{1+\cos \beta}{1-\cos \beta}$$

or

$$\frac{\tan (\theta+\frac{a}{2})}{\tan \frac{a}{2}} = \frac{1}{\tan^2 \frac{\beta}{2}}, \quad \text{by (59) and (101)}$$

whence

$$\tan (\theta+\frac{a}{2}) = \tan \frac{a}{2} \cot^2 \frac{\beta}{2}.$$

This determines $\theta+\frac{a}{2}$, and therefore θ becomes known by deducting $\frac{a}{2}$, thus $\theta = \tan^{-1} (\tan \frac{a}{2} \cot^2 \frac{\beta}{2}) - \frac{a}{2}$.

174. *Given* $\tan (\theta+a) = \sin \beta \tan \theta$, *to find* θ.

Putting the given equation in the form

$$\frac{\tan (\theta+a)}{\tan \theta} = \sin \beta,$$

we have, by composition and division,

$$\frac{\tan (\theta+a)+\tan \theta}{\tan (\theta+a)-\tan \theta} = \frac{1+\sin \beta}{1-\sin \beta},$$

whence
$$\frac{\sin(2\theta+a)}{\sin a}=\tan^2\left(45°+\frac{\beta}{2}\right),$$

therefore
$$\sin(2\theta+a)=\tan^2\left(45°+\frac{\beta}{2}\right)\sin a$$

and
$$\theta=\tfrac{1}{2}\sin^{-1}\left\{\tan^2\left(45°+\frac{\beta}{2}\right)\sin a\right\}-\frac{a}{2}.$$

175. *Given* $\tan(a+\theta)\tan\theta=\tan\beta$, *to find* θ.

Here we have
$$\frac{1-\tan(a+\theta)\tan\theta}{1+\tan(a+\theta)\tan\theta}=\frac{1-\tan\beta}{1+\tan\beta},$$

whence
$$\frac{\cos(a+2\theta)}{\cos a}=\tan(45°-\beta),$$

therefore
$$\cos(a+2\theta)=\tan(45°-\beta)\cos a$$

and
$$\theta=\tfrac{1}{2}\cos^{-1}\left\{\tan(45°-\beta)\cos a\right\}-\frac{a}{2}.$$

176. *Given* $x\sin\theta=a$ *and* $x\cos\theta=b$, *to find* θ *and* x.

By division we have
$$\tan\theta=\frac{a}{b},$$

which gives two values of θ, one less and the other greater than 180°, and also two values of x from the equation $x=a\operatorname{cosec}\theta$.

Limiting the values of θ to those less than 360°, the solution is determinate under the following restrictions:

1st. *When x is positive.*

The signs of $\sin\theta$ and $\cos\theta$ will be the same as those of a and b respectively, and therefore the quadrant in which θ must be taken is determined.

2nd. *When x is negative.*

The signs of $\sin\theta$ and $\cos\theta$ are the opposite of those of a and b, and therefore θ must be taken out accordingly.

3rd. *When $\theta < 180°$ or $> 180°$.*

Under either of these conditions the equation $\tan\theta=\frac{a}{b}$ gives only *one* value of θ (a and b being unrestricted as to sign). Under the former condition, x has the sign of a; and under the latter, the opposite sign to that of a.

4th. *When θ is limited to acute values, positive or negative.*

Under this condition x will always have the same sign as b.

Auxiliary Angles.

177. In the solution of equations by logarithms it is necessary to express the sum or difference of two quantities by means of a product. This can always be effected by introducing the sine, tangent or some other function of an angle chosen for that purpose. An angle which is thus introduced to assist in trigonometrical calculation, is called an *auxiliary angle*, and is of great utility and extensive application, particularly in Spherical Trigonometry and Spherical Astronomy.

Auxiliary angles have already been employed in a few examples. (See Arts. 120 and 144.)

Ex. 1.—Adapt $x = \sqrt{(a^2 \pm b^2)}$ to logarithms.

(1) $$x = \sqrt{(a^2 + b^2)} = a\sqrt{1 + \frac{b^2}{a^2}}.$$

Assume $\tan \theta = \dfrac{b}{a}$, an assumption always possible, since the tangent may be of any magnitude whatever, then

$$x = a\sqrt{(1 + \tan^2 \theta)} = a \sec \theta;$$

hence x will be found by the two equations

Log $\tan \theta = \log b - \log a + 10$;

$\log x = \log a + \text{Log sec } \theta - 10.$

(2) $$x = \sqrt{(a^2 - b^2)} = \sqrt{(a+b)(a-b)},$$

which is in a form adapted to logarithms; or we may proceed as follows:

$$x = a\sqrt{1 - \frac{b^2}{a^2}}.$$

Assume $\sin \theta = \dfrac{b}{a}$, since $\dfrac{b}{a}$ must be less than 1,

then $$x = a\sqrt{(1 - \sin^2 \theta)} = a \cos \theta.$$

AUXILIARY ANGLES.

Ex. 2.—*Adapt* $x = a \pm \sqrt{a^2 + b^2}$, *to logarithms.*

The equation may be written thus,

$$x = a\left\{1 \pm \sqrt{(1 + \frac{b^2}{a^2})}\right\}.$$

Assume $\tan \phi = \dfrac{b}{a}$,

then $\quad x = a\,(1 \pm \sec \phi)$

$$= a\frac{1 + \cos \phi}{\cos \phi} \text{ and } -a\,\frac{1 - \cos \phi}{\cos \phi}$$

$$= \frac{a \sin \phi}{\cos \phi} \cdot \frac{1 + \cos \phi}{\sin \phi} \text{ and } -\frac{a \sin \phi}{\cos \phi} \cdot \frac{1 - \cos \phi}{\sin \phi}$$

$$= a \tan \phi \cot \frac{\phi}{2} \text{ and } -a \tan \phi \tan \frac{\phi}{2}$$

$$= b \cot \frac{\phi}{2} \text{ and } -b \tan \frac{\phi}{2},$$

which are both adapted to logarithms.

Ex. 3.—*Adapt* $x = a \pm \sqrt{a^2 - b^2}$, *to logarithms.*

Assume $\sin \theta = \dfrac{b}{a}$, then we shall easily find

$$x = 2a \cos^2 \frac{\theta}{2} \text{ and } 2a \sin^2 \frac{\theta}{2}.$$

Ex. 4.—*Given* $\cos \phi \cos \delta \cos h + \sin \phi \sin \delta = \sin a$, *to find* ϕ *in a form adapted to logarithms.*

Assume $\qquad x \sin \theta = \cos \delta \cos h$

and $\qquad\qquad x \cos \theta = \sin \delta,$ $\qquad\qquad\qquad$ (1)

that is, $\qquad\qquad \tan \theta = \cot \delta \cos h;$ $\qquad\qquad$ (2)

then the given equation becomes

$$x\,(\cos \phi \sin \theta + \sin \phi \cos \theta) = \sin a,$$

or $\qquad\qquad x \sin (\phi + \theta) = \sin a$

and $\qquad\qquad\qquad x = \sin \delta \sec \theta;$

therefore $\qquad\qquad \sin (\phi + \theta) = \dfrac{\sin a}{\sin \delta \sec \theta}$

$$= \sin a\,\operatorname{cosec} \delta \cos \theta, \qquad (3)$$

which gives two values of $(\phi + \theta)$.

240   PLANE TRIGONOMETRY.

In this example, let $\delta = -30°\ 22'\ 47''.5$; $a = 29°\ 10'$; $h = -15°$, find ϕ.

By (1)
$$\text{Log cos } \delta = 9.935857$$
$$\text{Log cos } h = 9.984944$$
$$\text{Log } x \sin \theta = 9.920801$$
$$\text{Log } x \cos \theta = \text{Log sin } \delta = 9.703919n$$
$$\text{Log tan } \theta = 10.216882n$$
$$\theta = 121°\ 15'\ 13''.$$

By (3)
$$\text{Log sin } a = 9.687843$$
$$\text{Log cosec } \delta = 10.296081n$$
$$\text{Log cos } \theta = 9.715023n$$
$$\text{Log sin } (\phi+\theta) = 9.698947$$

$$\phi + \theta = \begin{cases} 29°\ 59'\ 53''.5 \\ \text{or } 150°\ 0'\ 6''.5, \end{cases}$$

therefore
$$\phi = \begin{cases} 28°\ 44'\ 53''.5 \\ \text{or } -91°\ 15'\ 19''.5. \end{cases}$$

If we take the acute value only of θ, we have
$$\theta = -58°\ 44'\ 47'',$$
which gives $\phi + \theta = -29°\ 59'\ 53''.5$ or $-150°\ 0'\ 6''.5$
and $\phi = 28°\ 44'\ 53''.5$ or $-91°\ 15'\ 19''.5$
as before.

Ex. 5.—Given $\tan \lambda = (\cos a \cos \beta - \cos \delta) \cos h - \sin a \cos \beta \sin h$, to find h in a form adapted to logarithms.

Writing the given equation in the form
$$\tan \lambda \text{ cosec } a \sec \beta = \frac{\cos a \cos \beta - \cos \delta}{\sin a \cos \beta} \cos h - \sin h,$$

and assuming $\cos \theta = \cos a \cos \beta$, since $\cos a \cos \beta$ is always less than 1, we have
$$\tan \lambda \text{ cosec } a \sec \beta = \frac{\cos \theta - \cos \delta}{\sin a \cos \beta} \cos h - \sin h$$
$$= \frac{2 \sin \frac{1}{2}(\delta+\theta) \sin \frac{1}{2}(\delta-\theta)}{\sin a \cos \beta} \cos h - \sin h.$$

Again, assuming $\cot \phi = \dfrac{2 \sin \frac{1}{2}(\delta+\theta) \sin \frac{1}{2}(\delta-\theta)}{\sin a \cos \beta}$, we have

$$\tan \lambda \operatorname{cosec} a \sec \beta = \cot \phi \cos h - \sin h$$

$$= \dfrac{\cos \phi \cos h - \sin \phi \sin h}{\sin \phi}$$

$$= \dfrac{\cos (\phi+h)}{\sin \phi},$$

whence $\cos (\phi+h) = \tan \lambda \operatorname{cosec} a \sec \beta \sin \phi$,

therefore $h = \cos^{-1} (\tan \lambda \operatorname{cosec} a \sec \beta \sin \phi) - \phi$.

Examples.

6. If $x = \dfrac{a^2 - b^2}{a^2 + b^2}$, shew that $x = \cos 2\theta$, where $\tan \theta = \dfrac{b}{a}$.

7. If $x = \sqrt{a+b} + \sqrt{a-b}$, shew that

$$x = 2\sqrt{a} \cos \left(45° - \dfrac{\theta}{2}\right), \text{ where } \cos \theta = \dfrac{b}{a}.$$

8. Given

$$\sin \theta \sqrt{(1+\tan^2 a \tan^2 \beta)} + \cos \theta \sqrt{(1-\tan^2 a \tan^2 \beta)} = \tan a + \tan \beta,$$

find θ in a form adapted to logarithms.

Ans. $\sin (\theta+\phi) = \dfrac{\sin (a+\beta)}{\sqrt{2} \cos a \cos \beta}$, where $\cos 2\phi = \tan^2 a \tan^2 \beta$.

9. If $\tan (\theta+45°) + \tan (\theta-45°) = 2 \tan 60°$, find θ.

Ans. $30°$ or $(3n+1) \dfrac{\pi}{6}$.

10. Shew that

$\sin A + \cos A + \sin B + \cos B = 2\sqrt{2} \cos \left\{45° - \frac{1}{2}(A+B)\right\} \cos \frac{1}{2}(A-B)$.

11. If $\cos (a+\phi) = \cos a \sin \phi + \sin \beta$, shew that

$\cos (\phi+\delta) = \sec a \sin \beta \cos \delta$, where $\tan \delta = \sqrt{2} \sin (45°+a) \sec a$.

12. Given $\sin x = \tan \delta \cot \omega,$
$\sin (x+2a) = \tan \delta' \cot \omega,$ find x and ω.

Ans. $\tan (x+a) = \dfrac{\sin (\delta'+\delta)}{\sin (\delta'-\delta)} \tan a,$

and $\cot \omega = \sin x \cot \delta$.

13. Given $\cot x \sin h = \tan \delta \cos \beta - \cos h \sin \beta$, find x in a form adapted to logarithms.

Ans. $\cot x = \cot h \operatorname{cosec} \phi \cos (\phi + \beta)$,
where $\tan \phi = \cos h \cot \delta$.

14. Given
$$\cos h = \cot \phi \tan \delta,$$
$$-\cos (a+h) = \tan \phi \tan \delta, \text{ to find } h \text{ and } \delta.$$

Ans. $\tan \left(h + \dfrac{a}{2}\right) = \cot \dfrac{a}{2} \sec 2\phi,$
$\tan \delta = \cos h \tan \phi.$

Quadratic Equations.

178. *To solve the equation*
$$x^2 + ax + b = 0, \tag{a}$$
where b is essentially positive and a either positive or negative.

By the ordinary algebraic method we have
$$x = -\frac{a}{2}\left(1 \pm \sqrt{1 - \frac{4b}{a^2}}\right),$$
which may be easily adapted to logarithms in a manner similar to that of *Exs.* 2 and 3 of the last Article.

The following method, however, will generally be found more convenient:

(1) $\sin \phi = 2 \sin \dfrac{\phi}{2} \cos \dfrac{\phi}{2} = \dfrac{2 \sin \dfrac{\phi}{2} \cos \dfrac{\phi}{2}}{\cos^2 \dfrac{\phi}{2} + \sin^2 \dfrac{\phi}{2}} = \dfrac{2 \tan \dfrac{\phi}{2}}{1 + \tan^2 \dfrac{\phi}{2}},$

whence $\tan^2 \dfrac{\phi}{2} - 2 \operatorname{cosec} \phi \tan \dfrac{\phi}{2} + 1 = 0.$ \hfill (225)

(2) $\tan \phi = \dfrac{\sin \phi}{\cos \phi} = \dfrac{2 \sin \dfrac{\phi}{2} \cos \dfrac{\phi}{2}}{\cos^2 \dfrac{\phi}{2} - \sin^2 \dfrac{\phi}{2}} = \dfrac{2 \tan \dfrac{\phi}{2}}{1 - \tan^2 \dfrac{\phi}{2}},$

whence $\tan^2 \dfrac{\phi}{2} + 2 \cot \phi \tan \dfrac{\phi}{2} - 1 = 0.$ \hfill (256)

QUADRATIC EQUATIONS.

In (a) let $x = y\sqrt{b}$, where the radical is taken with the *positive* sign, x and y having the same sign. We thus reduce (a) to

$$y^2 + \frac{a}{\sqrt{b}} y + 1 = 0,$$

which compared with (255) gives

$$-2 \csc \phi = \frac{a}{\sqrt{b}}, \qquad y = \tan \frac{\phi}{2},$$

or

$$\sin \phi = -\frac{2\sqrt{b}}{a}, \text{ and } x = \sqrt{b} \tan \frac{\phi}{2},$$

which gives two values of ϕ, and consequently two values of x. Let θ be the smaller of these two values of ϕ, then all the values of ϕ which have the same sine are

$$\theta, \ \pi - \theta, \ 2\pi + \theta, \ 3\pi - \theta, \ \&c.,$$

and all the values of $\tan \dfrac{\phi}{2}$ are

$$\tan \frac{\theta}{2}, \ \tan \tfrac{1}{2}(\pi - \theta), \ \tan \tfrac{1}{2}(2\pi + \theta), \ \tan \tfrac{1}{2}(3\pi - \theta), \ \&c.,$$

or

$$\tan \frac{\theta}{2}, \ \cot \frac{\theta}{2}, \ \tan \frac{\theta}{2}, \ \cot \frac{\theta}{2}, \ \&c.$$

Hence the roots (x_1, x_2) of (a) are found by the formulæ

$$\sin \theta = -\frac{2\sqrt{b}}{a}, \quad x_1 = \sqrt{b} \tan \frac{\theta}{2} \ \text{ and } \ x_2 = \sqrt{b} \cot \frac{\theta}{2},$$

where θ is always to be taken less than 90°, with the sign of its sine, $\sqrt{b}$ being regarded as a *positive* quantity, and a either *positive* or *negative*. When $2\sqrt{b}$ is greater than a, $\sin \theta$ is impossible, and both roots are imaginary.

179. *To solve the equation*

$$x^2 + ax - b = 0, \qquad (c)$$

where $-b$ *is essentially negative, a being either positive or negative.*

Let $x = y\sqrt{b}$, then (c) becomes

$$y^2 + \frac{a}{\sqrt{b}} y - 1 = 0,$$

which compared with (256) gives

$$2 \cot \phi = \frac{a}{\sqrt{b}}, \qquad y = \tan \frac{\phi}{2},$$

or
$$\tan \phi = \frac{2\sqrt{b}}{a}, \quad \text{and} \quad x = \sqrt{b} \tan \frac{\phi}{2},$$

which gives two values of ϕ and also two of x. If θ be the smaller of the two values of ϕ, the values of ϕ which have the same tangent are

$$\theta, \ \pi+\theta, \ 2\pi+\theta, \ 3\pi+\theta, \ \&c.,$$

and all the values of $\tan \frac{\phi}{2}$ are

$$\tan \frac{\theta}{2}, \ -\cot \frac{\theta}{2}, \ \tan \frac{\theta}{2}, \ -\cot \frac{\theta}{2}, \ \&c,$$

Therefore the roots of (c) are found by the formulæ

$$\tan \theta = \frac{2\sqrt{b}}{a}, \quad x_1 = \sqrt{b} \tan \frac{\theta}{2} \quad \text{and} \quad x_2 = -\sqrt{b} \cot \frac{\theta}{2},$$

where θ is to be taken less than 90°, with the sign of its tangent, the radical with the positive sign, and a either positive or negative. Here $\tan \theta$ is always possible, therefore both roots are real.

Ex.—Given $x^2 - 1.7246x + .72681 = 0$, find x.

Here we have $a = -1.7246$ and $b = .72681$.

$$\log(-2) = 0.3010300n$$
$$\log \sqrt{b} = \overline{1}.9307104$$
$$\text{ar. co. } \log a = 9.7633116n$$
$$\overline{\text{Log } \sin \theta = 9.9950520}$$
$$\theta = 81° \ 22' \ 3''$$
$$\frac{\theta}{2} = 40° \ 41' \ 1''.5$$

$$\log \sqrt{b} = \overline{1}.9307104$$
$$\text{Log } \tan \frac{\theta}{2} = 9.9343178$$
$$\overline{\log x_1 = \overline{1}.8650282}$$
$$x_1 = .732872$$
$$\text{Log } \cot \frac{\theta}{2} = 10.0656822$$
$$\overline{\log x_2 = \overline{1}.9963926}$$
$$x_2 = .991728$$

Cubic Equations.

180. Let the equation be transformed, if necessary, to another which wants the second term, so that it may be of the form

$$x^3 - qx - r = 0; \qquad (a)$$

if $x = \dfrac{y}{n}$, this becomes

$$y^3 - n^2 q y - n^3 r = 0. \tag{b}$$

From (104) we have, by writing ϕ for A,

$$\cos^3 \phi - \frac{3}{4} \cos \phi - \frac{\cos 3\phi}{4} = 0,$$

which compared with (b) gives

$$\cos \phi = y \quad \text{and} \quad x = \frac{\cos \phi}{n},$$

$$n^2 q = \frac{3}{4} \quad \text{or} \quad n = \frac{1}{2} \sqrt{\frac{3}{q}},$$

$$\frac{\cos 3\phi}{4} = n^3 r \quad \text{or} \quad \cos 3\phi = 4 n^3 r = \frac{r}{2} \sqrt{\frac{27}{q^3}},$$

where the radicals $\sqrt{\dfrac{3}{q}}$ and $\sqrt{\dfrac{27}{q^3}}$ are to be considered positive.

If θ be the circular measure of the least angle whose cosine is equal to $\cos 3\phi$, then by Art. 171 the three values of $\cos \phi$ are

$$\cos \frac{\theta}{3}, \quad \cos \frac{2\pi + \theta}{3} \quad \text{and} \quad \cos \frac{2\pi - \theta}{3},$$

and therefore the three values of x are

$$2\sqrt{\frac{q}{3}} \cdot \cos \frac{\theta}{3}, \quad 2\sqrt{\frac{q}{3}} \cdot \cos \frac{2\pi + \theta}{3} \quad \text{and} \quad 2\sqrt{\frac{q}{3}} \cdot \cos \frac{2\pi - \theta}{3}.$$

Since $\cos 3\phi < 1$, $\dfrac{r}{2}\sqrt{\dfrac{27}{q^3}} < 1$, or $\dfrac{r^2}{4} < \dfrac{q^3}{27}$.

Ex. Given $x^3 - 4x - \dfrac{8}{3} = 0$, find x.

Ans. $\dfrac{4}{\sqrt{3}} \cos 10°$, $-\dfrac{4}{\sqrt{3}} \cos 50°$, $-\dfrac{4}{\sqrt{3}} \cos 70°$.

The other forms of cubic equations can be solved by Trigonometry, but the solution is a matter more of curiosity than of utility. We shall therefore pursue the subject no further, but refer the student to the standard treatises on Algebra for a fuller elucidation of this subject.

Elimination of Trigonometrical Functions.

181. The elimination of the trigonometrical functions from a given number of equations, is frequently required in some of the higher branches of mathematics, and is generally effected by the aid of the various trigonometrical transformations given in the preceding chapters. The following examples illustrate the mode of proceeding in most cases:

Ex. 1.—Eliminate θ between the equations

$$\frac{x}{a \sin \theta} + \frac{y}{b \cos \theta} = 1 \tag{1}$$

$$\frac{x}{a \sin^2 \theta} + \frac{y}{b \cos^2 \theta} = 0. \tag{2}$$

Clearing of fractions gives us

$$bx \cos \theta + ay \sin \theta = ab \sin \theta \cos \theta, \tag{3}$$
$$bx \cos^2 \theta + ay \sin^2 \theta = 0. \tag{4}$$

Adding $bx \sin^2 \theta$ to both members of (4) we have

$$bx (\cos^2 \theta + \sin^2 \theta) = (bx - ay) \sin^2 \theta$$

or
$$\sqrt{bx} = \sqrt{bx - ay} \sin \theta,$$

whence
$$\sin \theta = \frac{\sqrt{bx}}{\sqrt{(bx - ay)}},$$

and
$$\cos \theta = \frac{\sqrt{ay}}{\sqrt{(ay - bx)}},$$

which substituted in (3) give after reduction

$$\sqrt{bx} \sqrt{(bx - ay)} + \sqrt{ay} \sqrt{(ay - bx)} = ab.$$

Ex. 2.—Eliminate ϕ between the equations

$$y \cos \phi - x \sin \phi = a \cos 2\phi \tag{1}$$
$$x \cos \phi + y \sin \phi = 2a \sin 2\phi. \tag{2}$$

From (1) and (2) we have by division

$$\frac{x \cos \phi + y \sin \phi}{y \cos \phi - x \sin \phi} = 2 \tan 2\phi$$

or
$$\frac{x + y \tan \phi}{y - x \tan \phi} = \frac{4 \tan \phi}{1 - \tan^2 \phi},$$

ELIMINATION OF TRIGONOMETRICAL FUNCTIONS. 247

whence
$$\tan^3 \phi - \frac{3x}{y} \tan^2 \phi + 3 \tan \phi - \frac{x}{y} = 0, \tag{3}$$

but
$$\tan^3 \phi \mp 3 \tan^2 \phi + 3 \tan \phi \mp 1 = (\tan \phi \mp 1)^3. \tag{4}$$

Subtracting (3) from (4), and using first the upper and then the lower sign, we get

$$3\left(\frac{x}{y} - 1\right) \tan^2 \phi + \frac{x}{y} - 1 = (\tan \phi - 1)^3,$$

whence
$$\frac{(x-y)^{\frac{2}{3}}}{y^{\frac{2}{3}}} = \frac{(\tan \phi - 1)^2}{(3 \tan^2 \phi + 1)^{\frac{2}{3}}}, \tag{5}$$

and likewise
$$\frac{(x+y)^{\frac{2}{3}}}{y^{\frac{2}{3}}} = \frac{(\tan \phi + 1)^2}{(3 \tan^2 \phi + 1)^{\frac{2}{3}}}. \tag{6}$$

Dividing (6) by (5) we have

$$\left(\frac{x+y}{x-y}\right)^{\frac{2}{3}} = \left(\frac{\tan \phi + 1}{\tan \phi - 1}\right)^2,$$

whence
$$\tan \phi = \frac{(x+y)^{\frac{1}{3}} + (x-y)^{\frac{1}{3}}}{(x+y)^{\frac{1}{3}} - (x-y)^{\frac{1}{3}}},$$

therefore
$$\sin \phi = \frac{(x+y)^{\frac{1}{3}} + (x-y)^{\frac{1}{3}}}{\sqrt{2} \{(x+y)^{\frac{2}{3}} + (x-y)^{\frac{2}{3}}\}^{\frac{1}{2}}},$$

$$\cos \phi = \frac{(x+y)^{\frac{1}{3}} - (x-y)^{\frac{1}{3}}}{\sqrt{2} \{(x+y)^{\frac{2}{3}} + (x-y)^{\frac{2}{3}}\}^{\frac{1}{2}}}$$

and
$$\sin 2\phi = 2 \sin \phi \cos \phi = \frac{(x+y)^{\frac{2}{3}} - (x-y)^{\frac{2}{3}}}{(x+y)^{\frac{2}{3}} + (x-y)^{\frac{2}{3}}},$$

which substituted in (2) give after reduction

$$(x+y)^{\frac{2}{3}} + (x-y)^{\frac{2}{3}} = 2a^{\frac{2}{3}}.$$

Examples.

3. Eliminate θ and ϕ between the equations
$$\sin \theta = m \cos \phi + n \sin \phi$$
$$\cos \theta = m \sin \phi - n \cos \phi.$$
Ans. $m^2 + n^2 = 1$.

4. Eliminate θ between the equations
$$\operatorname{cosec}^2 \theta = m \tan \theta$$
$$\sec^2 \theta = n \cot \theta.$$
Ans. $(mn)^{\frac{2}{3}} = (\sqrt{m} + \sqrt{n})^2$.

5. Eliminate B and C between the equations
$$a - b \cos C - c \cos B = 0$$
$$b - c \cos A - a \cos C = 0$$
$$c - a \cos B - b \cos A = 0.$$
Ans. $a^2 = b^2 + c^2 - 2bc \cos A$.

6. Eliminate θ and ϕ between the equations
$$\tan \theta + \tan \phi = a$$
$$\tan \theta \tan \phi (\operatorname{cosec} 2\theta + \operatorname{cosec} 2\phi) = b$$
$$\cos (\theta + \phi) = c \cos (\theta - \phi).$$
Ans. $a = b + bc$.

7. Eliminate θ between the equations
$$x = a (\cos \theta + \cos 2\theta)$$
$$y = b (\sin \theta + \sin 2\theta).$$
Ans. $\left(\dfrac{x^2}{a^2} + \dfrac{y^2}{b^2}\right)^2 - 3 \left(\dfrac{x^2}{a^2} + \dfrac{y^2}{b^2}\right) = \dfrac{2x}{a}$.

8. Eliminate α and β between the equations
$$\tan \alpha = m, \ \tan \beta = n, \ \tan \dfrac{\alpha}{2} \tan \dfrac{\beta}{2} = c.$$
Ans. $4c \left(\dfrac{1}{m} + \dfrac{c}{n}\right) \left(\dfrac{1}{n} + \dfrac{c}{m}\right) = (1 - c^2)^2$.

9. Eliminate θ between the equations
$$\sec \theta - \cos \theta = m$$
$$\operatorname{cosec} \theta - \sin \theta = n.$$
Ans. $(mn)^{\frac{2}{3}} (m^{\frac{2}{3}} + n^{\frac{2}{3}}) = 1$.

ELIMINATION OF TRIGONOMETRICAL FUNCTIONS. 249

10. Eliminate θ and ϕ between the equations

$$a \sin^2 \theta + b \cos^2 \theta = c$$
$$b \sin^2 \phi + a \cos^2 \phi = d$$
$$a \tan \theta - b \tan \phi = 0.$$

Ans. $\dfrac{1}{a} + \dfrac{1}{b} = \dfrac{1}{c} + \dfrac{1}{d}$.

11. Eliminate θ between the equations

$$a \sin \theta + b \cos \theta = m$$
$$a \cos \theta - b \sin \theta = n.$$

Ans. $a^2 + b^2 = m^2 + n^2$.

12. Eliminate θ and ϕ between the equations

$$x = a \cos^m \theta \cos^m \phi$$
$$y = b \cos^m \theta \sin^m \phi$$
$$z = c \sin^m \theta.$$

Ans. $\left(\dfrac{x}{a}\right)^{\frac{2}{m}} + \left(\dfrac{y}{b}\right)^{\frac{2}{m}} + \left(\dfrac{z}{c}\right)^{\frac{2}{m}} = 1$.

13. Eliminate θ between the equations

$$(a+b) \tan (\theta - \phi) = (a-b) \tan (\theta + \phi)$$
$$a \cos 2\phi + b \cos 2\theta = c.$$

Ans. $a^2 - b^2 + c^2 = 2 ac \cos 2\phi$.

14. Eliminate ϕ between the equations

$$n \sin \theta - m \cos \theta = 2m \sin \phi$$
$$n \sin 2\theta - m \cos 2\phi = n.$$

Ans. $(n \sin \theta + m \cos \theta)^2 = 2m(m+n)$.

15. Eliminate θ and r between the equations

$$r = 2a \cos 2\theta, \quad x = r \cos \theta, \quad y = r \sin \theta.$$

Ans. $(x^2 + y^2)^3 = 4a^2(x^2 - y^2)^2$.

16. Eliminate θ and r between the equations

$$r = \dfrac{a}{\sqrt{\theta}}, \quad x = r \cos \theta, \quad y = r \sin \theta.$$

Ans. $(x^2 + y^2) \tan^{-1} \dfrac{y}{x} = a^2$.

CHAPTER XIV.

ON THE COMPUTATION OF LOGARITHMS.

We will here prove all the formulæ necessary for the computation of Napierian and Common Logarithms; but before commencing this chapter the student should read carefully Articles 88–97, of Chapter VII.

The Exponential Theorem.

182. *To expand a^x in a series of ascending powers of x.*

$a^x = \{1+(a-1)\}^x$

$= 1+x(a-1)+\dfrac{x(x-1)}{1.2}(a-1)^2+\dfrac{x(x-1)(x-2)}{1.2.3}(a-1)^3+ \&\text{c.}$

$= 1+\left\{(a-1)-\tfrac{1}{2}(a-1)^2+\tfrac{1}{3}(a-1)^3-\tfrac{1}{4}(a-1)^4+\&\text{c.}\right\}x$

+terms in x^2+terms in x^3+ &c.

This shews that a^x can be expanded in a series beginning with unity and proceeding in ascending powers of x.

Let the coefficient of x, which is

$(a-1)-\tfrac{1}{2}(a-1)^2+\tfrac{1}{3}(a-1)^3-\tfrac{1}{4}(a-1)^4+ \&\text{c.},$

be represented by A, and the coefficients of x^2, x^3, &c., by B, C, &c. respectively, then we have

$a^x = 1+Ax+Bx^2+Cx^3+Dx^4+ \&\text{c.},$

where A, B, C, &c., are independent of x, and therefore remain unchanged however we may change x.

Assume also

$a^y = 1+Ay+By^2+Cy^3+Dy^4+ \&\text{c.}$

Now, since $a^{x+y}=a^x a^y$, and

$a^{x+y}=1+A(x+y)+B(x+y)^2+C(x+y)^3+ \&\text{c.},$

the last series is evidently equal to the product of the two former; therefore we have

$1+A(x+y)+B(x+y)^2+C(x+y)^3+ \&\text{c.}$
$=(1+Ax+Bx^2+Cx^3+ \&\text{c.})(1+Ay+By^2+Cy^3+ \&\text{c.}).$

THE EXPONENTIAL THEOREM.

Expanding both members of this equation we have

$$\left\{\begin{array}{l} 1+Ax+ Bx^2+ Cx^3 + Dx^4 + \&c. \\ Ay+2Bxy+3Cx^2y+4Dx^3y + \&c. \\ By^2+3Cxy^2+6Dx^2y^2 + \&c. \\ Cy^3 +4Dxy^3 + \&c. \\ Dy^4 + \&c. \end{array}\right.$$

$$= \left\{\begin{array}{l} 1+Ax+Bx^2 + Cx^3 + Dx^4 + \&c. \\ Ay+A^2xy+ABx^2y+ACx^3y + \&c. \\ By^2 +ABxy^2+B^2x^2y^2 + \&c. \\ Cy^3 +ACxy^3 + \&c. \\ Dy^4 + \&c. \end{array}\right.$$

Cancelling the terms common to both series, we have

$$\left\{\begin{array}{l} 2Bxy+3Cx^2y+4Dx^3y + \&c. \\ +3Cxy^2+6Dx^2y^2 + \&c. \\ +4Dxy^3 + \&c. \end{array}\right\} = \left\{\begin{array}{l} A^2xy+ABx^2y+ACx^3y + \&c. \\ +ABxy^2+B^2x^2y^2 + \&c. \\ +ACxy^3 + \&c. \end{array}\right\}$$

Equating the coefficients of xy, x^2y, x^3y, &c., we have

$$2B=A^2, \quad \text{or} \quad B=\frac{A^2}{1.2},$$

$$3C=AB, \quad \text{or} \quad C=\frac{A^3}{1.2.3},$$

$$4D=AC, \quad \text{or} \quad D=\frac{A^4}{1.2.3.4},$$

&c. &c.

Therefore $\quad a^x = 1+Ax+\dfrac{A^2x^2}{1.2}+\dfrac{A^3x^3}{1.2.3}+\dfrac{A^4x^4}{1.2.3.4}+ \&c.,$

where $\quad A=(a-1)-\tfrac{1}{2}(a-1)^2+\tfrac{1}{3}(a-)^3-\&c.$

Since this result is true for all values of x, take x such that

$$Ax=1, \quad \text{or} \quad x=\frac{1}{A}, \quad \text{then}$$

$$a^{\frac{1}{A}}=1+1+\frac{1}{1.2}+\frac{1}{1.2.3}+\frac{1}{1.2.3.4}+\&c.$$

$$=2.718281828459\ldots=e,$$

hence $\quad a=e^A \quad \text{and} \quad A=\log_e a.$

Therefore we have finally

$$a^x = 1 + (\log_e a)\frac{x}{1} + (\log_e a)^2 \frac{x^2}{1.2} + (\log_e a)^3 \frac{x^3}{1.2.3} + \&c. \qquad (257)$$

which is called the *Exponential Theorem*.

If $a = e$, we have

$$e^x = 1 + x + \frac{x^2}{1.2} + \frac{x^3}{1.2.3} + \frac{x^4}{1.2.3.4} + \&c. \qquad (258)$$

The Logarithmic Series.

183. *To express $\log_e (1+x)$ in a series of ascending powers of x.*

From the last Article we have

$$\log_e a = A$$
$$= (a-1) - \tfrac{1}{2}(a-1)^2 + \tfrac{1}{3}(a-1)^3 - \tfrac{1}{4}(a-1)^4 + \&c.,$$

in which write $1+x$ for a and we get

$$\log_e (1+x) = x - \frac{x^2}{2} + \frac{x^3}{3} - \frac{x^4}{4} + \frac{x^5}{5} - \&c. \qquad (259)$$

In the last series write $-x$ for x, then we have

$$\log_e (1-x) = -x - \frac{x^2}{2} - \frac{x^3}{3} - \frac{x^4}{4} - \&c. \qquad (260)$$

184. To prove the Logarithmic Series independently of the Exponential Theorem.

Assume $\quad Ax + Bx^2 + Cx^3 + Dx^4 + \ldots = \log_a(1+x)$
and $\quad\;\; Ay + By^2 + Cy^3 + Dy^4 + \ldots = \log_a(1+y)$

By subtraction we have

$$A(x-y) + B(x^2 - y^2) + C(x^3 - y^3) + \ldots = \log_a(1+x) - \log_a(1+y)$$
$$= \log_a\left(\frac{1+x}{1+y}\right)$$
$$= \log_a\left(1 + \frac{x-y}{1+y}\right)$$

We may consider $\dfrac{x-y}{1+y}$ as a simple quantity, and therefore $\log_a\left(1 + \dfrac{x-y}{1+y}\right)$ may be developed in the same manner as $\log_a(1+x)$.

THE LOGARITHMIC SERIES.

Thus,
$$\log_a\left(1+\frac{x-y}{1+y}\right)=A\left(\frac{x-y}{1+y}\right)+B\left(\frac{x-y}{1+y}\right)^2+C\left(\frac{x-y}{1+y}\right)^3+\ldots$$

Therefore
$$A(x-y)+B(x^2-y^2)+C(x^3-y^3)+\ldots$$
$$=A\left(\frac{x-y}{1+y}\right)+B\left(\frac{x-y}{1+y}\right)^2+C\left(\frac{x-y}{1+y}\right)^3+\ldots$$

Dividing by $x-y$ we have
$$A+B(x+y)+C(x^2+xy+y^2)+\ldots=A\left(\frac{1}{1+y}\right)+B\frac{(x-y)}{(1+y)^2}+C\frac{(x-y)^2}{(1+y)^3}+\ldots$$

Since this equation is true for all values of x and y, it must be true for $x=y$; hence, writing x for y, it becomes
$$A+2Bx+3Cx^2+\ldots=\frac{A}{1+x}$$
$$=A(1-x+x^2-x^3+\ldots).$$

Equating the coefficients of like powers of x, we have

$$A=A, \quad 2B=-A \quad \text{or} \quad B=-\frac{A}{2},$$
$$3C=A, \quad\quad \text{or} \quad C=\frac{A}{3},$$
$$4D=-A, \quad \text{or} \quad D=-\frac{A}{4},$$
$$\&c. \quad\quad\quad \&c.$$

Therefore
$$\log_a(1+x)=A\left(x-\frac{x^2}{2}+\frac{x^3}{3}-\frac{x^4}{4}+\ldots\right).$$

Dividing by x we have
$$\frac{1}{x}\log_a(1+x)=A\left(1-\frac{x}{2}+\frac{x^2}{3}-\frac{x^3}{4}+\ldots\right),$$

but $\quad \dfrac{1}{x}\log_a(1+x)=\log_a(1+x)^{\frac{1}{x}}, \quad$ (Art. 93.)
$$=\log_a\left(1+1+\frac{1-x}{1.2}+\frac{1-3x+2x^2}{1.2.3}+\ldots\right)$$

by the Binomial Theorem.

Therefore

$$A\left(1 - \frac{x}{2} + \frac{x^2}{3} - \frac{x^3}{4} + \ldots\right) = \log_a\left(1 + 1 + \frac{1-x}{1.2} + \frac{1-3x+2x^2}{1.2.3} + \ldots\right).$$

Since this equation is true for all values of x, it must be true when $x=0$, hence

$$A = \log_a\left(1 + 1 + \frac{1}{1.2} + \frac{1}{1.2.3} + \frac{1}{1.2.3.4} + \ldots\right)$$

$$= \log_a e = \frac{1}{\log_e a}, \quad \text{(Art. 97)}$$

= the modulus of the system whose base is a. (149)

Therefore we have generally

$$\log_a(1+x) = \frac{1}{\log_e a}\left(x - \frac{x^2}{2} + \frac{x^3}{3} - \frac{x^4}{4} + \ldots\right). \tag{261}$$

If $a = e$, we have

$$\log_e(1+x) = x - \frac{x^2}{2} + \frac{x^3}{3} - \frac{x^4}{4} + \ldots \text{ as before.}$$

185. To deduce the Exponential Theorem from the Logarithmic Series.

From (261) we have, by writing A for $\dfrac{1}{\log_e a}$,

$$a^{A\left(x - \frac{x^2}{2} + \frac{x^3}{3} - \&c.\right)} = 1 + x,$$

and raising both members of this equation to the power $\dfrac{y}{Ax}$, we have

$$a^{y\left(1 - \frac{x}{2} + \frac{x^2}{3} - \&c.\right)} = (1+x)^{\frac{y}{Ax}}$$

$$= 1 + \frac{y}{A} + \frac{y^2}{1.2.A^2} - \frac{xy}{1.2.A^2} + \frac{y^3}{1.2.3.A^3} - \frac{xy^2 + 2xy^2}{1.2.3.A^2} + \&c.,$$

and when $x = 0$, this becomes

$$a^y = 1 + \frac{y}{A} + \frac{y^2}{1.2.A^2} + \frac{y^3}{1.2.3.A^3} + \&c.$$

Restoring the value of A, we have

$$a^y = 1 + (\log_e a)\frac{y}{1} + (\log_e a)^2 \frac{y^2}{1.2} + (\log_e a)^3 \frac{y^3}{1.2.3} + \&c.$$

186. The Napierian Base.

The sum of the series

$$1 + 1 + \frac{1}{1.2} + \frac{1}{1.2.3} + \frac{1}{1.2.3.4} + \&c.,$$

which we have denoted by e, is the base of the Napierian system of logarithms. This base renders the logarithmic series simpler than any other base would, as is evident from a comparison of (259) and (261). Napierian logarithms are sometimes called *natural* logarithms, because they occur first in the investigation of formulæ for their calculation. The Napierian system is used for the most part in the higher *Analysis*, and, with the exception of the common system which is universally employed in arithmetical and trigonometrical calculations, it is the only one which we shall ever have to use.

187. The Napierian Base is incommensurable.

For, if possible, let $e = \frac{m}{n}$, where m and n are integers, then

$$\frac{m}{n} = 1 + 1 + \frac{1}{1.2} + \frac{1}{1.2.3} + \cdots \frac{1}{1.2.3\ldots n} + \frac{1}{1.2.3\ldots(n+1)} + \cdots$$

Multiply both members by $1.2.3\ldots n$, and we have

$1.2.3\ldots(n-1)m = 2.2.3.4\ldots n + 3.4.5\ldots n + \ldots$
$$+ \left\{ \frac{1}{n+1} + \frac{1}{(n+1)(n+2)} + \cdots \right\};$$

but the former member being integral, the latter must also be so, which is impossible since $\frac{1}{n+1} + \frac{1}{(n+1)(n+2)} + \cdots$ is greater than $\frac{1}{n+1}$ and less than the sum of the geometrical series $\frac{1}{n+1} + \frac{1}{(n+1)^2} + \frac{1}{(n+1)^3} + \cdots$, that is, less than $\frac{1}{n}$; therefore e is incommensurable.

188. Converging Series for the immediate calculation of Napierian Logarithms.

From (259) and (260) we have by subtraction

$$\log_e (1+x) - \log_e (1-x) = 2\left(x + \frac{x^3}{3} + \frac{x^5}{5} + \frac{x^7}{7} + \ldots\right)$$

or

$$\log_e \left(\frac{1+x}{1-x}\right) = 2\left(x + \frac{x^3}{3} + \frac{x^5}{5} + \frac{x^7}{7} + \ldots\right);$$

in which let $\frac{1+x}{1-x} = m$, and therefore $x = \frac{m-1}{m+1}$, thus we have

$$\log_e m = 2\left\{\left(\frac{m-1}{m+1}\right) + \frac{1}{3}\left(\frac{m-1}{m+1}\right)^3 + \ldots\right\}, \qquad (262)$$

which, however, converges too slowly to be of much utility in the calculation of the logarithms of integral numbers.

In (262) let $m = \frac{1+x}{x}$, and therefore $\frac{m-1}{m+1} = \frac{1}{2x+1}$,

thus we obtain

$$\log_e \frac{1+x}{x} = 2\left\{\frac{1}{2x+1} + \frac{1}{3}\frac{1}{(2x+1)^3} + \ldots\right\}$$

or

$$\log_e (1+x) = \log_e x + 2\left\{\frac{1}{2x+1} + \frac{1}{3}\frac{1}{(2x+1)^3} + \ldots\right\}, \qquad (263)$$

which converges very rapidly, especially when x is large.

In the last equation, let $1+x=y^2$, and therefore $x = y^2 - 1$ and $2x+1 = 2y^2 - 1$; then we have

$$\log_e y^2 = \log_e (y^2 - 1) + 2\left\{\frac{1}{2y^2-1} + \frac{1}{3}\frac{1}{(2y^2-1)^3} + \ldots\right\}$$

or $\log_e (y+1) = 2\log_e y - \log_e (y-1)$

$$-2\left\{\frac{1}{2y^2-1} + \frac{1}{3}\frac{1}{(2y^2-1)^3} + \ldots\right\}, \qquad (264)$$

which also converges very rapidly when x is large.

By judicious substitutions, many other series for the calculation of Napierian logarithms, may be deduced, but practically considered they are now unnecessary, as tables have been already computed to the highest attainable degree of accuracy. We shall therefore pursue this subject no farther, but refer the student to the third example at the end of this chapter for two other converging series, which may be advantageously used for the same purpose.

Calculation of Napierian Logarithms.

189. By the last three formulæ we are able to compute the Napierian logarithms of all numbers; but the properties established in Articles 91, 92 and 93, render it necessary to apply them to *prime* numbers only.

Thus, in (262) let $m=2$, then

$$\log_e 2 = 2 \left\{ \frac{1}{3} + \frac{1}{3}\left(\frac{1}{3}\right)^3 + \frac{1}{5}\left(\frac{1}{3}\right)^5 + \frac{1}{7}\left(\frac{1}{3}\right)^7 + \ldots \right\}$$

$$= .6931471 \ldots$$

In (263) let $x=2$, then

$$\log_e 3 = \log_e 2 + 2 \left\{ \frac{1}{5} + \frac{1}{3}\left(\frac{1}{5}\right)^3 + \frac{1}{5}\left(\frac{1}{5}\right)^5 + \ldots \right\}$$

$$= .6931471 + .4054650 = 1.098612.$$

$\log_e 4 = \log_e (2 \times 2) = 2 \log_e 2 = 1.386294.$

In (264) let $y=4$, then

$$\log_e 5 = 2 \log 4 - \log_e 3 - 2 \left\{ \frac{1}{31} + \frac{1}{3}\left(\frac{1}{31}\right)^3 + \frac{1}{5}\left(\frac{1}{31}\right)^5 + \ldots \right\}$$

$$= 2.772588 - 1.098612 - .064538$$

$$= 1.609438.$$

$\log_e 6 = \log_e 3 + \log_e 2 = 1.791759.$

In (264) let $y=6$, then

$$\log_e 7 = 2 \log_e 6 - \log_e 5 - 2 \left\{ \frac{1}{71} + \frac{1}{3}\left(\frac{1}{71}\right)^3 + \ldots \right\}$$

$$= 1.945910.$$

$\log_e 8 = \log_e 2^3 = 3 \log_e 2 = 2.079442.$
$\log_e 9 = \log_e 3^2 = 2 \log_e 3 = 2.197225.$
$\log_e 10 = \log_e (5 \times 2) = \log_e 5 + \log_e 2 = 2.302585.$

&c. &c.

Hence the modulus of the common system is

$$\frac{1}{\log_e 10} = \frac{1}{2.302585} = .4342944819\ldots$$

as was shewn in Art. 95.

Calculation of Common Logarithms.

190. Having computed the Napierian logarithms by the method of the last Article, we may convert them into common logarithms by (151), thus,

$$\text{Log } 2 = .43429448 \log_e 2 = .3010300$$
$$\log 3 = .43429448 \log_e 3 = .4771213$$
$$\&c. \qquad \&c.$$

By means of M, the modulus, we may adapt the series of Art. 188 to the calculation of common logarithms; thus, by (151)

$$\log m = 2M \left\{ \left(\frac{m-1}{m+1}\right) + \frac{1}{3}\left(\frac{m-1}{m+1}\right)^3 + \ldots \right\}. \tag{264}$$

$$\log (1+x) = \log x + 2M \left\{ \frac{1}{2x+1} + \frac{1}{3}\frac{1}{(2x+1)^3} + \ldots \right\}. \tag{265}$$

$$\log (y+1) = 2 \log y - \log (y-1) - 2M \left\{ \frac{1}{2y^2-1} + \frac{1}{3}\frac{1}{(2y^2-1)^3} + \ldots \right\}. \tag{266}$$

Having found the logarithms of prime numbers by the preceding series, the logarithms of composite numbers are easily found by the principle of Art. 91.

Thus, $\log 3360 = \log (2^5 \times 3 \times 5 \times 7)$
$$= 5 \log 2 + \log 3 + \log 5 + \log 7$$
$$= 3.5263393, \quad \&c.$$

Theory of Proportional Parts.

191. We shall now investigate how far the principle of proportional parts can be depended on in finding the logarithm of a number which is not found exactly in the tables. In the following investigation we will assume that the logarithms are calculated to seven decimal places, and that the table contains the logarithms of all whole numbers from 1 to 100000.

THEORY OF PROPORTIONAL PARTS.

To shew that in general the Increment of the Logarithm is approximately proportional to the Increment of the Number.

Let N and $N+h$ be two numbers, the former containing five digits and the latter six, the last (h) being after the decimal point. Then we have

$$\log (N+h) - \log N = \log \frac{N+h}{N} = \log \left(1 + \frac{h}{N}\right)$$

$$= M\left(\frac{h}{N} - \frac{h^2}{2N^2} + \ldots\right) \quad \text{by (261)}$$

where M is the modulus $.43429448\ldots$.

Now, if N is not less than 10000, the second term of this series is less than $.000,000,002,2$, which does not affect the seventh decimal place, and may therefore be neglected.

Hence, as far as seven places of decimals at least, we have

$$\log (N+h) - \log N = \frac{M}{N} h,$$

which shews that the change of the logarithm is approximately proportional to the change of the number.

We will now proceed to ascertain to what extent the principle of proportional parts can be applied in the case of the logarithmic trigonometrical functions.

192. To shew that in general the change of the Tabular Logarithmic Function of an Angle is approximately proportional to the change of the Angle.

Let θ denote any angle and h any small increment such as $1'$ or $10''$, then we have

$$\frac{\sin (\theta+h)}{\sin \theta} = \frac{\sin \theta \cos h + \sin h \cos \theta}{\sin \theta}$$

$$= 1 + h \cot \theta, \text{ approximately,}$$

since $\cos h = 1$ and $\sin h = h$ very nearly.

Therefore

$$\text{Log sin } (\theta+h) - \text{Log sin } \theta = \log(1 + h \cot \theta)$$
$$= M(h \cot \theta - \frac{h^2 \cot^2 \theta}{2} + \&c.)$$
$$= Mh \cot \theta, \text{ approximately.}$$

But when θ is very small $\cot \theta$ is very large, and therefore the second term $\dfrac{h^2 \cot^2 \theta}{2}$ may be too large to be neglected.

Thus, if $h = 1'$ and $\theta = 2°$, the value of the second term is .0000151, which is far too large to be disregarded. If, however, $h = 10''$ and $\theta = 2°$, the value of the second term is .0000004, which will affect the seventh figure but not generally the sixth; therefore we conclude that when θ is not very small,

$$\text{Log sin } (\theta+h) - \text{Log sin } \theta = Mh \cot \theta. \qquad (267)$$

that is, with the exception just stated, the change of the logarithmic sine is approximately proportional to the change of the angle.

193. In a similar manner we find

$$\text{Log cos } (\theta+h) - \text{Log cos } \theta = -Mh \tan \theta \qquad (268)$$

approximately.

When θ is near $90°$ the second term, omitted in (268), is too large to be neglected. Therefore, with this exception, the change of the logarithmic cosine is approximately proportional to the change of the angle.

194. In the case of the tangent we have

$$\tan (\theta+h) = \frac{\tan \theta + \tan h}{1 - \tan h \tan \theta}$$
$$= \frac{\tan \theta + h}{1 - h \tan \theta} = \tan \theta + h \sec^2 \theta$$

approximately,

therefore $\qquad \dfrac{\tan (\theta+h)}{\tan \theta} = 1 + 2h \operatorname{cosec} 2\theta.$

and $\qquad \text{Log tan } (\theta+h) - \text{Log tan } \theta = 2Mh \operatorname{cosec} 2\theta \qquad (269)$

approximately.

When θ is very small cosec 2θ is very large, and the second term is too large to be neglected; therefore, with this exception, the change of the logarithmic tangent is approximately proportional to the change in the angle.

195. A similar proof may be employed for each of the other Logarithmic functions. The following are the results which may be verified by the student:

$$\text{Log cot } (\theta+h) - \text{Log cot } \theta = -2Mh \text{ cosec } 2\theta. \qquad (270)$$

$$\text{Log sec } (\theta+h) - \text{Log sec } \theta = Mh \tan \theta. \qquad (271)$$

$$\text{Log cosec } (\theta-h) - \text{Log cosec } \theta = -Mh \cot \theta. \qquad (272)$$

196. From the preceding Articles it is seen that the change of the Logarithmic sine is equal to that of the Logarithmic cosecant, but with the opposite sign. Hence a column of "Differences for 1'" is printed in some tables *between* the columns of Logarithmic sines and cosecants, serving to the former as a column of *increments* for 1', and to the latter as a column of *decrements* for 1'.

In like manner the columns of cosines and secants have the same differences for 1', and so also have the tangents and cotangents; these columns serving as increments to the secants and tangents, and decrements to the cosines and cotangents.

197. From the preceding investigations, and from an inspection of the tables themselves, the student will see that the principle of proportional parts is not applicable to angles which are very small or nearly equal to a right angle. In the case of the Log sin and Log cosec, the differences are irregular for small angles and insensible for angles near 90°; for the Log cos and Log sec the differences are insensible for small angles and irregular for angles near 90°; for the Log tan and Log cot the differences are irregular both for small angles and for angles near 90°. For the methods of computing the Logarithmic functions of angles near the limits of the quadrant, the student is referred to Art. 112.

Examples.

1. Prove that
$$\log_e a = n\left\{(1-a^{-\frac{1}{n}}) + \tfrac{1}{2}(1-a^{-\frac{1}{n}})^2 + \tfrac{1}{3}(1-a^{-\frac{1}{n}})^3 + \ldots\right\}$$

2. Prove that
$$N = 1 + \frac{\log N}{M} + \frac{1}{1.2}\left\{\frac{\log N}{M}\right\}^2 + \frac{1}{1.2.3}\left\{\frac{\log N}{M}\right\}^3 + \ldots$$

where N is any number and M the modulus.

3. If a, b, c be three consecutive numbers, prove that
$$\log b = \tfrac{1}{2}(\log a + \log c) + M\left\{\frac{1}{2ac+1} + \tfrac{1}{3}\frac{1}{(2ac+1)^3} + \ldots\right\}$$
and
$$\log c = 2\log b - \log a - 2M\left\{\frac{1}{2b^2-1} + \tfrac{1}{3}\frac{1}{(2b^2-1)^3} + \ldots\right\}.$$

4. Prove that $\dfrac{1}{e} = \dfrac{1}{1.3} + \dfrac{1}{1.2.3.5} + \dfrac{1}{1.2.3.4.5.7} + \ldots$

5. Prove that $\dfrac{e}{2} = \dfrac{1}{1.2} + \dfrac{1+2}{1.2.3} + \dfrac{1+2+3}{1.2.3.4} + \dfrac{1+2+3+4}{1.2.3.4.5} + \ldots$

6. Prove that $2e = 1 + \dfrac{2}{1} + \dfrac{3}{1.2} + \dfrac{4}{1.2.3} + \dfrac{5}{1.2.3.4} + \ldots$

7. Find the modulus of the system whose base is $\dfrac{1}{3}$.

   Ans. $-.91024$.

8. Prove that $\log_e (\sqrt{-1})^{\frac{1}{2\sqrt{-1}}} = 1 - \dfrac{1}{3} + \dfrac{1}{5} - \dfrac{1}{7} + \dfrac{1}{9} - \dfrac{1}{11} + \ldots$

9. Prove that $\log_e 101 - \log_e 99 = \dfrac{1}{50}$ very nearly.

10. To what base is -5 the log of 32768?   *Ans.* $\dfrac{1}{8}$.

11. The log of a number to one base is the same as that of its reciprocal to another base; find the relation of the bases to each other.

CHAPTER XV.

DE MOIVRE'S THEOREM—EXPANSIONS OF CERTAIN TRIGONOMETRICAL FUNCTIONS.

De Moivre's Theorem.

198. *If m be any rational quantity, either integral or fractional, positive or negative, then*

$$(\cos\theta \pm \sqrt{-1}\sin\theta)^m = \cos m\theta \pm \sqrt{-1}\sin m\theta.$$

(1) Let the index be a positive whole number.

$$(\cos\theta \pm \sqrt{-1}\sin\theta)^2 = (\cos\theta \pm \sqrt{-1}\sin\theta)(\cos\theta \pm \sqrt{-1}\sin\theta)$$
$$= \cos^2\theta - \sin^2\theta \pm \sqrt{-1}\,2\sin\theta\cos\theta$$
$$= \cos 2\theta \pm \sqrt{-1}\sin 2\theta,$$

$$(\cos\theta \pm \sqrt{-1}\sin\theta)^3 = (\cos 2\theta \pm \sqrt{-1}\sin 2\theta)(\cos\theta \pm \sqrt{-1}\sin\theta)$$
$$= \cos 2\theta\cos\theta - \sin 2\theta\sin\theta$$
$$\pm \sqrt{-1}(\sin 2\theta\cos\theta + \cos 2\theta\sin\theta)$$
$$= \cos 3\theta \pm \sqrt{-1}\sin 3\theta, \quad \text{and so on.}$$

Suppose this law to hold for m factors, so that

$$(\cos\theta \pm \sqrt{-1}\sin\theta)^m = \cos m\theta \pm \sqrt{-1}\sin m\theta,$$

then

$$(\cos\theta \pm \sqrt{-1}\sin\theta)^{m+1} = (\cos m\theta \pm \sqrt{-1}\sin m\theta)(\cos\theta \pm \sqrt{-1}\sin\theta)$$
$$= \cos m\theta\cos\theta - \sin m\theta\sin\theta$$
$$\pm \sqrt{-1}(\sin m\theta\cos\theta + \cos m\theta\sin\theta)$$
$$= \cos(m+1)\theta \pm \sqrt{-1}\sin(m+1)\theta.$$

If, then, the law holds for m factors, it also holds for $m+1$ factors; but we have just shewn that it holds when $m=3$, therefore it holds when $m=4$, and by successive inductions we conclude that the formula is true for any positive integer.

Therefore we have

$$(\cos\theta \pm \sqrt{-1}\sin\theta)^m = \cos m\theta \pm \sqrt{-1}\sin m\theta, \qquad (273)$$

when m is any positive integer.

(2) Let the index be a negative whole number.

$$(\cos\theta \pm \sqrt{-1}\sin\theta)^{-m} = \frac{1}{(\cos\theta \pm \sqrt{-1}\sin\theta)^m}$$

$$= \frac{(\cos^2\theta + \sin^2\theta)^m}{(\cos\theta \pm \sqrt{-1}\sin\theta)^m}$$

$$= (\cos\theta \mp \sqrt{-1}\sin\theta)^m,$$

by actual division

$$= \cos m\theta \mp \sqrt{-1}\sin m\theta, \quad \text{by (273)}$$

$$= \cos(-m\theta) \pm \sqrt{-1}\sin(-m\theta),$$

which proves the theorem when m is a negative integer.

(3) Let the index be a fraction $\dfrac{m}{n}$, either positive or negative.

$$(\cos\theta \pm \sqrt{-1}\sin\theta)^m = \cos m\theta \pm \sqrt{-1}\sin m\theta$$

$$= \cos n\left(\frac{m}{n}\theta\right) \pm \sqrt{-1}\sin n\left(\frac{m}{n}\theta\right)$$

$$= (\cos\frac{m}{n}\theta \pm \sqrt{-1}\sin\frac{m}{n}\theta)^n, \quad \text{by (273)}$$

therefore

$$(\cos\theta \pm \sqrt{-1}\sin\theta)^{\frac{m}{n}} = \cos\frac{m}{n}\theta \pm \sqrt{-1}\sin\frac{m}{n}\theta, \qquad (274)$$

which proves the theorem for fractional indices.

199. As long as m is an integer, both members of (273) can have only one value; but in the case of (274), in which the index is a fraction, the first member has n different values in consequence of the nth root, while the second member has only one of these. The second member, however, may be transformed so as to exhibit the same number of values as the first.

DE MOIVRE'S THEOREM.

Thus, since $\cos\theta = \cos(2r\pi+\theta)$
and $\sin\theta = \sin(2r\pi+\theta)$, we have

$$(\cos\theta \pm \sqrt{-1}\sin\theta)^{\frac{m}{n}} = \{\cos(2r\pi+\theta) \pm \sqrt{-1}\sin(2r\pi+\theta)\}^{\frac{m}{n}}$$
$$= \cos\frac{m}{n}(2r\pi+\theta) \pm \sqrt{-1}\sin\frac{m}{n}(2r\pi+\theta), \quad (275)$$

in which the second member has n different values corresponding to the values $0, 1, 2, \ldots n-1$, of r, according to Art. 171. Therefore the theorem is entirely general under the form (275).

200. To express an imaginary quantity of the form $(a+b\sqrt{-1})^m$ by means of Trigonometrical Functions.

Assume $k\cos\theta = a$ and $k\sin\theta = b$,

whence $\tan\theta = \dfrac{b}{a}$ and $k = \sqrt{a^2+b^2}$.

Then, by De Moivre's Theorem,

$$(a+b\sqrt{-1})^m = k^m(\cos\theta + \sqrt{-1}\sin\theta)^m$$
$$= (a^2+b^2)^{\frac{m}{2}}(\cos m\theta + \sqrt{-1}\sin m\theta).$$

In a similar manner we find

$$(a+b\sqrt{-1})^{\frac{1}{m}} = (a^2+b^2)^{\frac{1}{2m}}\left(\cos\frac{2r\pi+\theta}{m} + \sqrt{-1}\sin\frac{2r\pi+\theta}{m}\right).$$

Ex.—Find the three values of

$$(\sqrt{5}+2\sqrt{-1})^{\frac{1}{3}}.$$

Here, $\tan\theta = \dfrac{2}{5}\sqrt{5}$ or $\theta = 41°\ 48'\ 37''$

and $k = 3$; then we have, by giving r the values $0, 1, 2$ in succession,

$$\sqrt[3]{3}\left(\cos\frac{\theta}{3} + \sqrt{-1}\sin\frac{\theta}{3}\right),$$

$$\sqrt[3]{3}\left(\cos\frac{2\pi+\theta}{3} + \sqrt{-1}\sin\frac{2\pi+\theta}{3}\right),$$

$$\sqrt[3]{3}\left(\cos\frac{4\pi+\theta}{3} + \sqrt{-1}\sin\frac{4\pi+\theta}{3}\right).$$

201. To express the sine and cosine of an Angle and of its multiple as Algebraic Binomials.

Since $(\cos\theta + \sqrt{-1}\sin\theta)(\cos\theta - \sqrt{-1}\sin\theta) = \cos^2\theta + \sin^2\theta = 1$,

if we assume $\cos\theta + \sqrt{-1}\sin\theta = x$,

then $\cos\theta - \sqrt{-1}\sin\theta = \dfrac{1}{x}$,

and by addition and subtraction we obtain

$$2\cos\theta = x + \frac{1}{x}, \quad \text{and } 2\sqrt{-1}\sin\theta = x - \frac{1}{x};$$

also, $\cos m\theta + \sqrt{-1}\sin m\theta = (\cos\theta + \sqrt{-1}\sin\theta)^m = x^m$,

and $\cos m\theta - \sqrt{-1}\sin m\theta = (\cos\theta - \sqrt{-1}\sin\theta)^m = \dfrac{1}{x^m}$,

by De Moivre's Theorem, whence as above

$$2\cos m\theta = x^m + \frac{1}{x^m}, \quad \text{and } 2\sqrt{-1}\sin m\theta = x^m - \frac{1}{x^m}. \qquad (276)$$

If the index be fractional, we have by Art. 199,

$$2\cos\frac{m}{n}(2r\pi + \theta) = x^{\frac{m}{n}} + \frac{1}{x^{\frac{m}{n}}}, \quad \text{and } 2\sqrt{-1}\sin\frac{m}{n}(2r\pi + \theta) = x^{\frac{m}{n}} - \frac{1}{x^{\frac{m}{n}}},$$

where r has the values $0, 1, 2, \ldots n-1$, so that each member has n different values.

This notation will be found very useful in subsequent investigations.

202. To express the sine and cosine of the multiple of an Angle in terms of the sine and cosine of the Simple Angle.

By De Moivre's Theorem we have

$\cos m\theta + \sqrt{-1}\sin m\theta$
$\qquad = (\cos\theta + \sqrt{-1}\sin\theta)^m$

$$= \cos^m \theta + m \cos^{m-1} \theta \sqrt{-1} \sin \theta - \frac{m(m-1)}{1.2} \cos^{m-2} \theta \sin^2 \theta$$

$$- \frac{m(m-1)(m-2)}{1.2.3} \cos^{m-3} \theta \sqrt{-1} \sin^3 \theta + \ldots$$

by the Binomial Theorem.

Equating the real and imaginary parts of this equation, we have

$$\cos m\theta = \cos^m \theta - \frac{m(m-1)}{1.2} \cos^{m-2} \theta \sin^2 \theta$$

$$+ \frac{m(m-1)(m-2)(m-3)}{1.2.3.4} \cos^{m-4} \theta \sin^4 \theta - \ldots \quad (277)$$

to $\tfrac{1}{2}(m+1)$ terms if m is odd, and to $\tfrac{1}{2}m+1$ terms if m is even.

$$\sin m\theta = m \cos^{m-1} \theta \sin \theta - \frac{m(m-1)(m-2)}{1.2.3} \cos^{m-3} \theta \sin^3 \theta$$

$$+ \frac{m(m-1)(m-2)(m-3)(m-4)}{1.2.3.4.5} \cos^{m-5} \theta \sin^5 \theta - \ldots \quad (278)$$

to $\tfrac{1}{2}(m+1)$ terms if m is odd, and to $\dfrac{m}{2}$ terms if m is even.

We may observe here that if m be *even*, the last term of the expansion of $(\cos \theta + \sqrt{-1} \sin \theta)^m$ is $(-1)^{\frac{m}{2}} \sin^m \theta$, which is real; and the last term but one is $m(-1)^{\frac{m-1}{2}} \cos \theta \sin^{m-1} \theta$, or $\sqrt{-1}\, m(-1)^{\frac{m-2}{2}} \cos \theta \sin^{m-1} \theta$, which is imaginary. Thus, when m is even, the last term of $\cos m\theta$ is $(-1)^{\frac{m}{2}} \sin^m \theta$, and the last term of $\sin m\theta$ is $m(-1)^{\frac{m-2}{2}} \cos \theta \sin^{m-1} \theta$.

Again, if m be *odd*, the last term of the expansion of $(\cos \theta + \sqrt{-1} \sin \theta)^m$ is $(-1)^{\frac{m}{2}} \sin^m \theta$ or $\sqrt{-1}\,(-1)^{\frac{m-1}{2}} \sin^m \theta$, which is imaginary; and the last term but one is $m(-1)^{\frac{m-1}{2}} \cos \theta \sin^{m-1} \theta$, which is real. Thus, when m is odd, the last term of $\cos m\theta$ is $m(-1)^{\frac{m-1}{2}} \cos \theta \sin^{m-1} \theta$, and the last term of $\sin m\theta$ is $(-1)^{\frac{m-1}{2}} \sin^m \theta$.

203. To express the tangent of the multiple of an Angle in terms of the tangent of the Simple Angle.

The quotient of (278) by (277) is

$$\tan m\theta = \frac{m\cos^{m-1}\theta \sin\theta - \dfrac{m(m-1)(m-2)}{1.2.3}\cos^{m-3}\theta \sin^3\theta + \ldots}{\cos^m\theta - \dfrac{m(m-1)}{1.2}\cos^{m-2}\theta \sin^2\theta + \ldots}$$

$$= \frac{m\tan\theta - \dfrac{m(m-1)(m-2)}{1.2.3}\tan^3\theta + \ldots}{1 - \dfrac{m(m-1)}{1.2}\tan^2\theta + \ldots}, \quad (279)$$

by dividing numerator and denominator by $\cos^m\theta$.

204. To express the sine and cosine of an Angle in terms of its Circular Measure.

In (278) and (277) assume $m\theta = a$ or $m = \dfrac{a}{\theta}$, then we have

$$\sin a = a\cos^{m-1}\theta \frac{\sin\theta}{\theta} - \frac{a(a-\theta)(a-2\theta)}{1.2.3}\cos^{m-3}\theta \left(\frac{\sin\theta}{\theta}\right)^3 + \ldots$$

$$\cos a = \cos^m\theta - \frac{a(a-\theta)}{1.2}\cos^{m-2}\theta\left(\frac{\sin\theta}{\theta}\right)^2 + \ldots$$

Now, when $m = \infty$, $\theta = 0$, $\cos\theta$ and all its powers $= 1$ and $\dfrac{\sin\theta}{\theta}$ and all its powers $= 1$. (Art. 74.)

Hence we have ultimately

$$\sin a = a - \frac{a^3}{1.2.3} + \frac{a^5}{1.2.3.4.5} - \ldots \qquad (280)$$

$$\cos a = 1 - \frac{a^2}{1.2} + \frac{a^4}{1.2.3.4} - \ldots \qquad (281)$$

By means of the last two series we may compute the sine and cosine of any angle. Thus, suppose we require the sine of 12°.

TRIGONOMETRICAL EXPANSIONS.   269

The circular measure of $12° = \dfrac{\pi}{15} = .20943951$,

then $\quad \sin 12° = .20943951 - \dfrac{(.20943951)^3}{1.2.3} + \dfrac{(.20943951)^5}{1.2.3.4.5} -$

$\quad\quad\quad = .20943951 - .00153117 + .00000335$
$\quad\quad\quad = .20791169\ldots ,$

agreeing with the tables which give .2079117.

205. To express the positive integral powers of the cosine of an Angle in terms of the cosines of its multiples.

Assume $\quad\quad 2\cos\theta = x + \dfrac{1}{x}$,

then $\quad\quad 2\cos n\theta = x^n + \dfrac{1}{x^n}\quad$ by (276).

$2^n \cos^n\theta = \left(x + \dfrac{1}{x}\right)^n$

$= x^n + nx^{n-2} + \dfrac{n(n-1)}{1.2}x^{n-4} + \ldots + \dfrac{n(n-1)}{1.2}\dfrac{1}{x^{n-4}} + n\dfrac{1}{x^{n-2}} + \dfrac{1}{x^n}$

$= \left(x^n + \dfrac{1}{x^n}\right) + n\left(x^{n-2} + \dfrac{1}{x^{n-2}}\right) + \dfrac{n(n-1)}{1.2}\left(x^{n-4} + \dfrac{1}{x^{n-4}}\right) + \ldots ,$

by placing together the first term and the last, the second and the last but one, and so on;

but $\quad x^n + \dfrac{1}{x^n} = 2\cos n\theta, \quad x^{n-2} + \dfrac{1}{x^{n-2}} = 2\cos(n-2)\theta, \ \&c.,$

therefore

$2^n \cos^n\theta = 2\cos n\theta + 2n\cos(n-2)\theta + 2.\dfrac{n(n-1)}{1.2}\cos(n-4)\theta + \ldots$

In the expansion of $\left(x + \dfrac{1}{x}\right)^n$ by the Binomial Theorem, there are $n+1$ terms, therefore when n is *even* there will be a middle term, viz., the $\left(\dfrac{n}{2}+1\right)^{\text{th}}$, which does not involve x, since $x^{\frac{n}{2}} \cdot \dfrac{1}{x^{\frac{n}{2}}} = x^0 = 1$.

This term is
$$\dfrac{n(n-1)(n-2)\ldots(\tfrac{1}{2}n+1)}{1.2.3\ldots\tfrac{1}{2}n}.$$

When n is *odd* there will be two middle terms of the expanded binomial, viz., the $\frac{1}{2}(n+1)^{\text{th}}$ and the $\frac{1}{2}(n+3)^{\text{th}}$, whose coefficients, however, will be equal, involving x and $\dfrac{1}{x}$ respectively; their sum is

$$\frac{n(n-1)(n-2)\ldots\frac{1}{2}(n+3)}{1.2.3\ldots\frac{1}{2}(n-1)}\left(x+\frac{1}{x}\right).$$

Therefore we have generally

$$2^{n-1}\cos^n\theta = \cos n\theta + n\cos(n-2)\theta + \frac{n(n-1)}{1.2}\cos(n-4)\theta + \ldots,$$

the last term being

$$\left.\begin{array}{l} +\frac{1}{2}n\cdot\dfrac{(n-1)(n-2)\ldots(\frac{1}{2}n+1)}{1.2.3\ldots\frac{1}{2}n}; \quad n \text{ even.} \\[1em] \text{or} \quad +\dfrac{n(n-1)(n-2)\ldots\frac{1}{2}(n+3)}{1.2.3\ldots\frac{1}{2}(n-1)}\cos\theta. \quad n \text{ odd.} \end{array}\right\} \quad (282)$$

206. To express the positive integral powers of the sine of an Angle in terms of the sines or cosines of its multiples.

Assume
$$2\sqrt{-1}\sin\theta = x - \frac{1}{x},$$

then
$$2\sqrt{-1}\sin n\theta = x^n - \frac{1}{x^n},$$

and $2^n(-1)^{\frac{n}{2}}\sin^n\theta = \left(x-\dfrac{1}{x}\right)^n$

$$= x^n - nx^{n-2} + \frac{n(n-1)}{1.2}x^{n-4} - \ldots$$

$$\pm \frac{n(n-1)}{1.2}\frac{1}{x^{n-4}} \mp n\frac{1}{x^{n-2}} \pm \frac{1}{x^n},$$

the upper or lower sign being taken as n is even or odd.

Arranging the terms as in the last Article we have

$$2^n(-1)^{\frac{n}{2}}\sin^n\theta = \left(x^n \pm \frac{1}{x^n}\right) - n\left(x^{n-2} \pm \frac{1}{x^{n-2}}\right) + \frac{n(n-1)}{1.2}\left(x^{n-4} \pm \frac{1}{x^{n-4}}\right) - \ldots$$

1. If n be even, it is of the form $4m$ or $4m+2$, since every *even*

number is either exactly divisible by 4, or divisible by 4 with a remainder 2.

If $n=4m$, $(\sqrt{-1})^{4m} = 1 = (-1)^{2m} = (-1)^{\frac{n}{2}}$.

If $n=4m+2$, $(\sqrt{-1})^{4m+2} = -1 = (-1)^{2m+1} = (-1)^{\frac{n}{2}}$.

The last term of the last series, which is the $\left(\dfrac{n}{2}+1\right)^{\text{th}}$ of the preceding one, does not involve x, and is

$$\pm \frac{n(n-1)(n-2)\ldots(\tfrac{1}{2}n+1)}{1.2.3\ldots\tfrac{1}{2}n},$$

according as $\dfrac{n}{2}$ is even or odd; it has therefore the same sign as $(-1)^{\frac{n}{2}}$. Hence we have, n being even,

$$2^{n-1}(-1)^{\frac{n}{2}}\sin^n\theta = \cos n\theta - n\cos(n-2)\theta + \frac{n(n-1)}{1.2}\cos(n-4)\theta - \ldots$$
$$+ (-1)^{\frac{n}{2}}\frac{\tfrac{1}{2}n(n-1)(n-2)\ldots(\tfrac{1}{2}n+1)}{1.2.3\ldots\tfrac{1}{2}n}. \quad (283)$$

(2) If n be odd, it is of the form $4m+1$ or $4m+3$, since every odd number is divisible by 4 with a remainder which is either 1 or 3.

If $n=4m+1$, $(\sqrt{-1})^{4m+1} = (-1)^{\frac{n-1}{2}}\sqrt{-1}$.

If $n=4m+3$, $(\sqrt{-1})^{4m+3} = (-1)^{\frac{n-1}{2}}\sqrt{-1}$.

The sum of the two middle terms of the first of the above series is

$$\pm \frac{n(n-1)(n-2)\ldots\tfrac{1}{2}(n+3)}{1.2.3\ldots\tfrac{1}{2}(n-1)}\left(x-\frac{1}{x}\right),$$

according as $\tfrac{1}{2}(n+1)$ is odd or even; it has therefore the same sign as $(-1)^{\frac{n-1}{2}}$. Hence we have, n being odd,

$$(-1)^{\frac{n-1}{2}}2^n\sqrt{-1}\sin^n\theta = 2\sqrt{-1}\sin n\theta - n2\sqrt{-1}\sin(n-2)\theta$$
$$+ \frac{n(n-1)}{1.2}2\sqrt{-1}\sin(n-4)\theta\ldots$$
$$+ (-1)^{\frac{n-1}{2}}\frac{n(n-1)\ldots\tfrac{1}{2}(n+3)}{1.2\ldots\tfrac{1}{2}(n-1)}2\sqrt{-1}\sin\theta;$$

or dividing by $2\sqrt{-1}$,

$$(-1)^{\frac{n-1}{2}} 2^{n-1} \sin^n \theta = \sin n\theta - n \sin (n-2)\theta + \frac{n(n-1)}{1.2} \sin (n-4)\theta + \ldots$$

$$+ (-1)^{\frac{n-1}{2}} \frac{n(n-1)(n-2)\ldots\frac{1}{2}(n+3)}{1.2.3\ldots\frac{1}{2}(n-1)} \sin \theta. \quad (284)$$

207. To express $\cos n\theta$ in a series of descending powers of $\cos \theta$, n being a positive integer.

Assume $2 \cos \theta = a + \dfrac{1}{a}$, then $2 \cos n\theta = a^n + \dfrac{1}{a^n}$.

Now, $(1 - ax)\left(1 - \dfrac{x}{a}\right) = 1 - \left(a + \dfrac{1}{a}\right)x + x^2$

$$= 1 - x(b - x),$$

if $b = a + \dfrac{1}{a} = 2 \cos \theta$.

Then $\log(1 - ax) + \log\left(1 - \dfrac{x}{a}\right) = \log\{1 - x(b-x)\}.$

Expanding both members of this equation by (260) we have

$$\left.\begin{array}{l} ax + \dfrac{a^2 x^2}{2} + \ldots \dfrac{a^n x^n}{n} + \ldots \\[6pt] \dfrac{x}{a} + \dfrac{x^2}{2a^2} + \ldots \dfrac{x^n}{na^n} + \ldots \end{array}\right\} = x(b-x) + \dfrac{x^2}{2}(b-x)^2 + \ldots$$

$$+ \dfrac{x^{n-2}}{n-2}(b-x)^{n-2} + \dfrac{x^{n-1}}{n-1}(b-x)^{n-1} + \dfrac{x^n}{n}(b-x)^n + \ldots$$

Equating the coefficients of x^n in this identity, we find on the left-hand side the coefficient of x^n to be $\dfrac{1}{n}\left(a^n + \dfrac{1}{a^n}\right)$ or $\dfrac{2}{n}\cos n\theta$, while the coefficient of x^n on the right-hand side is found by taking out the coefficient of x^n from the expansion of $\dfrac{x^n}{n}(b-x)^n$ and of all the terms that precede it. Thus, the coefficient of x^n

TRIGONOMETRICAL EXPANSIONS. 273

in $\dfrac{x^n}{n}(b-x)^n$ is $+\dfrac{b^n}{n}$,

in $\dfrac{x^{n-1}}{n-1}(b-x)^{n-1}$ is $-\dfrac{1}{n-1}\cdot(n-1)b^{n-2}$,

in $\dfrac{x^{n-2}}{n-2}(b-x)^{n-2}$ is $+\dfrac{1}{n-2}\cdot\dfrac{(n-2)(n-3)}{1.2}b^{n-4}$,

....

in $\dfrac{x^{n-r}}{n-r}(b-x)^{n-r}$ is $\dfrac{(-1)^r(n-r-1)\ldots(n-2r+1)}{1.2.3\ldots r}b^{n-2r}$.

Therefore we have

$$\dfrac{2}{n}\cos n\theta = \dfrac{b^n}{n} - b^{n-2} + \dfrac{n-3}{1.2}b^{n-4} - \ldots$$
$$+ \dfrac{(-1)^r(n-r-1)\ldots(n-2r+1)}{1.2.3\ldots r}b^{n-2r};$$

or writing $2\cos\theta$ for b and multiplying by n,

$$2\cos n\theta = (2\cos\theta)^n - n(2\cos\theta)^{n-2} + \dfrac{n(n-3)}{1.2}(2\cos\theta)^{n-4} - \ldots$$
$$+ \dfrac{(-1)^r n(n-r-1)\ldots(n-2r+1)}{1.2.3\ldots r}(2\cos\theta)^{n-2r} \quad (285)$$

Here we must observe that since none but *positive integral* powers of $(b-x)$ appear, the index of b in the general term must also be a positive integer, that is, r must not be greater than $\dfrac{n}{2}$.

208. In (285) write $\dfrac{\pi}{2}-\phi$ for θ, then we have when n is *even*,

$$(-1)^{\frac{n}{2}}2\cos n\phi = (2\sin\phi)^n - n(2\sin\phi)^{n-2} + \dfrac{n(n-3)}{1.2}(2\sin\phi)^{n-4} - \ldots$$
$$+(-1)^{n-r}\cdot\dfrac{n(n-r-1)\ldots(n-2r+1)}{1.2.3\ldots r}(2\sin\phi)^{n-2r}; \quad (286)$$

and when n is odd,

$$(-1)^{\frac{n-1}{2}}2\sin n\phi = (2\sin\phi)^n - n(2\sin\phi)^{n-2} + \dfrac{n(n-3)}{1.2}(2\sin\phi)^{n-4} - \ldots$$
$$+(-1)^{n-r-1}\cdot\dfrac{n(n-r-1)\ldots(n-2r+1)}{1.2.3\ldots r}(2\sin\phi)^{n-2r}. \quad (287)$$

19

209. To express $\cos n\theta$ in a series of ascending powers of $\cos \theta$, n being a positive integer.

(1) Let n be *even*.

In (285) r is limited to values not greater than $\dfrac{n}{2}$, therefore writing for r in the general term the values $\dfrac{n}{2}$, $\dfrac{n}{2}-1$, $\dfrac{n}{2}-2$, ...

....3, 2, 1, 0, in succession, the number of terms will be $\dfrac{n}{2}+1$, and as r is diminished successively by 1, the terms are alternately positive and negative, the first term having the same sign as $(-1)^{\frac{n}{2}}$.

Therefore we have

$$2\cos n\theta = (-1)^{\frac{n}{2}} \left\{ \frac{n(\tfrac{1}{2}n-1)(\tfrac{1}{2}n-2)\ldots 3.2.1}{1.2.3\ldots \tfrac{1}{2}n}(2\cos\theta)^0 \right.$$

$$-\frac{n.\tfrac{1}{2}n(\tfrac{1}{2}n-1)\ldots(n-2(\tfrac{1}{2}n-1)+1)}{1.2.3\ldots(\tfrac{1}{2}n-1)}(2\cos\theta)^2$$

$$+\frac{n(\tfrac{1}{2}n+1)\tfrac{1}{2}n(\tfrac{1}{2}n-1)\ldots(n-2(\tfrac{1}{2}n-2)+1)}{1.2.3\ldots(\tfrac{1}{2}n-2)}(2\cos\theta)^4$$

$$\left. -\frac{n(\tfrac{1}{2}n+2)(\tfrac{1}{2}n+1)\ldots(n-2(\tfrac{1}{2}n-3)+1)}{1.2.3\ldots(\tfrac{1}{2}n-3)}(2\cos\theta)^6 + \&c. \right\}$$

$$=(-1)^{\frac{n}{2}}\left\{ \frac{n(\tfrac{1}{2}n-1)(\tfrac{1}{2}n-2)\ldots 3.2.1}{1.2.3\ldots \tfrac{1}{2}n} \right.$$

$$-\frac{n.\tfrac{1}{2}n(\tfrac{1}{2}n-1)\ldots 4.3}{1.2.3\ldots(\tfrac{1}{2}n-1)}(2\cos\theta)^2$$

$$+\frac{n(\tfrac{1}{2}n+1)\tfrac{1}{2}n(\tfrac{1}{2}n-1)\ldots 6.5}{1.2.3\ldots(\tfrac{1}{2}n-2)}(2\cos\theta)^4$$

$$\left. -\frac{n(\tfrac{1}{2}n+2)(\tfrac{1}{2}n+1)\ldots 8.7}{1.2.3\ldots(\tfrac{1}{2}n-3)}(2\cos\theta)^6 + \&c. \right\}$$

$$=(-1)^{\frac{n}{2}}\left\{ 2 - \frac{n.\tfrac{1}{2}n}{1.2}.(2\cos\theta)^2 + \frac{n(\tfrac{1}{2}n+1)\tfrac{1}{2}n(\tfrac{1}{2}n-1)}{1.2.3.4}(2\cos\theta)^4 \right.$$

$$\left. -\frac{n(\tfrac{1}{2}n+2)(\tfrac{1}{2}n+1)\tfrac{1}{2}n(\tfrac{1}{2}n-1)(\tfrac{1}{2}n-2)}{1.2.3.4.5.6}(2\cos\theta)^6 + \&c. \right\};$$

dividing by 2, and reducing coefficients we have

$$\cos n\theta = (-1)^{\frac{n}{2}} \left\{ 1 - \frac{n^2}{1.2} \cos^2\theta + \frac{n^2(n^2-2^2)}{1.2.3.4} \cos^4\theta \right.$$
$$\left. - \frac{n^2(n^2-2^2)(n^2-4^2)}{1.2.3.4.5.6} \cos^6\theta + \&c. \right\}. \quad (288)$$

(2) Let n be *odd*.

Since r cannot be greater than $\frac{n}{2}$, it may be $\frac{1}{2}(n-1)$, the integer next less than $\frac{n}{2}$; the terms are alternately positive and negative, the first term having the same sign as $(-1)^{\frac{1}{2}(n-1)}$, and the number of terms will be $\frac{1}{2}(n-1)+1$ or $\frac{1}{2}(n+1)$. Writing for r, $\frac{1}{2}(n-1)$, $\frac{1}{2}(n-3)$, $\frac{1}{2}(n-5)\ldots\ldots 3, 2, 1, 0$ successively in the general term of (285), we have as before, by reducing the coefficients and dividing by 2,

$$\cos n\theta = (-1)^{\frac{n-1}{2}} \left\{ n\cos\theta - \frac{n(n^2-1^2)}{1.2.3}\cos^3\theta + \frac{n(n^2-1^2)(n^2-3^2)}{1.2.3.4.5}\cos^5\theta \right.$$
$$\left. - \frac{n(n^2-1)(n^2-3^2)(n^2-5^2)}{1.2.3.4.5.6.7}\cos^7\theta + \&c. \right\}. \quad (289)$$

210. To expand sin x and cos x in series containing the ascending powers of x, independently of De Moivre's Theorem.

The series for $\sin x$ must vanish when $x=0$, therefore it can contain no term independent of x, nor can the even powers of x enter into the series; for suppose

$$\sin x = Ax + Bx^2 + Cx^3 + Dx^4 + \&c.;$$

substitute $-x$ for x and this becomes

$$\sin(-x) = -Ax + Bx^2 - Cx^3 + Dx^4 - \&c.;$$

but $\sin(-x) = -\sin x,$ (Art. 41)
$$= -Ax - Bx^2 - Cx^3 - Dx^4 - \&c.,$$

therefore $B = -B, \; D = -D, \; \&c.,$ which is absurd
unless $B = 0, \; D = 0, \; \&c.,$

therefore $\quad \sin x = Ax + Cx^3 + Ex^5 + \&c.,$

and $\quad \dfrac{\sin x}{x} = A + Cx^2 + Ex^4 + \&c.;$

but if $\quad x = 0, \dfrac{\sin x}{x} = 1,$ (Art. 74)

therefore $\quad A = 1,$ and we have

$$\sin x = x + Cx^3 + Ex^5 + \&c. \qquad (a)$$

Again, the series for $\cos x$ must $= 1$ when $x = 0$, therefore its first term is 1, and it can contain no odd powers of x; for suppose

$$\cos x = 1 + Ax + Bx^2 + Cx^3 + Dx^4 + \&c.,$$

then $\quad \cos(-x) = 1 - Ax + Bx^2 - Cx^3 + Dx^4 - \&c.,$

but $\quad \cos(-x) = \cos x \qquad$ (Art. 41)

$\quad = 1 + Ax + Bx^2 + Cx^3 + Dx^4 + \&c.,$

therefore $\quad A = -A, \ C = -C, \ \&c., \qquad$ which is absurd

unless $\quad A = 0, \ C = 0, \ \&c.,$

therefore $\quad \cos x = 1 + Bx^2 + Dx^4 + \&c. \qquad (b)$

Adding and subtracting (a) and (b) we get

$$\cos x + \sin x = 1 + x + Bx^2 + Cx^3 + Dx^4 + Ex^5 + \&c. \qquad (c)$$

$$\cos x - \sin x = 1 - x + Bx^2 - Cx^3 + Dx^4 - Ex^5 + \&c. \qquad (d)$$

In (c) write $x + h$ for x, then it becomes

$$\cos(x+h) + \sin(x+h) = 1 + (x+h) + B(x+h)^2 + C(x+h)^3 + \ldots$$

but

$\cos(x+h) + \sin(x+h) = \cos x \cos h - \sin x \sin h + \sin x \cos h + \cos x \sin h$

$\quad = \cos h (\cos x + \sin x) + \sin h (\cos x - \sin x)$

$\quad = (1 + Bh^2 + Dh^4 + \ldots)(1 + x + Bx^2 + Cx^3 + \ldots)$

$\quad + (h + Ch^3 + Eh^5 + \ldots)(1 - x + Bx^2 - Cx^3 + \ldots).$

Equating the second members of the last two equations and expanding we have

$$\left.\begin{array}{l} 1 + x + Bx^2 + Cx^3 + Dx^4 + \\ h + 2Bhx + 3Chx^2 + 4Dhx^3 + \\ Bh^2 + 3Ch^2x + 6Dh^2x^2 + \\ \qquad Ch^3 + 4Dh^3x + \\ \qquad\qquad Dh^4 + \end{array}\right\} = \left\{\begin{array}{l} 1 + x + Bx^2 + Cx^3 \quad + Dx^4 + \\ h - hx + Bhx^2 - Chx^3 + \\ Bh^2 + Bh^2x + B^2h^2x^2 + \\ \qquad Ch^3 \quad - Ch^3x + \\ \qquad\qquad Dh^4 + \end{array}\right.$$

Cancelling the terms common to both members of this equation and equating coefficients of hx, hx^2, &c., we have

$$2B = -1 \qquad B = -\frac{1}{1.2}$$

$$3C = B \qquad C = -\frac{1}{1.2.3}$$

$$4D = -C \qquad D = \frac{1}{1.2.3.4}$$

$$5E = D \qquad E = \frac{1}{1.2.3.4.5},$$

&c., &c.

Hence we have by substituting in (a) and (b)

$$\sin x = x - \frac{x^3}{1.2.3} + \frac{x^5}{1.2.3.4.5} - \ldots$$

$$\cos x = 1 - \frac{x^2}{1.2} + \frac{x^4}{1.2.3.4} - \ldots$$

211. Sines and tangents of small Angles.

The last two formulæ furnish us with the means of finding the sine and tangent of a small angle, and conversely, of finding a small angle from its sine or tangent.

Thus, when x is small

$$\sin x = x - \frac{x^3}{1.2.3}, \text{ very nearly,}$$

$$= x \left(1 - \frac{x^2}{1.2} + \frac{x^4}{1.2.3.4}\right)^{\frac{1}{3}} = x \cos^{\frac{1}{3}} x.$$

Let x be an angle containing n'',

then $\qquad n = \dfrac{x}{\sin 1''} \quad$ or $\quad x = n \sin 1''$,

since $\sin 1'' =$ circular measure of $1''$ very nearly;

therefore $\quad \sin x = n \sin 1'' \cos^{\frac{1}{3}} x$

or $\quad$ Log $\sin n'' = \log n + $ Log $\sin 1'' + \frac{1}{3}($Log $\cos x - 10)$

$\qquad = \log n + $ Log $\sin 1'' - \frac{1}{3}(10 - $ Log $\cos x)$

$\qquad = \log n + 4.6855749 - \frac{1}{3}($Log $\sec x - 10).$ $\qquad$ (290)

and also $\quad \log n = \text{Log} \sin n'' - \text{Log} \sin 1'' + \frac{1}{3}(\text{Log} \sec x - 10)$
$\qquad\qquad = \text{Log} \sin n'' + \text{Log} \csc 1'' + \frac{1}{3}(\text{Log} \sec x - 10)$
$\qquad\qquad = \text{Log} \sin n'' + \overline{5}.3144251 + \frac{1}{3}(\text{Log} \sec x - 10).\qquad (291)$

That is, to find the sine of a small angle we have the following rule:

"To the logarithm of the angle reduced to seconds add 4.6855749, and from the sum subtract $\frac{1}{3}$ of its logarithmic secant, the characteristic of the latter logarithm being previously diminished by 10; the remainder is the logarithmic sine."—(Chambers's Logarithmic Tables, Art. 27.)

To find a small angle from Log sin, we have this rule:

"To the given Log sin add $\overline{5}.3144251$ and $\frac{1}{3}$ of the corresponding Log sec, the characteristic of the latter logarithm being previously diminished by 10, and the sum will be the logarithm of the number of seconds in the angle."

In like manner a formula may be established for finding the tangent of a small angle, and conversely.

Thus $\qquad\qquad \tan x = \dfrac{\sin x}{\cos x} = \dfrac{x \cos^{\frac{2}{3}} x}{\cos x} = \dfrac{x}{\cos^{\frac{1}{3}} x},$

and if the angle x contain n'', we have as before $x = n \sin 1''$, and

$\quad\text{Log} \tan n'' = \log n + \text{Log} \sin 1'' - \frac{2}{3}(\text{Log} \cos x - 10),$
$\qquad\qquad = \log n + \text{Log} \sin 1'' + \frac{2}{3}(\text{Log} \sec x - 10).\qquad (292)$

and $\quad \log n = \text{Log} \tan n'' - \text{Log} \sin 1'' - \frac{2}{3}(\text{Log} \sec x - 10),$
$\qquad\qquad = \text{Log} \tan n'' + \text{Log} \csc 1'' - \frac{2}{3}(\text{Log} \sec x - 10).\qquad (293)$

This method of finding the sine and tangent of a small angle is generally known as Maskelyne's method, and was first given by him in his Introduction to Taylor's Logarithms. It is not as convenient as that of Art. 112, which is known as Delambre's method.

Examples.

1. Prove that

$\qquad \sin 5x = 5 \cos^4 x \sin x - 10 \cos^2 x \sin^3 x + \sin^5 x.$
$\qquad \cos 5x = \cos^5 x - 10 \cos^3 x \sin^2 x + 5 \cos x \sin^4 x.$

EXAMPLES.

2. Prove that
$$2^4 \sin^5 x = 10 \sin x - 5 \sin 3x + \sin 5x.$$
$$2^4 \cos^5 x = \cos 5x + 5 \cos 3x + 10 \cos x.$$

3. If $\tan \theta = \dfrac{b}{a}$, shew that
$$(a+b\sqrt{-1})^{\frac{m}{n}} + (a-b\sqrt{-1})^{\frac{m}{n}} = 2(a^2+b^2)^{\frac{m}{2n}} \cos \frac{m}{n} \theta.$$

4. Prove that
$$\cos 4x = 1 - 8 \cos^2 x + 8 \cos^4 x,$$
$$\sin 5x = 16 \sin^5 x - 20 \sin^3 x + 5 \sin x.$$

5. By means of (280) prove the following rule for finding the length of a small arc: "From eight times the chord of half the arc subtract the chord of the whole arc: one-third of the remainder is equal to the arc very nearly."

6. Prove that when θ is small
$$2 \sin \theta + \tan \theta = 3\theta, \text{ nearly.}$$

7. Shew that
$$\sin^3 x = \frac{3}{4}\left(\frac{3^2-1}{1.2.3} x^3 - \frac{3^4-1}{1.2.3.4.5} x^5 + \ldots \pm \frac{3^{2n}-1}{1.2\ldots(2n+1)} x^{2n+1} \mp \ldots\right).$$

8. Find the three values of $(-1)^{\frac{1}{3}}$ by De Moivre's Theorem.

CHAPTER XVI.

EXPONENTIAL FORMULÆ—COMPUTATION OF THE NUMERICAL VALUE OF π—TRIGONOMETRICAL SERIES.

Exponential values of the sine, cosine and tangent.

212. From (258) we have

$$e^x = 1 + x + \frac{x^2}{1.2} + \frac{x^3}{1.2.3} + \frac{x^4}{1.2.3.4} + \&c.,$$

in which let $\theta\sqrt{-1}$ and $-\theta\sqrt{-1}$ be successively substituted for x, then we find

$$e^{\theta\sqrt{-1}} = 1 + \theta\sqrt{-1} - \frac{\theta^2}{1.2} - \frac{\theta^3}{1.2.3}\sqrt{-1} + \frac{\theta^4}{1.2.3.4} + \frac{\theta^5}{1.2.3.4.5}\sqrt{-1} - \&c.,$$

$$e^{-\theta\sqrt{-1}} = 1 - \theta\sqrt{-1} - \frac{\theta^2}{1.2} + \frac{\theta^3}{1.2.3}\sqrt{-1} + \frac{\theta^4}{1.2.3.4} - \frac{\theta^5}{1.2.3.4.5}\sqrt{-1} - \&c.,$$

the sum and difference of which are

$$e^{\theta\sqrt{-1}} + e^{-\theta\sqrt{-1}} = 2(1 - \frac{\theta^2}{1.2} + \frac{\theta^4}{1.2.3.4} - \&c.)$$
$$= 2\cos\theta, \quad \text{by (281)}. \tag{294}$$

$$e^{\theta\sqrt{-1}} - e^{-\theta\sqrt{-1}} = 2\sqrt{-1}\,(\theta - \frac{\theta^3}{1.2.3} + \frac{\theta^5}{1.2.3.4.5} - \&c.)$$
$$= 2\sqrt{-1}\sin\theta, \quad \text{by (280)}. \tag{295}$$

The quotient of (295) by (294) is

$$\sqrt{-1}\tan\theta = \frac{e^{\theta\sqrt{-1}} - e^{-\theta\sqrt{-1}}}{e^{\theta\sqrt{-1}} + e^{-\theta\sqrt{-1}}} = \frac{e^{2\theta\sqrt{-1}} - 1}{e^{2\theta\sqrt{-1}} + 1}, \tag{296}$$

by multiplying numerator and denominator by $e^{\theta\sqrt{-1}}$.

EXPONENTIAL FORMULÆ.

These formulæ, which are due to Euler, are reckoned amongst the most useful in Modern Analysis.

By the addition and subtraction of (294) and (295) we get

$$e^{\theta\sqrt{-1}} = \cos\theta + \sqrt{-1}\sin\theta \qquad (297)$$

and
$$e^{-\theta\sqrt{-1}} = \cos\theta - \sqrt{-1}\sin\theta; \qquad (298)$$

or introducing the notation of Art. 201, we have

$$x = e^{\theta\sqrt{-1}} = \cos\theta + \sqrt{-1}\sin\theta \qquad (299)$$

and
$$\frac{1}{x} = e^{-\theta\sqrt{-1}} = \cos\theta - \sqrt{-1}\sin\theta. \qquad (300)$$

Hence (294) and (295) may be written

$$\left.\begin{array}{r}2\cos\theta = x + \dfrac{1}{x}\\[4pt] 2\sqrt{-1}\sin\theta = x - \dfrac{1}{x}\end{array}\right\}, \qquad (301)$$

and if we substitute $m\theta$ for θ in (299) and (300) we find by addition and subtraction

$$\left.\begin{array}{r}2\cos m\theta = x^m + \dfrac{1}{x^m}\\[4pt] 2\sqrt{-1}\sin m\theta = x^m - \dfrac{1}{x^m}\end{array}\right\}. \qquad (302)$$

213. To express the Circular Measure of an Angle in terms of its tangent.

From (297) we have

$$e^{\theta\sqrt{-1}} = \cos\theta + \sqrt{-1}\sin\theta$$
$$= \cos\theta\,(1 + \sqrt{-1}\tan\theta).$$

Express in Napierian logarithms, thus

$$\theta\sqrt{-1} = \log_e\cos\theta + \log_e(1 + \sqrt{-1}\tan\theta)$$
$$= \log_e\cos\theta + \sqrt{-1}\tan\theta + \frac{\tan^2\theta}{2} - \sqrt{-1}\,\frac{\tan^3\theta}{3} - \&c., \text{ by (259)}.$$

Equating the real and imaginary parts of this equation we have

$$0 = \log_e \cos\theta + \frac{\tan^2\theta}{2} - \frac{\tan^4\theta}{4} + \frac{\tan^6\theta}{6} - \&c., \qquad (303)$$

$$\theta = \tan\theta - \frac{\tan^3\theta}{3} + \frac{\tan^5\theta}{5} - \frac{\tan^7\theta}{7} + \&c., \qquad (304)$$

the last of which is known as Gregory's series, and is convergent for all angles whose tangent is not greater than 1.

If $\tan\theta = x$, so that $\theta = \tan^{-1} x$, the series may be written

$$\tan^{-1} x = x - \frac{x^3}{3} + \frac{x^5}{5} - \frac{x^7}{7} + \frac{x^9}{9} - \&c. \qquad (305)$$

214. To find the numerical value of π.

In Gregory's series let $\theta = \dfrac{\pi}{4}$, then since $\tan\dfrac{\pi}{4} = 1$ we have

$$\frac{\pi}{4} = 1 - \frac{1}{3} + \frac{1}{5} - \frac{1}{7} + \frac{1}{9} - \&c.,$$

$$= 2\left(\frac{1}{1\cdot 3} + \frac{1}{5\cdot 7} + \frac{1}{9\cdot 11} + \&c.\right). \qquad (306)$$

This series converges too slowly to be of much use. To obtain a rapidly converging series, let

$$\frac{\pi}{4} = \tan^{-1} m + \tan^{-1} n$$

or
$$\tan^{-1} 1 = \tan^{-1}\frac{m+n}{1-mn}, \quad \text{by (253)}$$

therefore
$$1 = \frac{m+n}{1-mn} \text{ and } m = \frac{1-n}{1+n}.$$

If we make $n = \dfrac{1}{3}$, we find $m = \dfrac{1}{2}$.

therefore $\dfrac{\pi}{4} = \tan^{-1}\dfrac{1}{2} + \tan^{-1}\dfrac{1}{3}$

$$= \left\{\begin{array}{l} \dfrac{1}{2} - \dfrac{1}{3}\left(\dfrac{1}{2}\right)^3 + \dfrac{1}{5}\left(\dfrac{1}{2}\right)^5 - \dfrac{1}{7}\left(\dfrac{1}{2}\right)^7 + \&c. \\ \dfrac{1}{3} - \dfrac{1}{3}\left(\dfrac{1}{3}\right)^3 + \dfrac{1}{5}\left(\dfrac{1}{3}\right)^5 - \dfrac{1}{7}\left(\dfrac{1}{3}\right)^7 + \&c. \end{array}\right\}, \qquad (307)$$

which are Euler's series for the computation of π. They converge much more rapidly than (306).

215. Machin's Series for computing π.

$$\tan^{-1} 1 - \tan^{-1} \frac{1}{5} = \tan^{-1} \frac{1 - \frac{1}{5}}{1 + \frac{1}{5}} = \tan^{-1} \frac{2}{3}$$

$$\tan^{-1} \frac{2}{3} - \tan^{-1} \frac{1}{5} = \tan^{-1} \frac{\frac{2}{3} - \frac{1}{5}}{1 + \frac{2}{3.5}} = \tan^{-1} \frac{7}{17}$$

$$\tan^{-1} \frac{7}{17} - \tan^{-1} \frac{1}{5} = \tan^{-1} \frac{\frac{7}{17} - \frac{1}{5}}{1 + \frac{7}{17.5}} = \tan^{-1} \frac{9}{46}$$

$$\tan^{-1} \frac{9}{46} - \tan^{-1} \frac{1}{5} = \tan^{-1} \frac{\frac{9}{46} - \frac{1}{5}}{1 + \frac{9}{46.5}} = \tan^{-1} \left(\frac{-1}{239}\right)$$

$$= -\tan^{-1} \frac{1}{239}.$$

Adding these equations and cancelling the terms common to both sides, we have

$$\tan^{-1} 1 - 4 \tan^{-1} \frac{1}{5} = -\tan^{-1} \frac{1}{239}$$

$$\tan^{-1} 1 = 4 \tan^{-1} \frac{1}{5} - \tan^{-1} \frac{1}{239},$$

therefore

$$\frac{\pi}{4} = \left\{ \begin{array}{l} 4\left(\frac{1}{5} - \frac{1}{3}\left(\frac{1}{5}\right)^3 + \frac{1}{5}\left(\frac{1}{5}\right)^5 - \&c. \right) \\ -\frac{1}{239} + \frac{1}{3}\left(\frac{1}{239}\right)^3 - \frac{1}{5}\left(\frac{1}{239}\right)^5 + \&c. \end{array} \right\}. \quad (308)$$

These series converge quite rapidly, especially the latter. If we take eight terms of the first and three of the second we find

$$\pi = 3.141592653589793\ldots.$$

For other series for computing π, see the examples at the end of this chapter.

Expansion of certain Trigonometrical Equations into Series.

216. *Given $\sin p = \sin P \sin (z+p)$, it is required to express p in a series of multiples of z.* [See (a), Art. 151.]

Multiplying both members of the given equation by $2\sqrt{-1}$, we have

$$2\sqrt{-1} \sin p = \sin P \cdot 2\sqrt{-1} \sin (z+p)$$

or $\quad e^{p\sqrt{-1}} - e^{-p\sqrt{-1}} = \sin P(e^{(z+p)\sqrt{-1}} - e^{-(z+p)\sqrt{-1}})$, by (295)

whence $\quad e^{2p\sqrt{-1}} = \dfrac{1 - \sin P e^{-z\sqrt{-1}}}{1 - \sin P e^{z\sqrt{-1}}},$

and taking the Napierian logarithms of both members we have

$$2p\sqrt{-1} = \log(1 - \sin P e^{-z\sqrt{-1}}) - \log(1 - \sin P e^{z\sqrt{-1}})$$

$$= \begin{cases} -\sin P e^{-z\sqrt{-1}} - \tfrac{1}{2}\sin^2 P e^{-2z\sqrt{-1}} - \tfrac{1}{3}\sin^3 P e^{-3z\sqrt{-1}} - \ldots \\ +\sin P e^{z\sqrt{-1}} + \tfrac{1}{2}\sin^2 P e^{2z\sqrt{-1}} + \tfrac{1}{3}\sin^3 P e^{3z\sqrt{-1}} + \ldots \end{cases}$$

by (260)

$$= \sin P(e^{z\sqrt{-1}} - e^{-z\sqrt{-1}}) + \tfrac{1}{2}\sin^2 P(e^{2z\sqrt{-1}} - e^{-2z\sqrt{-1}})$$

$$+ \tfrac{1}{3}\sin^3 P(e^{3z\sqrt{-1}} - e^{-3z\sqrt{-1}}) + \ldots$$

$$= \sin P \cdot 2\sqrt{-1} \sin z + \tfrac{1}{2}\sin^2 P \cdot 2\sqrt{-1} \sin 2z$$

$$+ \tfrac{1}{3}\sin^3 P \cdot 2\sqrt{-1} \sin 3z + \ldots$$

therefore

$$p = \sin P \sin z + \tfrac{1}{2}\sin^2 P \sin 2z + \tfrac{1}{3}\sin^3 P \sin 3z + \ldots \qquad (309)$$

Here, p is expressed in circular measure; to find p in *seconds* we must divide both members by $\sin 1''$, according to (128), and since

2 sin 1″=sin 2″, 3 sin 1″=sin 3″, &c., approximately, the last equation may be written thus,

$$p'' = \frac{\sin P \sin z}{\sin 1''} + \frac{\sin^2 P \sin 2z}{\sin 2''} + \frac{\sin^3 P \sin 3z}{\sin 3''} + \ldots \quad (310)$$

Ex.—Given $P=58' 10''$ and $z=40°$, to find p. (Same as Ex. 64, Chapter X.)

sin P=8.228380	sin² P=6.4567	sin³ P=4.685
sin z=9.808067	sin $2z$=9.9933	sin $3z$=9.937
cosec 1″=5.314425	cosec 2″=5.0134	cosec 3″=4.837
2243″.23=3.350872	29″.07=1.4634	0″.29=$\overline{1}$.460

$$p = 2243''.23 + 29''.07 + 0''.29 = 37' 52.6.$$

Here we omit the symbol Log for the sake of brevity.

217. *Given* $\tan x = n \tan y$, *to express x in a series of multiples of y.*

Multiplying both members of the given equation by $\sqrt{-1}$, we have by (296)

$$\frac{e^{2x\sqrt{-1}}-1}{e^{2x\sqrt{-1}}+1} = n \frac{e^{2y\sqrt{-1}}-1}{e^{2y\sqrt{-1}}+1},$$

whence
$$e^{2x\sqrt{-1}} = \frac{(1+n)e^{2y\sqrt{-1}}+(1-n)}{(1-n)e^{2y\sqrt{-1}}+(1+n)}$$

$$= \frac{e^{2y\sqrt{-1}} + \frac{1-n}{1+n}}{\frac{1-n}{1+n}e^{2y\sqrt{-1}}+1}$$

$$= e^{2y\sqrt{-1}} \left(\frac{1 + \frac{1-n}{1+n} e^{-2y\sqrt{-1}}}{1 + \frac{1-n}{1+n} e^{2y\sqrt{-1}}} \right)$$

$$= e^{2y\sqrt{-1}} \left(\frac{1 + me^{-2y\sqrt{-1}}}{1 + me^{2y\sqrt{-1}}} \right),$$

by putting $m = \dfrac{1-n}{1+n}$

or $e^{(x-y)2\sqrt{-1}} = \dfrac{1 + me^{-2y\sqrt{-1}}}{1 + me^{2y\sqrt{-1}}}$, and in logarithms

$(x-y)2\sqrt{-1} = \log(1 + me^{-2y\sqrt{-1}}) - \log(1 + me^{2y\sqrt{-1}})$

$$= \begin{cases} me^{-2y\sqrt{-1}} - \tfrac{1}{2} m^2 e^{-4y\sqrt{-1}} + \tfrac{1}{3} m^3 e^{-6y\sqrt{-1}} - \ldots \\ -me^{2y\sqrt{-1}} + \tfrac{1}{2} m^2 e^{4y\sqrt{-1}} - \tfrac{1}{3} m^3 e^{6y\sqrt{-1}} + \ldots \end{cases}$$

$= -m(e^{2y\sqrt{-1}} - e^{-2y\sqrt{-1}}) + \tfrac{1}{2} m^2 (e^{4y\sqrt{-1}} - e^{-4y\sqrt{-1}})$

$\qquad\qquad - \tfrac{1}{3} m^3 (e^{6y\sqrt{-1}} - e^{-6y\sqrt{-1}}) + \ldots$

$= -m2\sqrt{-1} \sin 2y + \tfrac{1}{2} m^2 2\sqrt{-1} \sin 4y$

$\qquad\qquad - \tfrac{1}{3} m^3 2\sqrt{-1} \sin 6y + \ldots$

therefore $\quad x = y - m \sin 2y + \tfrac{1}{2} m^2 \sin 4y - \tfrac{1}{3} m^3 \sin 6y + \ldots \quad (311)$

or $\quad x'' = y'' - \dfrac{m \sin 2y}{\sin 1''} + \dfrac{m^2 \sin 4y}{\sin 2''} - \dfrac{m^3 \sin 6y}{\sin 3''} + \ldots \quad (312)$

218. *Given* $\tan x = \dfrac{n \sin y}{1 - n \cos y}$, *to express x in a series of multiples of y.*

The given equation may be easily reduced to

$$\sin x = n \sin(x + y), \qquad (a)$$

which is of the same form as the equation of Art. 216. Hence writing x for p, n for $\sin P$, and y for z in (309), we have at once

$$x = n \sin y + \tfrac{1}{2} n^2 \sin 2y + \tfrac{1}{3} n^3 \sin 3y + \ldots \qquad (313)$$

Formulæ (309)–(313) are very useful in Spherical Astronomy. The left-hand members are limited to acute angles, but entire generality may be conferred upon them by observing that the equation $\tan x = n \tan y$ is true when we write $n'\pi + x$ for x and $m'\pi + y$ for y, and therefore for $x - y$ we may write $x - y - (m' - n')\pi$ or $x - y - p\pi$

where p is, like n' and m', any integer or zero. Therefore (311) may be written thus:

$$x = p\pi + y - m \sin 2y + \tfrac{1}{2} m^2 \sin 4y - \ldots.$$

In the same manner the other series of Articles 216-218 may be generalized.

219. *In a triangle ABC, given two sides a and b and the included angle C, to express either of the other angles by a series of multiples of C.*

Since
$$\frac{b}{a} = \frac{\sin B}{\sin A} = \frac{\sin (A+C)}{\sin A}$$

we have
$$\sin A = \frac{a}{b} \sin (A+C),$$

which, compared with (a) of the last Article, gives by (313)

$$A = \frac{a}{b} \sin C + \frac{a^2}{b^2} \cdot \frac{\sin 2C}{2} + \frac{a^3}{b^3} \cdot \frac{\sin 3C}{3} + \ldots.$$

or
$$A'' = \frac{a}{b} \frac{\sin C}{\sin 1''} + \frac{a^2}{b^2} \cdot \frac{\sin 2C}{\sin 2''} + \frac{a^3}{b^3} \cdot \frac{\sin 3C}{\sin 3''} + \ldots. \qquad (314)$$

which will be convergent when $\dfrac{a}{b}$ is a proper fraction.

220. *In a triangle, given two sides a and b and the included angle C, to express c by a series of multiples of C.*

From Art. 122 we have

$$c^2 = a^2 + b^2 - 2ab \cos C$$

$$= a^2 \left(1 - 2\frac{b}{a} \cos C + \frac{b^2}{a^2}\right)$$

$$= a^2 \left(1 - \frac{b}{a}\left(x + \frac{1}{x}\right) + \frac{b^2}{a^2}\right), \quad \text{by (301)}$$

$$= a^2 \left(1 - \frac{b}{a} x\right)\left(1 - \frac{b}{ax}\right)$$

$$2 \log_e c = 2 \log_e a + \log_e \left(1 - \frac{b}{a} x\right) + \log_e \left(1 - \frac{b}{ax}\right)$$

$$= \begin{cases} 2\log_e a - \dfrac{b}{a}x - \dfrac{b^2}{2a^2}x^2 - \dfrac{b^3}{3a^3}x^3 - \ldots \\ - \dfrac{b}{ax} - \dfrac{b^2}{2a^2x^2} - \dfrac{b^3}{2a^3x^3} - \ldots, \quad \text{by (260)} \end{cases}$$

$$= 2\log_e a - \dfrac{b}{a}\left(x+\dfrac{1}{x}\right) - \dfrac{b^2}{2a^2}\left(x^2+\dfrac{1}{x^2}\right) - \dfrac{b^3}{3a^3}\left(x^3+\dfrac{1}{x^3}\right) - \ldots$$

$$= 2\log_e a - \dfrac{b}{a}\,2\cos C - \dfrac{b^2}{2a^2}\,2\cos 2C - \dfrac{b^3}{3a^3}\,2\cos 3C - \ldots$$

$$\log_e c = \log_e a - \left(\dfrac{b}{a}\cos C + \dfrac{b^2}{2a^2}\cos 2C + \dfrac{b^3}{3a^3}\cos 3C + \ldots\right)$$

or $\log c = \log a - M\left(\dfrac{b}{a}\cos C + \dfrac{b^2}{2a^2}\cos 2C + \dfrac{b^3}{3a^3}\cos 3C + \ldots\right)$, (315)

where M is the modulus.

221. Given $r = \dfrac{a(1-e^2)}{1+e\cos\theta}$, *to express r in a series of multiples of θ; e being less than 1.* (*The Polar Equation to the Ellipse.*)

Assume $e = \dfrac{2b}{1+b^2}$, which is always possible, since $1+b^2 > 2b$.

then $\qquad 1-e^2 = \left(\dfrac{1-b^2}{1+b^2}\right)^2$ and $b = \dfrac{e}{1+\sqrt{1-e^2}}$.

Let $\quad 2\cos\theta = x + \dfrac{1}{x}$,

then $\quad 1 + e\cos\theta = 1 + \dfrac{b}{1+b^2}\left(x+\dfrac{1}{x}\right)$

$$= \dfrac{1}{1+b^2}\left(1 + b\left(x+\dfrac{1}{x}\right) + b^2\right)$$

$$= \dfrac{1}{1+b^2}(1+bx)\left(1+\dfrac{b}{x}\right),$$

therefore $\quad r = a\left(\dfrac{1-b^2}{1+b^2}\right)^2 (1+b^2) \cdot \dfrac{1}{1+bx} \cdot \dfrac{1}{1+\dfrac{b}{x}}$

$$= a\,\dfrac{(1-b^2)^2}{1+b^2}\begin{cases}(1 - bx + b^2x^2 - b^3x^3 + b^4x^4 - \ldots) \\ \times\left(1 - \dfrac{b}{x} + \dfrac{b^2}{x^2} - \dfrac{b^3}{x^3} + \dfrac{b^4}{x^4} - \ldots\right)\end{cases}$$

$$= a\,\frac{(1-b^2)^2}{1+b^2}\Big\{(1+b^2+b^4+b^6+b^8+\ldots)$$
$$-(b+b^3+b^5+b^7+b^9+\ldots)\left(x+\frac{1}{x}\right)$$
$$+(b^2+b^4+b^6+b^8+b^{10}+\ldots)\left(x^2+\frac{1}{x^2}\right)$$
$$-(b^3+b^5+b^7+b^9+b^{11}+\ldots)\left(x^3+\frac{1}{x^3}\right)$$
$$+(\ldots\ldots\ldots\ldots\ldots)\ldots\Big\}.$$

But
$$(1+b^2+b^4+b^6+\ldots)=\frac{1}{1-b^2}$$
$$-(b+b^3+b^5+b^7+\ldots)=-\frac{b}{1-b^2}$$
$$+(b^2+b^4+b^6+b^8+\ldots)=\frac{b^2}{1-b^2}$$
$$-(b^3+b^5+b^7+b^9+\ldots)=-\frac{b^3}{1-b^2}$$
$$+(\ldots\ldots\ldots\ldots)=\ldots$$

Therefore
$$r = a\,\frac{(1-b^2)^2}{1+b^2}\Big\{\frac{1}{1-b^2}-\frac{b}{1-b^2}\left(x+\frac{1}{x}\right)+\frac{b^2}{1-b^2}\left(x^2+\frac{1}{x^2}\right)$$
$$-\frac{b^3}{1-b^2}\left(x^3+\frac{1}{x^3}\right)+\ldots\Big\}$$
$$= a\,\frac{1-b^2}{1+b^2}\{1-2b\cos\theta+2b^2\cos 2\theta-2b^3\cos 3\theta+\ldots\}$$
$$= a\sqrt{1-e^2}\,(1-2b\cos\theta+2b^2\cos 2\theta-2b^3\cos 3\theta+\ldots).\quad(316)$$

222. *To shew that*
$$\cos\frac{x}{2}\cos\frac{x}{2^2}\cos\frac{x}{2^3}\ldots ad\ inf. = \frac{\sin x}{x}.$$

$$\sin x = 2\sin\frac{x}{2}\cos\frac{x}{2}$$
$$= 2^2\cos\frac{x}{2}\sin\frac{x}{2^2}\cos\frac{x}{2^2},\text{ since }\sin\frac{x}{2}=2\sin\frac{x}{2^2}\cos\frac{x}{2^2},$$

$$= 2^3 \cos \frac{x}{2} \cos \frac{x}{2^2} \sin \frac{x}{2^3} \cos \frac{x}{2^3}.$$

$$\cdots\cdots\cdots\cdots$$

$$= 2^n \cos \frac{x}{2} \cos \frac{x}{2^2} \cos \frac{x}{2^3} \cdots \cos \frac{x}{2^n} \sin \frac{x}{2^n}.$$

Therefore $\quad \cos \dfrac{x}{2} \cos \dfrac{x}{2^2} \cos \dfrac{x}{2^3} \cdots \cos \dfrac{x}{2^n} = \dfrac{\sin x}{2^n \sin \dfrac{x}{2^n}},$

but $\quad 2^n \sin \dfrac{x}{2^n} = x \dfrac{\sin \dfrac{x}{2^n}}{\dfrac{x}{2^n}} = x,$ when $n = \infty,$ by Art. 74.

Hence $\quad \cos \dfrac{x}{2} \cos \dfrac{x}{2^2} \cos \dfrac{x}{2^3} \cdots \text{ad inf.} = \dfrac{\sin x}{x}.$ (317)

Whence we also get

$$x = \sin x \sec \frac{x}{2} \sec \frac{x}{2^2} \sec \frac{x}{2^3} \cdots \text{ad inf.} \qquad (318)$$

Multiplying (317) by $\cos x$ and expressing in logarithms we have

$$\log \cos x + \log \cos \frac{x}{2} + \log \cos \frac{x}{2^2} + \cdots \text{ad inf.} = \log \left(\frac{\sin 2x}{2x}\right)$$

Summation of Trigonometrical Series.

223. *To find the sum of the sines and cosines of a series of angles in arithmetical progression.*

(1) $\quad \sin x + \sin (x+y) + \sin (x+2y) + \&c.,$ to n terms.

By (54) we have

$$\cos \left(x - \frac{y}{2}\right) - \cos \left(x + \frac{y}{2}\right) = 2 \sin \frac{y}{2} \sin x,$$

$$\cos \left(x + \frac{y}{2}\right) - \cos \left(x + \frac{3}{2} y\right) = 2 \sin \frac{y}{2} \sin (x+y),$$

$$\cos \left(x + \frac{3}{2} y\right) - \cos \left(x + \frac{5}{2} y\right) = 2 \sin \frac{y}{2} \sin (x+2y),$$

$$\cdots\cdots\cdots\cdots\cdots\cdots\cdots\cdots\cdots$$

$$\cos\left(x+\frac{2n-3}{2}y\right) - \cos\left(x+\frac{2n-1}{2}y\right) = 2\sin\frac{y}{2}\sin\left(x+(n-1)y\right),$$

Let S denote the sum of the series, then we have by addition

$$\cos\left(x-\frac{y}{2}\right) - \cos\left(x+\frac{2n-1}{2}y\right)$$

$$= 2\sin\frac{y}{2}\left(\sin x + \sin(x+y) + \sin(x+2y) + \ldots\right)$$

$$= 2\sin\frac{y}{2}\cdot S$$

whence
$$S = \frac{\cos\left(x-\frac{y}{2}\right) - \cos\left(x+\frac{2n-1}{2}y\right)}{2\sin\frac{y}{2}}$$

$$= \frac{\sin\left(x+\frac{n-1}{2}y\right)\sin\frac{ny}{2}}{\sin\frac{y}{2}}. \tag{319}$$

(2) $\quad \cos x + \cos(x+y) + \cos(x+2y) + \&c.$, to n terms.

By (52) we have

$$\sin\left(x+\frac{y}{2}\right) - \sin\left(x-\frac{y}{2}\right) = 2\sin\frac{y}{2}\cos x,$$

$$\sin\left(x+\frac{3}{2}y\right) - \sin\left(x+\frac{y}{2}\right) = 2\sin\frac{y}{2}\cos(x+y),$$

$$\sin\left(x+\frac{5}{2}y\right) - \sin\left(x+\frac{3}{2}y\right) = 2\sin\frac{y}{2}\cos(x+2y),$$

$$\ldots\ldots\ldots\ldots\ldots\ldots\ldots$$

$$\sin\left(x+\frac{2n-1}{2}y\right) - \sin\left(x+\frac{2n-3}{2}y\right) = 2\sin\frac{y}{2}\cos\left(x+(n-1)y\right).$$

Let S denote the sum of the proposed series, then we have by addition

$$\sin\left(x+\frac{2n-1}{2}y\right) - \sin\left(x-\frac{y}{2}\right) = 2\sin\frac{y}{2}\left(\cos x + \cos(x+y) + \&c.\ldots\right)$$

$$= 2 \sin \frac{y}{2} \cdot S$$

whence
$$S = \frac{\sin\left(x + \frac{2n-1}{2} y\right) - \sin\left(x - \frac{y}{2}\right)}{2 \sin \frac{y}{2}}$$

$$= \frac{\cos\left(x + \frac{n-1}{2} y\right) \sin \frac{ny}{2}}{\sin \frac{y}{2}}. \tag{320}$$

224. If $y = x, 2x$, &c., in succession, we find from (319) and (320)

$$\sin x + \sin 2x + \sin 3x + \&c \ldots + \sin nx \quad = \frac{\sin \frac{n+1}{2} x \sin \frac{nx}{2}}{\sin \frac{x}{2}}. \tag{321}$$

$$\sin x + \sin 3x + \sin 5x + \&c \ldots + \sin (2n-1)x = \frac{\sin^2 nx}{\sin x}. \tag{322}$$

&c., &c., &c.

$$\cos x + \cos 2x + \cos 3x + \&c \ldots + \cos nx \quad = \frac{\cos \frac{n+1}{2} x \sin \frac{nx}{2}}{\sin \frac{x}{2}}. \tag{323}$$

$$\cos x + \cos 3x + \cos 5x + \&c \ldots + \cos (2n-1)x = \frac{\sin 2nx}{2 \sin x}. \tag{324}$$

&c., &c., &c.

225. The formulæ of the last two Articles enable us to find the sum of the squares of the sines or cosines of a series of angles in arithmetical progression; thus, let

$$\sin^2 x + \sin^2 (x+y) + \sin^2 (x+2y) + \&c.,$$

to n terms be the proposed series.

Since $2\sin^2 x = 1 - \cos 2x$, $2\sin^2(x+y) = 1 - \cos(2x+2y)$ &c., we have by substitution

$2S = 1 - \cos 2x + 1 - \cos(2x+2y) + 1 - \cos(2x+4y) + 1 - \cos(2x+6y) + \&c.$,
to n terms

$= n - \big(\cos 2x + \cos(2x+2y) + \cos(2x+4y) + \&c.$, to n terms$\big)$

$= n - \dfrac{\cos\big(2x+(n-1)y\big)\sin ny}{\sin y}$, by (320)

and

$S = \dfrac{n}{2} - \dfrac{\cos\big(2x+(n-1)y\big)\sin ny}{2\sin y}$.

In a similar manner we may find the sum of the series

$$\cos^2 x + \cos^2(x+y) + \cos^2(x+2y) + \&c.,$$

to n terms by using $2\cos^2 x = 1 + \cos 2x$, &c.

226. *To find the sum of n terms of a series of the form*

$$\sin x \cos y + \sin 2x \cos 3y + \&c. \ldots + \sin nx \cos(2n-1)y.$$

By (51) we have

$\sin(x+y) + \sin(x-y) = 2\sin x \cos y$
$\sin(2x+3y) + \sin(2x-3y) = 2\sin 2x \cos 3y$,
&c., &c.

Thus by addition the proposed series is resolved into two others which can be summed by the method of Art. 223.

227. *To find the sum of n terms of the series*

$$\tan x + 2\tan 2x + 4\tan 4x + \&c.$$

By (119) we have

$\tan x = \cot x - 2\cot 2x$
$2\tan 2x = 2\cot 2x - 4\cot 4x$
$4\tan 4x = 4\cot 4x - 8\cot 8x$
$\ldots\ldots \quad \ldots\ldots$
$2^{n-1}\tan 2^{n-1}x = 2^{n-1}\cot 2^{n-1}x - 2^n \cot 2^n x$;

therefore by addition $\quad S = \cot x - 2^n \cot 2^n x.$ \hfill (325)

228. *To find the sum of n terms of the series*
$$\csc x + \csc 2x + \csc 4x + \&c.$$

By (120) we have
$$\csc x = \cot \frac{x}{2} - \cot x$$
$$\csc 2x = \cot x - \cot 2x$$
$$\csc 4x = \cot 2x - \cot 4x$$
$$\cdots \cdots \cdots \cdots$$
$$\csc 2^{n-1} x = \cot 2^{n-2} x - \cot 2^{n-1} x,$$

therefore by addition
$$S = \cot \frac{x}{2} - \cot 2^{n-1} x. \qquad (326)$$

229. *To find the sum of n terms of the series*
$$a \sin \theta + a^2 \sin 2\theta + a^3 \sin 3\theta + \&c.$$

Putting S equal to the sum and multiplying both members by $2\sqrt{-1}$, we have

$$2\sqrt{-1}\, S = a \cdot 2\sqrt{-1} \sin \theta + a^2 \cdot 2\sqrt{-1} \sin 2\theta + a^3 \cdot 2\sqrt{-1} \sin 3\theta + \&c.$$

$$= a\left(x - \frac{1}{x}\right) + a^2\left(x^2 - \frac{1}{x^2}\right) + a^3\left(x^3 - \frac{1}{x^3}\right) + \&c., \text{ to } n \text{ terms}$$
by (302)

$$= \begin{cases} ax + a^2 x^2 + a^3 x^3 + \&c., \text{ to } n \text{ terms} \\ -\left(\dfrac{a}{x} + \dfrac{a^2}{x^2} + \dfrac{a^3}{x^3} + \&c., \text{ to } n \text{ terms.}\right) \end{cases}$$

$$= \frac{ax(a^n x^n - 1)}{ax - 1} - \frac{ax^{-1}(a^n x^{-n} - 1)}{ax^{-1} - 1}, \text{ (Colenso's Alg., Art. 157)}$$

$$= \frac{a^{n+2}(x^n - x^{-n}) - a^{n+1}(x^{n+1} + x^{-n-1}) + a(x - x^{-1})}{a^2 - a(x + x^{-1}) + 1},$$

therefore
$$S = \frac{a^{n+2} \sin n\theta - a^{n+1} \sin (n+1)\theta + a \sin \theta}{a^2 - 2a \cos \theta + 1}. \qquad (327)$$

If a be a proper fraction and $n = \infty$, we shall have
$$S = \frac{a \sin \theta}{a^2 - 2a \cos \theta + 1}$$

as the sum of the series continued to infinity.

230. *To find the sum of the series*

$$\cos\theta\sin\theta - \tfrac{1}{2}\cos^2\theta\sin 2\theta + \tfrac{1}{3}\cos^3\theta\sin 3\theta - \&c., \text{ ad inf.}$$

By (302) we have

$$2\sqrt{-1}\,S = \cos\theta\,(x - x^{-1}) - \frac{\cos^2\theta}{2}(x^2 - x^{-2}) + \frac{\cos^3\theta}{3}(x^3 - x^{-3}) - \&c.$$

$$= \left\{\begin{array}{l} x\cos\theta - \dfrac{x^2\cos^2\theta}{2} + \dfrac{x^3\cos^3\theta}{3} - \&c. = \log(1 + x\cos\theta) \\ -x^{-1}\cos\theta + \dfrac{x^{-2}\cos^2\theta}{2} - \dfrac{x^{-3}\cos^3\theta}{3} + \&c. = -\log(1 + x^{-1}\cos\theta) \end{array}\right\}$$

$$= \log\frac{1 + x\cos\theta}{1 + x^{-1}\cos\theta}, \text{ therefore } e^{2S\sqrt{-1}} = \frac{1 + x\cos\theta}{1 + x^{-1}\cos\theta},$$

and by composition and division,

$$\frac{e^{2S\sqrt{-1}} - 1}{e^{2S\sqrt{-1}} + 1} = \frac{\cos\theta\,(x - x^{-1})}{2 + \cos\theta\,(x + x^{-1})} = \frac{2\cos\theta\sin\theta\sqrt{-1}}{2 + 2\cos^2\theta},$$

therefore by (296) $\sqrt{-1}\tan S = \dfrac{2\cos\theta\sin\theta\sqrt{-1}}{2(1 + \cos^2\theta)}$

and $\qquad\qquad\qquad S = \tan^{-1}\left(\dfrac{\sin 2\theta}{2(1 + \cos^2\theta)}\right).\qquad(328)$

231. The investigation of Trigonometrical Series cannot be fully carried on without the aid of the Differential Calculus. The student must therefore consult the treatises on that branch of mathematics for further information on this subject. (See Todhunter's Diff. Cal., chaps. vi., vii.; Clark's Cal., chaps. v., vi.; Williamson's Diff. Cal., chap. iii.)

Examples.

1. Prove by the aid of (294), (295) and (296) the following identities:

(1) $\tan\dfrac{\theta}{2} = \dfrac{1 - \cos\theta}{\sin\theta}.$

(2) $\cos 2\theta = \cos^2\theta - \sin^2\theta.$

(3) $\sin\theta = 2\sin\dfrac{\theta}{2}\cos\dfrac{\theta}{2}.$

(4) $\tan 2\theta = \dfrac{\sin\theta + \sin 3\theta}{\cos\theta + \cos 3\theta}.$

(5) $\tan(45° - \theta)\tan(45° - 3\theta) = \dfrac{1 - 2\sin 2\theta}{1 + 2\sin 2\theta}$

(6) $2\cos^2\theta = 1 + \cos 2\theta.$

2. If $2\cos\theta = x+x^{-1}$, and $2\cos\phi = y+y^{-1}$, then will
$$2\sqrt{-1}\sin(m\theta+n\phi) = x^m y^n - x^{-m} y^{-n}.$$

3. Shew that $e^{\pi\sqrt{-1}} = -1$

and $\pi - \dfrac{\pi^3}{1.2.3} + \dfrac{\pi^5}{1.2.3.4.5} - \&c.,$ ad inf. $= 0$.

4. Shew that $\pi = -\sqrt{-1}.\log(-1)$

and $(\sqrt{-1})^{\sqrt{-1}} = 1 - \dfrac{\pi}{2} + \dfrac{1}{1.2}\left(\dfrac{\pi}{2}\right)^2 - \dfrac{1}{1.2.3}\left(\dfrac{\pi}{2}\right)^3 + \&c.,$ ad inf.

5. Prove that versin $x = -\frac{1}{2}(e^{\frac{x}{2}\sqrt{-1}} - e^{-\frac{x}{2}\sqrt{-1}})^2$.

6. Prove that
$$(1+\cos\theta+\sqrt{-1}\sin\theta)^n + (1+\cos\theta-\sqrt{-1}\sin\theta)^n = 2^{n+1}\cos^n\dfrac{\theta}{2}\cos\dfrac{n\theta}{2}.$$

7. If $a^2+b^2=2$, and $ab = \sqrt{-1}\tan 2\phi$, shew that $a = \cos\phi\sqrt{2\sec 2\phi}$.

8. If $2\cos a_n = x_n + \dfrac{1}{x_n}$, and $a_1+a_2+a_3+\ldots+a_n = 2\pi$,

then will $x_1 x_2 x_3 \ldots x_n = 1$.

9. Prove that
$$2\sin(x\pm y\sqrt{-1}) = (e^y+e^{-y})\sin x \pm (e^y-e^{-y})\sqrt{-1}\cos x.$$

10. If $S_1 = \tan A + \tan B + \tan C + \&c. =$ sum of the tangents,
$S_2 =$ sum of their products taken 2 and 2,
$S_3 =$ sum of their products taken 3 and 3,
 &c., &c.,

shew that
$$\tan(A+B+C+\&c.) = \dfrac{S_1 - S_3 + S_5 - \&c.}{1 - S_2 + S_4 - \&c.}.$$

11. In any triangle shew that
$$\log\dfrac{b}{a} = M\left\{(\cos 2A - \cos 2B) + \tfrac{1}{2}(\cos 4A - \cos 4B) + \tfrac{1}{3}(\cos 6A - \cos 6B) + \&c.\right\}.$$

EXAMPLES.

12. Prove that
$$\log_e \sec\theta = \tfrac{1}{2}\tan^2\theta - \tfrac{1}{4}\tan^4\theta + \tfrac{1}{6}\tan^6\theta - \tfrac{1}{8}\tan^8\theta + \&c.\ldots$$

13. Prove by Gregory's series that
$$\sec^{-1}\tfrac{1}{2}\left(\tfrac{m}{n}+\tfrac{n}{m}\right) = \tfrac{1}{2}\left(\tfrac{m}{n}-1\right)\left(\tfrac{n}{m}+1\right) - \tfrac{1}{3.2^3}\left(\tfrac{m}{n}-1\right)^3\left(\tfrac{n}{m}+1\right)^3 + \&c.\ldots$$

14. Prove that $\log\tan^{\frac{1}{2}}\left(\tfrac{\pi}{4}+\theta\right) = \tan\theta + \tfrac{1}{3}\tan^3\theta + \tfrac{1}{5}\tan^5\theta + \&c.\ldots$

15. Shew that $\dfrac{\pi}{8} = \dfrac{1}{1.3} + \dfrac{1}{5.7} + \dfrac{1}{9.11} + \ldots$

16. Prove that $\dfrac{\pi}{4} = 4\tan^{-1}\dfrac{1}{5} - \tan^{-1}\dfrac{1}{70} + \tan^{-1}\dfrac{1}{99}$

and $\dfrac{\pi}{4} = 4\tan^{-1}\dfrac{1}{5} - 2\tan^{-1}\dfrac{1}{408} + \tan^{-1}\dfrac{1}{1393}.$

17. If $\tan 2\theta = \sin 2\phi$, shew that
$$\theta = \cos 2\phi \tan\phi - \tfrac{1}{3}\cos 6\phi \tan^3\phi + \tfrac{1}{5}\cos 10\phi \tan^5\phi - \ldots$$

Sum to n terms the following four series:

18. $\tan\dfrac{\theta}{2}\sec\theta + \tan\dfrac{\theta}{4}\sec\dfrac{\theta}{2} + \tan\dfrac{\theta}{8}\sec\dfrac{\theta}{4} + \ldots$

$\qquad\qquad\qquad\qquad\qquad$ Ans. $\tan\theta - \tan\dfrac{\theta}{2^n}$.

19. $\dfrac{1}{2}\sec\theta + \dfrac{1}{2^2}\sec\theta\sec 2\theta + \dfrac{1}{2^3}\sec\theta\sec 2\theta\sec 2^2\theta + \ldots$

$\qquad\qquad\qquad\qquad\qquad$ Ans. $\cos\theta - \sin\theta\cot 2^n\theta$.

20. $\sin(x-y) + \sin(2x-3y) + \sin(3x-5y) + \ldots$

$\qquad\qquad$ Ans. $\dfrac{\sin\left(\tfrac{1}{2}(n+1)x - ny\right)\sin(\tfrac{1}{2}nx - ny)}{\sin\tfrac{1}{2}(x-2y)}.$

21. $\dfrac{1}{2}\log\tan 2x + \dfrac{1}{2^2}\log\tan 2^2 x + \dfrac{1}{2^3}\log\tan 2^3 x + \ldots$

$\qquad\qquad\qquad$ Ans. $\log 2\sin 2x - \dfrac{\log 2\sin 2^{n+1}x}{2^n}.$

Sum to infinity the following twelve series:

22. $\dfrac{\cos\phi}{1} + \dfrac{\cos^2\phi}{2} + \dfrac{\cos^3\phi}{3} + \ldots \qquad$ Ans. $\log\left(\cot\dfrac{\phi}{2}\text{cosec}\,\phi\right).$

23. $1 + \dfrac{\cos 2\theta}{1.2} + \dfrac{\cos 4\theta}{1.2.3.4} + \dfrac{\cos 6\theta}{1.2.3.4.5.6} + \ldots$

Ans. $\frac{1}{2} \cos(\sin\theta)(e^{\cos\theta} + e^{-\cos\theta})$.

24. $\sin a \sin a + \frac{1}{3} \sin^3 a \sin 3a + \frac{1}{5} \sin^5 a \sin 5a + \ldots$

Ans. $\frac{1}{2} \tan^{-1}(2 \tan^2 a)$.

25. $x \cos(a+\beta) + \dfrac{x^2}{1.2} \cos(a+2\beta) + \dfrac{x^3}{1.2.3} \cos(a+3\beta) + \ldots$

Ans. $e^{x \cos \beta} \cos(a + x \sin \beta) - \cos a$.

26. $\sin a - \frac{1}{2} \sin 2a + \frac{1}{3} \sin 3a - \frac{1}{4} \sin 4a + \ldots$ *Ans.* $\dfrac{a}{2}$.

27. $\sin x \sin x - \frac{1}{2} \sin 2x \sin^2 x + \frac{1}{3} \sin 3x \sin^3 x - \ldots$

Ans. $\cot^{-1}(1 + \cot x + \cot^2 x)$.

28. $\cos 2\theta + \frac{1}{3} \cos^3 2\theta + \frac{1}{5} \cos^5 2\theta + \ldots$ *Ans.* $\log \cot \theta$.

29. $(1+2) \log 2 + \dfrac{1+2^2}{1.2}(\log 2)^2 + \dfrac{1+2^3}{1.2.3}(\log 2)^3 + \ldots$ *Ans.* 4.

30. $2\{\sin^2 x - \frac{1}{2} \sin^2 2x + \frac{1}{3} \sin^2 3x - \frac{1}{4} \sin^2 4x + \ldots\}$.

Ans. $\log \sec x$.

31. If $\cos x + \frac{1}{2} \cos^2 x + \frac{1}{3} \cos^3 x + \ldots = S$,
and $\cos x + \frac{1}{2} \cos 2x + \frac{1}{3} \cos 3x + \ldots = s$,
shew that $S - 2s = \log 2$.

32. $\dfrac{\sin \theta}{1.2} + \dfrac{\sin 2\theta}{2.2^2} + \dfrac{\sin 3\theta}{3.2^3} + \dfrac{\sin 4\theta}{4.2^4} + \ldots$ *Ans.* $\tan^{-1} \dfrac{\sin \theta}{2 - \cos \theta}$.

33. $2 \cos \theta + \dfrac{3}{2} \cos^2 \theta + \dfrac{4}{3} \cos^3 \theta + \dfrac{5}{4} \cos^4 \theta + \ldots$

Ans. $\dfrac{\cos \theta}{1 - \cos \theta} - \log(1 - \cos \theta)$.

34. Prove by Art. 222 that

$$\left(1 - \tan^2 \dfrac{x}{2}\right)\left(1 - \tan^2 \dfrac{x}{2^2}\right)\left(1 - \tan^2 \dfrac{x}{2^3}\right) \ldots \text{ad inf.} = x \cot x.$$

35. Shew that

$1 + \dfrac{n}{1} a \cos mx + \dfrac{n(n-1)}{1.2} a^2 \cos 2mx + \dfrac{n(n-1)(n-2)}{1.2.3} a^3 \cos 3mx$

$+ \&c. = r^n \cos n\theta$,

where $r^2 = 1 + 2a \cos mx + a^2$, and $\tan \theta = \dfrac{a \sin mx}{1 + a \cos mx}$.

CHAPTER XVII.

RESOLUTION OF CERTAIN EQUATIONS INTO THEIR SIMPLE AND QUADRATIC FACTORS.

232. To resolve the Equation
$$x^{2n} - 2x^n \cos \phi + 1 = 0. \tag{329}$$
into its Simple and Quadratic Factors.

By completing the square and extracting the square root we have
$$x^n = \cos \phi \pm \sqrt{-1} \sin \phi = \cos (2r\pi + \phi) \pm \sqrt{-1} \sin (2r\pi + \phi)$$
whence
$$x = \{\cos (2r\pi + \phi) \pm \sqrt{-1} \sin (2r\pi + \phi)\}^{\frac{1}{n}}$$
$$= \cos \frac{2r\pi + \phi}{n} \pm \sqrt{-1} \sin \frac{2r\pi + \phi}{n}, \tag{330}$$
by De Moivre's Theorem.

Therefore the $2n$ values of x will be found by assigning to r the values $0, 1, 2, \ldots n-1$, in succession; thus, using first the upper sign, we have

(1) if $r = 0$, $\quad x = \cos \dfrac{\phi}{n} + \sqrt{-1} \sin \dfrac{\phi}{n}$,

(2) $\quad r = 1$, $\quad x = \cos \dfrac{2\pi + \phi}{n} + \sqrt{-1} \sin \dfrac{2\pi + \phi}{n}$,

(3) $\quad r = 2$, $\quad x = \cos \dfrac{4\pi + \phi}{n} + \sqrt{-1} \sin \dfrac{4\pi + \phi}{n}$,

....

(n) $\quad r = n-1$, $\quad x = \cos \dfrac{2(n-1)\pi + \phi}{n} + \sqrt{-1} \sin \dfrac{2(n-1)\pi + \phi}{n}$;

and using the lower sign,

(1) if $r=0$, $\quad x = \cos\dfrac{\phi}{n} - \sqrt{-1}\sin\dfrac{\phi}{n}$,

(2) $\quad r=1$, $\quad x = \cos\dfrac{2\pi+\phi}{n} - \sqrt{-1}\sin\dfrac{2\pi+\phi}{n}$,

(3) $\quad r=2$, $\quad x = \cos\dfrac{4\pi+\phi}{n} - \sqrt{-1}\sin\dfrac{4\pi+\phi}{n}$,

....

(n) $\quad r=n-1$, $\quad x = \cos\dfrac{2(n-1)\pi+\phi}{n} - \sqrt{-1}\sin\dfrac{2(n-1)\pi+\phi}{n}$.

Now, from the first of each of these groups we have the first two simple factors

$$\left(x - \cos\dfrac{\phi}{n} - \sqrt{-1}\sin\dfrac{\phi}{n}\right) \text{ and } \left(x - \cos\dfrac{\phi}{n} + \sqrt{-1}\sin\dfrac{\phi}{n}\right),$$

the product of which is $x^2 - 2x\cos\dfrac{\phi}{n} + 1$, the first quadratic factor; from the second of each we have the second two simple factors

$$\left(x - \cos\dfrac{2\pi+\phi}{n} - \sqrt{-1}\sin\dfrac{2\pi+\phi}{n}\right) \text{ and } \left(x - \cos\dfrac{2\pi+\phi}{n} + \sqrt{-1}\sin\dfrac{2\pi+\phi}{n}\right),$$

the product of which is $x^2 - 2x\cos\dfrac{2\pi+\phi}{n} + 1$, the second quadratic factor; and so on, there being $2n$ simple factors in all.

Therefore we have for the n quadratic factors

$$x^{2n} - 2x^n\cos\phi + 1 = (x^2 - 2x\cos\dfrac{\phi}{n} + 1)$$
$$\times (x^2 - 2x\cos\dfrac{2\pi+\phi}{n} + 1)$$
$$\times (x^2 - 2x\cos\dfrac{4\pi+\phi}{n} + 1)$$
$$\times \quad \ldots\ldots\ldots$$
$$\times (x^2 - 2x\cos\dfrac{2(n-1)\pi+\phi}{n} + 1). \qquad (331)$$

RESOLUTION OF EQUATIONS. 301

Since the simple factors of $x^{2n} - 2x^n \cos\phi + 1$ are obtained from (330) by making $r = 0, 1, 2, \ldots n-1$ in succession, it is evident that they are all included in the form

$$x - \left(\cos\frac{2r\pi + \phi}{n} \pm \sqrt{-1} \sin\frac{2r\pi + \phi}{n}\right) \quad (332)$$

Ex. Resolve the equation $x^4 - x^2 + 1 = 0$ into its quadratic and simple factors.

Here $n = 2$, $2\cos\phi = 1$, therefore $\phi = \dfrac{\pi}{3}$, and by (331)

$$\begin{aligned}
x^4 - x^2 + 1 &= \left(x^2 - 2x\cos\frac{\pi}{6} + 1\right)\left(x^2 - 2x\cos\frac{7\pi}{6} + 1\right) \\
&= (x^2 - \sqrt{3}\,x + 1)(x^2 + \sqrt{3}\,x + 1), \\
&= \left(x - \cos\frac{\pi}{6} - \sqrt{-1}\sin\frac{\pi}{6}\right) \\
&\quad \times \left(x - \cos\frac{\pi}{6} + \sqrt{-1}\sin\frac{\pi}{6}\right) \\
&\quad \times \left(x + \cos\frac{\pi}{6} + \sqrt{-1}\sin\frac{\pi}{6}\right) \\
&\quad \times \left(x + \cos\frac{\pi}{6} - \sqrt{-1}\sin\frac{\pi}{6}\right), \quad \text{by (332)} \\
&= (x - \tfrac{1}{2}\sqrt{3} - \tfrac{1}{2}\sqrt{-1}) \\
&\quad \times (x - \tfrac{1}{2}\sqrt{3} + \tfrac{1}{2}\sqrt{-1}) \\
&\quad \times (x + \tfrac{1}{2}\sqrt{3} + \tfrac{1}{2}\sqrt{-1}) \\
&\quad \times (x + \tfrac{1}{2}\sqrt{3} - \tfrac{1}{2}\sqrt{-1}).
\end{aligned}$$

Hence the four roots of $x^4 - x^2 + 1 = 0$, are

$$\tfrac{1}{2}(\sqrt{3} \pm \sqrt{-1}) \text{ and } \tfrac{1}{2}(-\sqrt{3} \pm \sqrt{-1}).$$

233. In (331) let $x = 1$, then we shall have

$$2 - 2\cos\phi = \left(2 - 2\cos\frac{\phi}{n}\right)\left(2 - 2\cos\frac{2\pi + \phi}{n}\right)\left(2 - 2\cos\frac{4\pi + \phi}{n}\right), \&c.,$$

or to n factors,

$$2^2 \sin^2\frac{\phi}{2} = 2^{2n} \sin^2\frac{\phi}{2n} \sin^2\frac{2\pi + \phi}{2n} \sin^2\frac{4\pi + \phi}{2n}, \&c.,$$

and writing $2n\theta$ for ϕ, dividing by 2^2 and extracting the square root, we have

$$\sin n\theta = 2^{n-1} \sin \theta \sin \left(\theta + \frac{\pi}{n}\right) \sin \left(\theta + \frac{2\pi}{n}\right) \sin \left(\theta + \frac{3\pi}{n}\right), \&c., \quad (333)$$
$$\text{to } n \text{ factors.}$$

Let $n\theta = \frac{\pi}{2}$ or $\theta = \frac{\pi}{2n}$, then we have

$$1 = 2^{n-1} \sin \frac{\pi}{2n} \sin \frac{3\pi}{2n} \sin \frac{5\pi}{2n}, \&c., \text{ to } n \text{ factors.} \quad (334)$$

Again, in (331) let $x = -1$, then we shall have

$$\left. \begin{array}{l} (n \text{ even}) \; 2 - 2 \cos \phi \\ (n \text{ odd}) \; 2 + 2 \cos \phi \end{array} \right\}$$

$$= \left(2 + 2 \cos \frac{\phi}{n}\right) \left(2 + 2 \cos \frac{2\pi + \phi}{n}\right) \left(2 + 2 \cos \frac{4\pi + \phi}{n}\right), \&c.,$$

whence, putting $2n\theta$ for ϕ and proceeding as before, we have

$$\left. \begin{array}{l} (n \text{ even}) \pm \sin n\theta \\ (n \text{ odd}) \pm \cos n\theta \end{array} \right\} = 2^{n-1} \cos \theta \cos \left(\theta + \frac{\pi}{n}\right) \cos \left(\theta + \frac{2\pi}{n}\right), \&c., \quad (335)$$
$$\text{to } n \text{ factors.}$$

Dividing (333) by (335) we get

$$\left. \begin{array}{l} (n \text{ even}) \pm 1 \\ (n \text{ odd}) \pm \tan n\theta \end{array} \right\} = \tan \theta \tan \left(\theta + \frac{\pi}{n}\right) \tan \left(\theta + \frac{2\pi}{n}\right), \&c., \quad (336)$$
$$\text{to } n \text{ factors.}$$

In (336) let $n\theta = \frac{\pi}{4}$, then we have, whether n be even or odd,

$$1 = \tan \frac{\pi}{4n} \tan \frac{5\pi}{4n} \tan \frac{9\pi}{4n}, \&c., \text{ to } n \text{ factors.} \quad (337)$$

234. To resolve the Equation $x^n - 1 = 0$ into its Quadratic Factors, n being odd.

In (331) let $\phi = 0$ and it becomes

$$(x^n - 1)^2 = (x - 1)^2 \times \left(x^2 - 2x \cos \frac{2\pi}{n} + 1\right)$$
$$\times \left(x^2 - 2x \cos \frac{4\pi}{n} + 1\right)$$
$$\times \quad \ldots\ldots\ldots$$
$$\times \left(x^2 - 2x \cos \frac{(n-1)\pi}{n} + 1\right). \quad (338)$$

RESOLUTION OF EQUATIONS. 303

Now, as n is odd, and as there are in all n factors, the number of quadratic factors in (338), *exclusive* of $(x-1)^2$, is *even;* and since

$$\cos\frac{2(n-1)\pi}{n} = \cos\left(2\pi - \frac{2\pi}{n}\right) = \cos\frac{2\pi}{n};$$

$$\cos\frac{2(n-2)\pi}{n} = \cos\left(2\pi - \frac{4\pi}{n}\right) = \cos\frac{4\pi}{n}, \text{ and so on;}$$

the first and the last of these factors are equal; the second and the last but one are equal, and so on; hence uniting these equal factors and extracting the square root, we have, when n is odd,

$$x^n - 1 = (x-1) \times (x^2 - 2x\cos\frac{2\pi}{n} + 1)$$

$$\times (x^2 - 2x\cos\frac{4\pi}{n} + 1)$$

$$\times \quad \ldots\ldots$$

$$\times (x^2 - 2x\cos\frac{(n-1)\pi}{n} + 1). \quad (339)$$

235. To resolve the Equation $x^n - 1 = 0$ into its Quadratic Factors, n being even.

When n is *even* the number of quadratic factors in (338), *exclusive* of $(x-1)^2$, is *odd,* and there is a middle factor, the $\frac{n}{2}^{\text{th}}$, which will not combine with any other. This factor is

$$x^2 - 2x\cos\frac{2\frac{n}{2}\pi}{n} + 1 = x^2 + 2x + 1 = (x+1)^2.$$

Hence uniting the other factors and extracting the square root, we have, when n is even,

$$x^n - 1 = (x-1)(x+1) \times (x^2 - 2x\cos\frac{2\pi}{n} + 1)$$

$$\times (x^2 - 2x\cos\frac{4\pi}{n} + 1)$$

$$\times \quad \ldots\ldots$$

$$\times (x^2 - 2x\cos\frac{(n-2)\pi}{n} + 1). \quad (340)$$

236. To resolve the Equation $x^n + 1 = 0$ into its Quadratic Factors, n being odd.

In (331) let $\phi = \pi$ and it becomes

$$(x^n + 1)^2 = (x^2 - 2x \cos \frac{\pi}{n} + 1)$$
$$\times (x^2 - 2x \cos \frac{3\pi}{n} + 1)$$
$$\times (x^2 - 2x \cos \frac{5\pi}{n} + 1)$$
$$\times \quad \ldots\ldots\ldots$$
$$\times (x^2 - 2x \cos \frac{(2n-1)\pi}{n} + 1). \quad (341)$$

Since there are n quadratic factors and n is odd, there is a middle factor which will not combine with any other. This factor is evidently

$$x^2 - 2x \cos \frac{n\pi}{n} + 1 = (x+1)^2,$$

and it is easily shewn, as in Art. 234, that the factors equally distant from the first and last, are equal; hence uniting the equal factors and extracting the square root, we have, when n is odd,

$$x^n + 1 = (x+1) \times (x^2 - 2x \cos \frac{\pi}{n} + 1)$$
$$\times (x^2 - 2x \cos \frac{3\pi}{n} + 1)$$
$$\times (x^2 - 2x \cos \frac{5\pi}{n} + 1)$$
$$\times \quad \ldots\ldots$$
$$\times (x^2 - 2x \cos \frac{(n-2)\pi}{n} + 1). \quad (342)$$

237. To resolve the Equation $x^n + 1 = 0$ into its Quadratic Factors, n being even.

When n is even the number of quadratic factors in (341) is even; therefore we have

RESOLUTION OF EQUATIONS.   305

$$x^n+1 = (x^2 - 2x\cos\frac{\pi}{n}+1)$$
$$\times (x^2 - 2x\cos\frac{3\pi}{n}+1)$$
$$\times (x^2 - 2x\cos\frac{5\pi}{n}+1)$$
$$\times \quad \ldots\ldots\ldots$$
$$\times (x^2 - 2x\cos\frac{n-1}{n}\pi+1). \qquad (343)$$

238. The simple factors in each case may be found by resolving each of the quadratic factors; or, they may be found from (332) by putting $\phi=0$ for the form $x^n-1=0$, and $\phi=\pi$ for the form $x^n+1=0$.

Ex. 1.—Find the quadratic factors of $x^5+1=0$.

Here $n=5$, then by (342) we have

$$x^5+1=(x+1)(x^2-2x\cos 36°+1)(x^2-2x\cos 108°+1)$$
$$=(x+1)\left(x^2-\frac{\sqrt{5}+1}{2}x+1\right)\left(x^2+\frac{\sqrt{5}-1}{2}x+1\right).$$

By putting each of the quadratic factors equal to 0 and solving the equation we obtain the five simple factors.

Ex. 2.—Find the roots of $x^4+1=0$.

By (343) we have

$$x^4+1=(x^2-2x\cos 45°+1)(x^2-2x\cos 135°+1)$$
$$=(x^2-\sqrt{2}\,x+1)(x^2+\sqrt{2}\,x+1).$$

Hence $\quad x=\dfrac{1\pm\sqrt{-1}}{\sqrt{2}}\ $ and $\ \dfrac{-1\pm\sqrt{-1}}{\sqrt{2}}$.

239. In (339), divide both sides by $x-1$, thus

$$x^{n-1}+x^{n-2}+\ldots x+1 = (x^2-2x\cos\frac{2\pi}{n}+1)$$
$$\times (x^2-2x\cos\frac{4\pi}{n}+1)$$
$$\times \quad \ldots\ldots\ldots$$
$$\times (x^2-2x\cos\frac{n-3}{n}\pi+1)$$
$$\times (x^2-2x\cos\frac{n-1}{n}\pi+1),$$

there being $\dfrac{n-1}{2}$ quadratic factors.

Put $x=1$, then we have

$$n = 2^{\frac{n-1}{2}}\left(1-\cos\frac{2\pi}{n}\right)\left(1-\cos\frac{4\pi}{n}\right)\ldots\left(1-\cos\frac{n-3}{n}\pi\right)\left(1-\cos\frac{n-1}{n}\pi\right),$$

$$= 2^{n-1}\sin^2\frac{2\pi}{2n}\sin^2\frac{4\pi}{2n}\ldots\sin^2\frac{n-3}{2n}\pi\sin^2\frac{n-1}{2n}\pi,$$

or

$$n^{\frac{1}{2}} = 2^{\frac{n-1}{2}}\sin\frac{2\pi}{2n}\sin\frac{4\pi}{2n}\ldots\sin\frac{n-3}{2n}\pi\sin\frac{n-1}{2n}\pi. \quad (344)$$

240. In (340), divide by $x-1$ and put $x=1$; then we have, there being $\frac{n-2}{2}$ quadratic factors,

$$n = 2 \cdot 2^{\frac{n-2}{2}}\left(1-\cos\frac{2\pi}{n}\right)\left(1-\cos\frac{4\pi}{n}\right)\ldots\left(1-\cos\frac{n-4}{n}\pi\right)\left(1-\cos\frac{n-2}{n}\pi\right);$$

$$= 2 \cdot 2^{n-2}\sin^2\frac{2\pi}{2n}\sin^2\frac{4\pi}{2n}\ldots\sin^2\frac{n-4}{2n}\pi\sin^2\frac{n-2}{2n}\pi,$$

or

$$n^{\frac{1}{2}} = 2^{\frac{n-1}{2}}\sin\frac{2\pi}{2n}\sin\frac{4\pi}{2n}\ldots\sin\frac{n-4}{2n}\pi\sin\frac{n-2}{2n}\pi. \quad (345)$$

241. In (342), divide by $x+1$ and put $x=1$; then we have, there being $\frac{n-1}{2}$ quadratic factors,

$$1 = 2^{\frac{n-1}{2}}\left(1-\cos\frac{\pi}{n}\right)\left(1-\cos\frac{3\pi}{n}\right)\ldots\left(1-\cos\frac{n-4}{n}\pi\right)\left(1-\cos\frac{n-2}{n}\pi\right),$$

$$= 2^{n-1}\sin^2\frac{\pi}{2n}\sin^2\frac{3\pi}{2n}\ldots\sin^2\frac{n-4}{2n}\pi\sin^2\frac{n-2}{2n}\pi,$$

or

$$1 = 2^{\frac{n-1}{2}}\sin\frac{\pi}{2n}\sin\frac{3\pi}{2n}\ldots\sin\frac{n-4}{2n}\pi\sin\frac{n-2}{2n}\pi. \quad (346)$$

242. In (343), put $x=1$ and we have

$$2 = 2^{\frac{n}{2}}\left(1-\cos\frac{\pi}{n}\right)\left(1-\cos\frac{3\pi}{n}\right)\ldots\left(1-\cos\frac{n-3}{n}\pi\right)\left(1-\cos\frac{n-1}{n}\pi\right)$$

or

$$1 = 2^{\frac{n-1}{2}}\sin\frac{\pi}{2n}\sin\frac{3\pi}{2n}\ldots\sin\frac{n-3}{2n}\pi\sin\frac{n-1}{2n}\pi. \quad (347)$$

243. To resolve sin x and cos x into factors.

The series for sin x, Art. 210, may be written thus

$$\sin x = x\left(1 - \frac{x^2}{1.2.3} + \frac{x^4}{1.2.3.4.5} - \&c.\right), \qquad (348)$$

from which we see that x is one factor, and the factors of the series within the parenthesis must evidently be of the form

$$1 - \frac{x^2}{a},$$

where a is constant but has a different value in each factor.

For suppose the factors to be

$$1 - \frac{x^2}{a}, \quad 1 - \frac{x^2}{a_1}, \quad 1 - \frac{x^2}{a_2}, \ \&c.,$$

the product of these will give a series of the same form as that within the parenthesis. Now, the required factors must reduce the second member to zero when the first member is zero, that is, when $x = 0$ or $= \pm n\pi$; therefore the general form of the factor is

$$1 - \frac{n^2\pi^2}{a};$$

and since this must be equal to zero when $x = \pm n\pi$, we have

$$1 - \frac{n^2\pi^2}{a} = 0,$$

whence $a = n^2\pi^2$,

therefore the general factor is $\quad 1 - \dfrac{x^2}{n^2\pi^2};$

then making $n = 1, 2, 3, \&c.$, in succession, (348) becomes

$$\sin x = x\left(1 - \frac{x^2}{1^2\pi^2}\right)\left(1 - \frac{x^2}{2^2\pi^2}\right)\left(1 - \frac{x^2}{3^2\pi^2}\right)\ldots \qquad (349)$$

Again, the factors of the series for cos x, Art. 210, must also be of the form

$$1 - \frac{x^2}{a};$$

but the first member is zero for $x = \pm(2n+1)\dfrac{\pi}{2}$, where n is any integer including zero; therefore we have

308 PLANE TRIGONOMETRY.

$$1-\frac{(2n+1)^2\pi^2}{2^2 a}=0,$$

whence
$$a=\frac{(2n+1)^2\pi^2}{2^2},$$

therefore the general factor is $1-\dfrac{2^2 x^2}{(2n+1)^2\pi^2}$;

and making $n=0, 1, 2,$ &c., in succession, we have

$$\cos x = \left(1-\frac{2^2 x^2}{1^2\pi^2}\right)\left(1-\frac{2^2 x^2}{3^2\pi^2}\right)\left(1-\frac{2^2 x^2}{5^2\pi^2}\right)\cdots \qquad (350)$$

Logarithmic sines and cosines.

244. The last two series furnish us with the means of calculating the logarithmic sines and cosines without first computing the natural sines and cosines.

In (349) and (350) put $x=m\dfrac{\pi}{2}$, then

$$\sin m\frac{\pi}{2}=\frac{m\pi}{2}\left(1-\frac{m^2}{2^2}\right)\left(1-\frac{m^2}{4^2}\right)\left(1-\frac{m^2}{6^2}\right)\cdots$$

$$\cos m\frac{\pi}{2}=\left(1-\frac{m^2}{1^2}\right)\left(1-\frac{m^2}{3^2}\right)\left(1-\frac{m^2}{5^2}\right)\cdots$$

and expressing these by logarithms we have

$$\log \sin m\frac{\pi}{2}=\log\frac{\pi}{2}+\log m+\log\left(1-\frac{m^2}{2^2}\right)+\log\left(1-\frac{m^2}{4^2}\right)+\text{&c.},$$

$$\log \cos m\frac{\pi}{2}=\log\left(1-\frac{m^2}{1^2}\right)+\log\left(1-\frac{m^2}{3^2}\right)+\log\left(1-\frac{m^2}{5^2}\right)+\text{&c.}$$

Developing the second members of these equations by the logarithmic series, and arranging according to the powers of m, we obtain

$$\left.\begin{aligned}\log \sin \frac{m\pi}{2}=\log\frac{\pi}{2}+\log m &-m^2\cdot\frac{M}{1}\left(\frac{1}{2^2}+\frac{1}{4^2}+\frac{1}{6^2}+\text{&c.}\right)\\ &-m^4\cdot\frac{M}{2}\left(\frac{1}{2^4}+\frac{1}{4^4}+\frac{1}{6^4}+\text{&c.}\right)\\ &-m^6\cdot\frac{M}{3}\left(\frac{1}{2^6}+\frac{1}{4^6}+\frac{1}{6^6}+\text{&c.}\right)\\ &-\ldots\ldots\ldots\ldots\text{&c.}\end{aligned}\right\}\cdot\quad(351)$$

$$\begin{aligned}
\log \cos \frac{m\pi}{2} = &-m^2 \cdot \frac{M}{1}\left(\frac{1}{1^2}+\frac{1}{3^2}+\frac{1}{5^2}+\&c.\right) \\
&-m^4 \cdot \frac{M}{2}\left(\frac{1}{1^4}+\frac{1}{3^4}+\frac{1}{5^4}+\&c.\right) \\
&-m^6 \cdot \frac{M}{3}\left(\frac{1}{1^6}+\frac{1}{3^6}+\frac{1}{5^6}+\&c.\right) \\
&- \quad \ldots\ldots\ldots\ldots \&c.
\end{aligned} \quad (352)$$

By summing the constant numerical series, substituting the value of the modulus M and giving m different values, the logarithmic sine and cosine of any angle may be easily computed. Of course 10 must be added to these expressions to give the tabular logarithmic sine, &c.

Examples.

1. Prove that
$$e^x + e^{-x} = 2\left(1+\frac{2^2 x^2}{\pi^2}\right)\left(1+\frac{2^2 x^2}{3^2 \pi^2}\right)\ldots$$
and
$$e^x - e^{-x} = 2x\left(1+\frac{x^2}{\pi^2}\right)\left(1+\frac{x^2}{2^2 \pi^2}\right)\ldots$$

2. Shew that $\sin 20° \sin 40° \sin 60° \sin 80° = \dfrac{3}{16}$.

3. If $\theta = \cos \theta$, shew that $\theta = 42° \ 20'$ very nearly.

4. Eliminate θ between the equations
$$a^2 \cos^2 \theta = \frac{1-a^2}{3}, \quad \tan a = \tan^3 \frac{\theta}{2}.$$

Ans. $\sin^{\frac{2}{3}} a + \cos^{\frac{2}{3}} a = (2a)^{\frac{2}{3}}$.

MISCELLANEOUS EXAMPLES.

1. Shew that $\cos^2(45° - \theta) = \tfrac{1}{2}(1 + \operatorname{cosec} 2\theta)\sin 2\theta$.

2. If $m \tan(a - \theta)\sec^2\theta = n \tan\theta \sec^2(a - \theta)$, find θ.

   Ans. $\theta = \tfrac{1}{2}\left(a - \tan^{-1}\dfrac{n-m}{n+m}\tan a\right)$.

3. In any triangle prove that

   $$\cos\frac{A}{2} + \cos\frac{B}{2} + \cos\frac{C}{2} = 4\cos\left(45° - \frac{A}{4}\right)\cos\left(45° - \frac{B}{4}\right)\cos\left(45° - \frac{C}{4}\right).$$

4. From the figure of Art. 121 prove formulæ (186) and (187) geometrically.

5. In any triangle prove that $\tan\dfrac{A}{2} + \tan\dfrac{B}{2} = \cos\dfrac{C}{2}\sec\dfrac{B}{2}\sec\dfrac{A}{2}$.

6. Solve the equation $\sin 5x = \sin x + \cos 3x$.

   Ans. $x = (2n+1)\dfrac{\pi}{6}$.

7. If a, b, c, d be the sides of a quadrilateral circumscribed about a circle, shew that the area of the quadrilateral is $\sqrt{abcd}\sin\theta$, where 2θ is the sum of two opposite angles.

8. In any triangle shew that $\sin\dfrac{A}{2} + \sin\dfrac{B}{2} > \cos\dfrac{C}{2}$.

9. In the *ambiguous* case, a, b and A being given, if c_1, c_2 are the third sides of the two triangles, shew that the distance between the centres of the circumscribing circles is $\tfrac{1}{2}(c_1 - c_2)\operatorname{cosec} A$.

10. Prove that

    $$\left(1 + \tan\frac{x}{2} + \sec\frac{x}{2}\right)\left(1 + \tan\frac{x}{2} - \sec\frac{x}{2}\right) = 2\tan\frac{x}{2}.$$

11. Shew that $\cos 33° 45' = \tfrac{1}{2}\left\{2 + (2 - \sqrt{2})^{\frac{1}{2}}\right\}^{\frac{1}{2}}$.

MISCELLANEOUS EXAMPLES. 311

12. If $\tan(\theta-a)\tan(\theta+a) = \dfrac{1-2\cos 2a}{1+2\cos 2a}$, shew that

$$\theta = \tfrac{1}{2}\cos^{-1}(2\cos^2 2a).$$

13. If two circles whose radii are r, r_1, touch each other externally, and if θ is the angle contained by the two common tangents to these circles, shew that $\sin\theta = \dfrac{4(r-r_1)\sqrt{rr_1}}{(r+r_1)^2}$.

14. If r_1, r_2, r_3 be the radii of the circles inscribed in the quadrilaterals AO, BO, CO of the figure of Art. 152, prove that

$$\dfrac{r_1}{r-r_1}+\dfrac{r_2}{r-r_2}+\dfrac{r_3}{r-r_3}=\dfrac{s}{r},$$

where r is the radius of the inscribed circle.

15. In any triangle prove that

$$\tan^2\dfrac{A}{2}+\tan^2\dfrac{B}{2}+\tan^2\dfrac{C}{2}$$

$$=4R^2\left\{\left(\dfrac{\sin^2\dfrac{A}{2}}{\dfrac{a}{2}}\right)^2+\left(\dfrac{\sin^2\dfrac{B}{2}}{\dfrac{b}{2}}\right)^2+\left(\dfrac{\sin^2\dfrac{C}{2}}{\dfrac{c}{2}}\right)^2\right\}.$$

16. Shew that $\sec 72° - \sec 36° = \sec 60°$.

17. In the ambiguous case, prove that the circles which pass respectively through the middle points of the sides of the two triangles, are equal, and that their common chord is equal to half the side which is common to the triangles.

18. If the hypothenuse AB of a right-angled triangle be divided in D by a line which bisects the right angle, prove that

$$AD:BD :: 1-\tan\tfrac{1}{2}(A-B) : 1+\tan\tfrac{1}{2}(A-B).$$

19. If $a+\beta+\gamma=180°$, prove that

$$\cos\dfrac{a}{2}+\cos\dfrac{\beta}{2}+\cos\dfrac{\gamma}{2}=4\cos\dfrac{a+\beta}{4}\cos\dfrac{\beta+\gamma}{4}\cos\dfrac{a+\gamma}{4}.$$

20. Prove that $\sin 5° \sin 15° \sin 25° \ldots \sin 85° = 2^{-\tfrac{17}{2}}$.

21. In any triangle prove that
$$\frac{b^2-c^2}{a^2}\sin 2A + \frac{c^2-a^2}{b^2}\sin 2B + \frac{a^2-b^2}{c^2}\sin 2C = 0.$$

22. If r_1, r_2, r_3 denote the radii of the circles inscribed between the inscribed circle and the sides containing the angles A, B, C respectively, prove that
$$\sqrt{r_1 r_2} + \sqrt{r_1 r_3} + \sqrt{r_2 r_3} = r.$$

23. If $\sin^{-1}\theta + \sin^{-1}\dfrac{\theta}{2} = \dfrac{\pi}{4}$, then $\theta = \dfrac{2}{17}(5 - 2\sqrt{2})$.

24. In any triangle if $\cos A$, $\cos B$, $\cos C$ be in arithmetical progression, $s-a$, $s-b$ and $s-c$ are in harmonical progression.

25. Shew that $\tan^{-1} a - \tan^{-1} b = \tan^{-1}\dfrac{a-1}{a+1} - \tan^{-1}\dfrac{b-1}{b+1}$.

26. Solve the equation $\sin\left(\dfrac{a}{2}+\theta\right)\sin\dfrac{a}{2} = \cos^2\dfrac{\theta}{2}$.

    Ans. $\theta = n\pi - a$, where n is any odd integer.

27. Given $\tan\dfrac{\theta}{2} = \operatorname{cosec}\theta - \sin\theta$, then $\cos\theta = \tfrac{1}{2}(\sqrt{5}-1)$.

28. If the sines of the angles A, B, C of a triangle are in arithmetical progression, shew that
$$\sin\frac{A}{2} : \sin\frac{C}{2} :: \sin\frac{A-B}{2} : \sin\frac{B-C}{2}.$$

29. Given $\log 2$ and $\log 3$, find the logs of $\sqrt[3]{\dfrac{1}{6}}$ and 135.

30. If D is the middle point of the base BC of a triangle ABC, prove that
$$\sin BAD = \frac{b\sin A}{\sqrt{(b^2+c^2+2bc\cos A)}}.$$

31. If the centres of the escribed circles of a triangle be joined, prove that the distances of the centres of the escribed circles of this new triangle from the centre of its inscribed circle are
$$8R\sin\tfrac{1}{4}(B+C),\ 8R\sin\tfrac{1}{4}(A+C),\ 8R\sin\tfrac{1}{4}(A+B),$$
where R is the radius of the circle circumscribing the original triangle.

32. Through the centre O of the circumscribing circle of a triangle ABC, AOD is drawn to meet BC in D, prove that
$$OD : BD : CD = \cos A : \sin 2C : \sin 2B.$$

33. In any triangle prove that $\cot A = \dfrac{R(b^2 + c^2 - a^2)}{abc}$.

34. Prove that $\sin^{-1} \dfrac{2mn}{m^2 + n^2} + \sin^{-1} \dfrac{m^2 - n^2}{m^2 + n^2} = \dfrac{\pi}{2}$.

35. If p, q, r be the perpendiculars from the angles of a triangle ABC, upon the opposite sides a, b, c respectively, shew that
$$a \sin A + b \sin B + c \sin C = 2(p \cos A + q \cos B + r \cos C).$$

36. Shew that
$$\cos \theta - \tfrac{1}{2} \cos 2\theta + \tfrac{1}{3} \cos 3\theta - \&c., \text{ ad inf.} = \log_e \left(2 \cos \dfrac{\theta}{2}\right).$$

37. If $\cos^2 \beta \tan (a+\theta) = \sin^2 \beta \cot (a-\theta)$, find θ.
$$\textit{Ans. } \tan \theta = \sqrt{\tan (a+\beta) \tan (a-\beta)}.$$

38. If $\log_a b = m$ and $\log_b a = n$, shew that $\dfrac{\log_a m}{\log_b n} = \sqrt{\dfrac{m}{n}}$.

39. Prove that
$$\cos \left(\dfrac{\pi}{4} + x\sqrt{-1}\right) = \dfrac{1}{2\sqrt{2}} \left\{(e^x + e^{-x}) - \sqrt{-1}\,(e^x - e^{-x})\right\}.$$

40. Shew that $\log (x + y\sqrt{-1}) = \tfrac{1}{2} \log (x^2 + y^2) + \tan^{-1} \dfrac{y}{x} \sqrt{-1}$.

41. If c be the hypothenuse of a right-angled triangle whose sides are a and b, prove that
$$\log c = \tfrac{1}{2}(\log 2 + \log a + \log b) + M\left\{\left(\dfrac{a-b}{a+b}\right)^2 + \dfrac{1}{3}\left(\dfrac{a-b}{a+b}\right)^6 + \dfrac{1}{5}\left(\dfrac{a-b}{a+b}\right)^{10} + ..\right\}$$

42. Solve the equation $\cos^{-1} \dfrac{\sqrt{3}}{4} + \sec^{-1} x = \dfrac{\pi}{3}$.
$$\textit{Ans. } x = \dfrac{\sqrt{38}}{36} (\sqrt{39} - \sqrt{3}).$$

43. In any triangle prove that
$$\sin^2 A + \sin^2 B + \sin^2 C - 2 \cos A \cos B \cos C = 2.$$

44. If the tangents of the angles of a triangle are in a geometrical progression whose ratio is n, prove that

$$\sin 2C = n \sin 2A.$$

45. The shadows of two vertical walls at right angles to each other, whose height are a and b feet, are observed at noon to be c and d feet respectively in breadth; shew that if a be the sun's altitude and β the inclination of the first wall to the meridian,

$$\cot a = \sqrt{\left(\frac{c^2}{a^2} + \frac{d^2}{b^2}\right)}, \quad \cot \beta = \frac{ad}{bc}.$$

46. Sum to n terms $\cos x + \cos 2x + \cos 3x + \ldots$

$$Ans. \ s = \frac{\sin \dfrac{nx}{2}}{\sin \dfrac{x}{2}} \cos \frac{(n+1)x}{2}.$$

47. Given the perimeter $=2s$, the area $= \Delta$, and the angle A of a triangle, find a.

$$Ans. \ a = \frac{s^2 - \Delta \cot \dfrac{A}{2}}{s}.$$

48. Given the area $= \Delta$, the angle C and $a+b=m$, find the sides of the triangle.

$$Ans. \ a = \tfrac{1}{2}(m + \sqrt{m^2 - 8\Delta \operatorname{cosec} C}).$$

49. Given R, r and p the perpendicular from C on c, to solve the triangle.

$$Ans. \ \cos \tfrac{1}{2}(A+B) = \frac{r \sec \theta}{\sqrt{2Rp}}, \ \cos \tfrac{1}{2}(A-B) = \frac{(p-r)\sec \theta}{\sqrt{2Rp}}.$$

$$\text{where } \sin \theta = \sqrt{\frac{2r}{p}}.$$

50. Given the perimeter $=2s$, C and p the perpendicular from C on c, to solve the triangle.

$$Ans. \ c = s \cos^2 \theta, \text{ where } \tan^2 \theta = \frac{p}{2s} \cot \frac{C}{2}.$$

51. If a, β, γ be the distances between the centres of the escribed

MISCELLANEOUS EXAMPLES. 315

circles, and a', β', γ' the distances between the centre of the inscribed circle and the centres of the escribed circles, prove that

$$\frac{a'\beta'\gamma'}{a\beta\gamma} = \frac{r}{s}, \text{ and that}$$

$$a' : \beta' : \gamma' :: \frac{r_1 - r}{a} \cos \frac{A}{2} : \frac{r_2 - r}{b} \cos \frac{B}{2} : \frac{r_3 - r}{c} \cos \frac{C}{2}.$$

52. If z and z' be the meridional zenith distances of the moon or a planet as seen from two observatories on the same meridian, and whose latitudes are ϕ and ϕ' respectively, one being N and the other S, r the earth's radius, D the distance of the moon or planet and P the horizontal parallax, shew that

$$D = r \frac{\sin z + \sin z'}{(z + z') - (\phi + \phi')},$$

$$P = \frac{(z + z') - (\phi + \phi')}{\sin z + \sin z'}.$$

53. If a quadrilateral can be inscribed in a circle and can also have a circle described about it, the area of the quadrilateral is equal to the square root of the product of the four sides.

54. Shew that, if Q and G be the centre of the circumscribed circle and the intersection of the perpendiculars of a triangle,

$$QG^2 = R^2(1 - 8 \cos A \cos B \cos C).$$

55. If a, β, γ, δ be the angles of a quadrilateral, prove that

$\cos a + \cos \beta + \cos \gamma + \cos \delta = 4 \cos \frac{1}{2}(a + \beta) \cos \frac{1}{2}(\beta + \gamma) \cos \frac{1}{2}(a + \gamma).$

56. Shew that

$(1 + \sec 2x)(1 + \sec 4x)(1 + \sec 8x) \ldots (1 + \sec 2^n x) = \tan 2^n x \cot x.$

57. If $\tan^{-1}(x+1)\sqrt{2} - \tan^{-1}\frac{x-1}{\sqrt{2}} = \cot^{-1} 4\sqrt{2}$, find x.

Ans. 6 or -2.

58. In any triangle if a^2, b^2, c^2 be in arithmetical progression, shew that

$$\frac{\sin 3B}{\sin B} = \left(\frac{a^2 - c^2}{2ac}\right)^2.$$

59. Prove that $2^n \cos \theta \cos 2\theta \cos 2^2\theta \ldots \cos 2^n \theta$
$= \cos \theta + \cos 3\theta + \cos 5\theta + \ldots + \cos (2^{n+1} - 1)\theta.$

60. If
$$\tan A \sec A + \tan B \sec B + \tan C \sec C + 2 \tan A \tan B \tan C = 0,$$
prove that
$$\sec^2 A + \sec^2 B + \sec^2 C = 1 \pm 2 \sec A \sec B \sec C.$$

61. Straight lines AO, BO, CO are drawn bisecting the sides BC, AC, AB of a triangle in the points D, E, F respectively; if r_1, r_2, r_3 are the radii of the circles circumscribed about the triangles EOF, FOD, DOE, shew that
$$\frac{OA^2 \cdot r_1^2}{a^2} = \frac{OB^2 \cdot r_2^2}{b^2} = \frac{OC^2 \cdot r_3^2}{c^2} = \frac{1}{3} \cdot \frac{a^2+b^2+c^2}{\dfrac{a^2}{r_1^2}+\dfrac{b^2}{r_2^2}+\dfrac{c^2}{r_3^2}}.$$

62. If $\cos \theta = \dfrac{a}{\sqrt{a^2+b^2}}$, shew that
$$\sqrt[3]{(a+b\sqrt{-1})} + \sqrt[3]{(a-b\sqrt{-1})} = 2\sqrt[6]{(a^2+b^2)} \cos \frac{\theta}{3}.$$

63. Sum to infinity $\dfrac{\sin \theta}{2} + \dfrac{\sin 2\theta}{2^2} + \dfrac{\sin 3\theta}{2^3} + \ldots$

$$Ans. \quad \frac{4}{9 \tan \dfrac{\theta}{2} + \cot \dfrac{\theta}{2}}.$$

64. If A, B, C, D are the angles of a quadrilateral, shew that
$$\frac{\tan A + \tan B + \tan C + \tan D}{\cot A + \cot B + \cot C + \cot D} = \tan A \cdot \tan B \cdot \tan C \cdot \tan D.$$

65. If a and β be the roots of the equation $x^2 - px + q = 0$, prove that
$$\tan^{-1} a + \tan^{-1} \beta = \tan^{-1} \frac{p}{1-q}.$$

66. Find θ from the equation
$$\sin (3\theta + a) + \cos (3\theta - a) = \cos (45° + \theta)$$
$$Ans. \quad \theta = 45° \text{ and } \sin 2\theta = \tfrac{1}{2}\left(\tfrac{1}{2} \operatorname{cosec}(45°+a) - 1\right).$$

67. If $\phi = \dfrac{\pi}{13}$, shew that
$$\cos \phi + \cos 3\phi + \cos 5\phi + \ldots + \cos 11\phi = \frac{1}{2}.$$

MISCELLANEOUS EXAMPLES. 317

68. Prove the formulæ,

$$1 \pm \sin\theta = 2\sin^2\left(\frac{\pi}{4} \pm \frac{\theta}{2}\right), \quad \sin^2\theta\frac{\cos 3\theta}{3} + \cos^2\theta\frac{\sin 3\theta}{3} = \frac{\sin 4\theta}{4}.$$

69. If $\sin(x+y\sqrt{-1}) = \beta(\cos a + \sin a\sqrt{-1})$, shew that

$$\tan a = \frac{e^y - e^{-y}}{e^y + e^{-y}}\cot x \text{ and } \beta^2 = \tfrac{1}{4}(e^{2y} + e^{-2y} - 2\cos 2x).$$

70. The area of a regular polygon circumscribed about a circle is a harmonic mean between the area of an inscribed regular polygon of the same number of sides and a circumscribed polygon of half the number of sides.

71. If $3\sin\theta = \sin(a-\theta)$, shew that

$$\theta = \frac{1}{3}\sin a - \frac{1}{2}\cdot\frac{1}{3^2}\sin 2a + \frac{1}{3}\cdot\frac{1}{3^3}\sin 3a - \&c.$$

72. Shew that

$$e^{-x}\cos\theta - \tfrac{1}{3}e^{-3x}\cos 3\theta + \tfrac{1}{5}e^{-5x}\cos 5\theta - \ldots \text{ ad inf.} = \tfrac{1}{2}\tan^{-1}\frac{2\cos\theta}{e^x - e^{-x}}.$$

73. Prove that $\log_a m = \log_a b \cdot \log_b c \cdot \log_c d \ldots \log_l m$.

74. Shew that

$$\left(\tan\frac{\theta}{3} + \tan\frac{2\pi+\theta}{3} + \tan\frac{4\pi+\theta}{3}\right)\left(\cot\frac{\theta}{3} + \cot\frac{2\pi+\theta}{3} + \cot\frac{4\pi+\theta}{3}\right) = 9.$$

75. A common tangent is drawn to two circles which touch each other and whose radii are r and $3r$; shew that the area of the curvilinear triangle bounded by the common tangent and the two circles is

$$\left(4\sqrt{3} - \frac{11}{6}\pi\right)r^2.$$

76. Shew that $\sin^2(x+y) + \cos^2(x-y) = 1 + \sin 2x \sin 2y$.

77. Given $\tan(\pi\cot\theta) = \cot(\pi\tan\theta)$, find θ.

Ans. $\tan\theta = \tfrac{1}{2}\left(2n+1 \pm \sqrt{4n(n+1)-15}\right).$

78. Shew that when n is an odd integer,

$$\cos n\left(\frac{\pi}{2} - x\right) = (-1)^{\frac{n-1}{2}}\sin nx.$$

79. Shew that

$$(\sec x - \tan x)^2 \left(\sin \frac{x}{2} + \cos \frac{x}{2}\right)^2 = \left(\sin \frac{x}{2} - \cos \frac{x}{2}\right)^2.$$

80. In a right-angled triangle, given $a+b-c=m$ and the angle A, shew that
$$a = \frac{m}{\sqrt{2}} \cos \frac{A}{2} \cosec \left(45° - \frac{A}{2}\right).$$

81. If straight lines be drawn from any point in the circumference of a circle to the angular points of an inscribed regular polygon of n sides, shew that the sum of the squares of these lines $= 2n \times (\text{radius})^2$.

82. Solve the equation $x^3 - \frac{3}{2}x = \frac{\sqrt{5}+1}{4\sqrt{2}}$.

Ans. $\sqrt{2} \cos 12°$, $-\sqrt{2} \cos 48°$, $-\sqrt{2} \cos 72°$.

83. A circle is inscribed in an equilateral triangle; an equilateral triangle in the circle; a circle again in the latter triangle, and so on: if r, r_1, r_2, r_3, &c., be the radii of the circles, prove that
$$r = r_1 + r_2 + r_3 + \&c., \text{ ad inf.}$$

84. The radii of two circles which intersect are r, r_1, and a the distance between their centres; prove that the common chord
$$= \frac{1}{a} \left\{ (r+r_1+a)(r+r_1-a)(r-r_1+a)(-r+r_1+a) \right\}^{\frac{1}{2}}.$$

85. Prove that in any triangle $\tan \left(B + \frac{A}{2}\right) = \frac{c+b}{c-b} \tan \frac{A}{2}$.

86. Find the value of $\cos^{-1} \frac{1}{2}(-1)^m$, m being any integer, and express the different values by one formula.

Ans. $\left\{6n \pm \frac{1}{2}(3-(-1)^m)\right\} \frac{\pi}{3}$.

87. If p be the perpendicular from the angle A of a triangle upon the opposite side, shew that
$$p = \frac{bc}{2} \cdot \frac{a \sin A + b \sin B + c \sin C}{bc \cos A + ac \cos B + ab \cos C}.$$

88. The straight lines which bisect the angles A and B of a tri-

angle ABC, meet the opposite sides in D and E respectively; shew that the area of the triangle CED is

$$\Delta \sin\frac{A}{2}\sin\frac{B}{2}\sec\tfrac{1}{2}(C-A)\sec\tfrac{1}{2}(C-B),$$

where Δ is the area of the triangle ABC.

89. In any triangle prove that

$$\frac{a^2\cos\tfrac{1}{2}(B-C)}{\cos\tfrac{1}{2}(B+C)}+\frac{b^2\cos\tfrac{1}{2}(C-A)}{\cos\tfrac{1}{2}(C+A)}+\frac{c^2\cos\tfrac{1}{2}(A-B)}{\cos\tfrac{1}{2}(A+B)}=2(ab+bc+ac).$$

90. If the straight lines which bisect the angles A, B, C of a triangle ABC meet the opposite sides in D, E, F respectively, shew that the area of the triangle DEF is

$$2\Delta\sin\frac{A}{2}\sin\frac{B}{2}\sin\frac{C}{2}\sec\frac{B-C}{2}\sec\frac{C-A}{2}\sec\frac{A-B}{2},$$

where Δ is the area of the triangle ABC.

91. If p_1, p_2, p_3 denote the perpendiculars drawn from the centre of the circumscribed circle of a triangle to the sides a, b, c respectively, shew that

$$4\left(\frac{a}{p_1}+\frac{b}{p_2}+\frac{c}{p_3}\right)=\frac{abc}{p_1 p_2 p_3},$$

and $\quad 4\left(p_1^2+p_2^2+p_3^2\right)=a^2\cot^2 A+b^2\cot^2 B+c^2\cot^2 C.$

92. Shew by means of a trigonometrical formula that if $a+b+c=abc$, then

$$\frac{2a}{1-a^2}+\frac{2b}{1-b^2}+\frac{2c}{1-c^2}=\frac{2a}{1-a^2}\cdot\frac{2b}{1-b^2}\cdot\frac{2c}{1-c^2}.$$

93. Solve the equation

$$\tan x+2\cot 2x=\sin x\left(1+\tan x\tan\frac{x}{2}\right).$$

Ans. $x=(2n+1)\dfrac{\pi}{4}.$

94. If x be the length of the line which bisects the angle A of a triangle, and is terminated by the base, ϕ the angle it makes with the base, shew that

$$x=\frac{(a+b+c)\left(\sin\phi-\sin\dfrac{A}{2}\right)}{2\sin\phi\cos\dfrac{A}{2}}.$$

95. In the Figure of Art. 153, shew that the area of the triangle $O_1 O_2 O_3$ is

$$\frac{abc}{2} \left\{ \left(\frac{1}{a}+\frac{1}{b}\right) \tan \frac{C}{2} + \left(\frac{1}{a}+\frac{1}{c}\right) \tan \frac{B}{2} + \left(\frac{1}{b}+\frac{1}{c}\right) \tan \frac{A}{2} \right\}.$$

96. In the Figure of Art. 153, join OO_3 and shew that the area of the triangle OO_1O_2 is $\dfrac{abc}{a+b+c} \cot \dfrac{C}{2}$.

97. Shew that $\dfrac{1}{\pi} = \dfrac{1}{2^2} \tan \dfrac{\pi}{2^2} + \dfrac{1}{2^3} \tan \dfrac{\pi}{2^3} + \dfrac{1}{2^4} \tan \dfrac{\pi}{2^4} + \ldots$

98. If a, b, c, d denote the sides of a quadrilateral, and ϕ the sum of two opposite angles, shew that

$$(\text{area})^2 = (s-a)(s-b)(s-c)(s-d) - abcd \cos^2 \frac{\phi}{2}.$$

where $2s = a+b+c+d$.

99. If P denote the point of intersection of the perpendiculars from the angles of a triangle on the opposite sides, prove that

$$PA^2 = 4R^2 - a^2.$$

100. Shew that the area of the triangle formed by joining the points of contact of the inscribed circle, or an escribed circle of a triangle is $\dfrac{R_c \Delta}{2R}$, where R_c is the radius of the circle.

101. If a, b, c, d be the sides of a trapezium taken in order, and a the acute angle between the diagonals, shew that the area of the trapezium is $\frac{1}{4}\{(a^2+c^2)\sim(b^2+d^2)\} \tan a$, and examine this result when $a = \dfrac{\pi}{2}$.

102. If p, q, r be the perpendiculars drawn from the angles A, B, C to the opposite sides, shew that

$$a \cdot \frac{\frac{1}{q}+\frac{1}{r}}{\frac{1}{b}+\frac{1}{c}} = b \cdot \frac{\frac{1}{p}+\frac{1}{r}}{\frac{1}{a}+\frac{1}{c}} = c \cdot \frac{\frac{1}{p}+\frac{1}{q}}{\frac{1}{a}+\frac{1}{b}}.$$

103. If a and b be the opposite and parallel sides of a trapezium, a being the greater, and θ and ϕ the acute angles at the extremity of a, shew that the area is $\frac{1}{2}(a^2 - b^2) \sin \theta \sin \phi \csc(\theta+\phi)$.

104. Shew that $\dfrac{1}{1^2}+\dfrac{1}{2^2}+\dfrac{1}{3^2}+\dfrac{1}{4^2}+$ &c., ad inf. $=\dfrac{\pi^2}{6}$.

105. In any triangle prove that
$$\Delta=\sin\frac{A}{2}\sin\frac{B}{2}\sin\frac{C}{2}\left(\frac{a^2}{\sin A}+\frac{b^2}{\sin B}+\frac{c^2}{\sin C}\right).$$

106. If a and β be two different values of θ which satisfy the equation $\dfrac{\cos\theta}{a}+\dfrac{\sin\theta}{b}=\dfrac{1}{c}$, shew that

$$a\cos\tfrac{1}{2}(a+\beta)=b\sin\tfrac{1}{2}(a+\beta)=c\cos\tfrac{1}{2}(a-\beta).$$

107. The angles A, B, C of a triangle are in descending order of magnitude; if another triangle be constructed having two of its angles $(A-B), (B-C)$ and sides m, n, p, then will $an+cm=bp$.

108. In any right-angled triangle, C being the right-angle, prove that $a^3\cos A+b^3\cos B=abc$.

109. Three circles are so inscribed in a triangle that each touches the other two and two sides of the triangle; prove that the radius of that circle which touches the sides AB, AC, is

$$\frac{r}{2}\left\{\frac{\left(1+\tan\dfrac{B}{4}\right)\left(1+\tan\dfrac{C}{4}\right)}{1+\tan\dfrac{A}{4}}\right\},$$

where r is the radius of the inscribed circle of the triangle.

110. If all the angular points of a regular polygon be joined, and r be the radius of the circumscribed circle, prove that the sum of all the lines including the perimeter of the polygon is $nr\cot\dfrac{\pi}{2n}$.

111. If R, r be the radii of the circumscribed and inscribed circles of a regular polygon of n sides, and R', r' the corresponding radii for a regular polygon of $2n$ sides, and having the same perimeter as the former, shew that $R+r=2r'$ and $Rr'=R'^2$.

112. Solve the equation $\cos 7\theta+7\cos\theta=0$.

Ans. $\theta=(2n+1)\dfrac{\pi}{2}$.

113. In any triangle ABC, if the lines which bisect the angles A, B, C and terminate in the opposite sides be denoted by α, β, γ respectively, prove that

$$\frac{a+b+c}{\alpha\beta\gamma} = R\left(\frac{1}{a}+\frac{1}{b}\right)\left(\frac{1}{a}+\frac{1}{c}\right)\left(\frac{1}{b}+\frac{1}{c}\right),$$

and
$$\frac{\alpha\beta\gamma}{abc} = 4\Delta \cdot \frac{a+b+c}{(a+b)(b+c)(a+c)}.$$

114. Sum to infinity

$$(3^{\frac{1}{2}}-1)\,3^{-\frac{1}{2}} - \frac{1}{3}(3^{\frac{3}{2}}-1)\,3^{-\frac{3}{2}} + \frac{1}{5}(3^{\frac{5}{3}}-1)\,3^{-\frac{5}{3}} - \&c.$$

Ans. $\dfrac{\pi}{12}$.

115. In any triangle prove that

$$a^2+b^2+c^2+r^2+r_1^2+r_2^2+r_3^2 = 4R^2.$$

116. In any triangle, the square of the distance between the centre of the inscribed circle and the intersection of the perpendiculars is $4R^2 - 2Rr - \dfrac{a^3+b^3+c^3}{a+b+c}$.

117. If h and k be the diagonals of a quadrilateral and ϕ their angle of intersection, shew that the area is $\frac{1}{2}hk \sin\phi$.

118. Shew that $\tan^{-1}\dfrac{\sqrt{2}+1}{\sqrt{2}-1} - \tan^{-1}\dfrac{1}{\sqrt{2}} = \dfrac{\pi}{4}$.

119. Eliminate θ from the equations

$$3a\cos\theta + a\cos 3\theta = 4m$$
$$3a\sin\theta - a\sin 3\theta = 4n$$

Result, $a^2 - m^2 - n^2 = 3a^{\frac{2}{3}} m^{\frac{2}{3}} n^{\frac{2}{3}}$

120. Solve the equation $\tan^{-1}\left(\dfrac{1}{x-1}\right) - \tan^{-1}\left(\dfrac{1}{x+1}\right) = \dfrac{\pi}{2}$.

Ans. $x = \pm \dfrac{2}{\sqrt{3}-1}$.

121. Prove that $\theta + \tan^{-1}(\cot 2\theta) = \tan^{-1}(\cot\theta)$.

MISCELLANEOUS EXAMPLES.

122. In any triangle prove that

$$\cos^2\frac{A}{2}+\cos^2\frac{B}{2}+\cos^2\frac{C}{2}=2+\frac{r}{2R}.$$

123. Sum to n terms

$$\tan^{-1} x+\tan^{-1}\frac{x}{1+1.2x^2}+\tan^{-1}\frac{x}{1+2.3x^2}+\&c.$$

Ans. $\tan^{-1} nx.$

124. Shew that $\sin^3\theta\cos^2\theta = \frac{1}{8}\sin\theta - \frac{1}{16}\sin 5\theta + \frac{1}{16}\sin 3\theta.$

125. Shew that $\sin 9° = \frac{1}{4}\{\sqrt{3+\sqrt{5}}-\sqrt{5-\sqrt{5}}\}.$

126. Given $\sin 2(a+\theta)+\sin 2a = 2\sin 2\theta$, find θ.

Ans. $\theta = \tan^{-1}(3\tan a) - a.$

127. In any triangle prove that

$$a\cos^2\frac{A}{2}+b\cos^2\frac{B}{2}+c\cos^2\frac{C}{2}=\Delta\left(\frac{1}{R}+\frac{1}{r}\right).$$

128. In the figure of Art. 152, prove that
$$OA \cdot OB \cdot OC \text{(...)} = 4\text{(...)} \cdot BD \cdot CE.$$

129. In the figure of Art. 153, prove that if R_1, R_2, R_3 denote the radii of the circles described about the triangles O_1BC, O_2AC, O_3AB respectively, $R_1R_2R_3 = 2R^2r$.

130. A circle whose radius is r is inscribed in a quadrilateral; if t_1, t_2, t_3, t_4 denote the tangents into which the sides of the quadrilateral are divided at the points of contact, prove that

$$\frac{t_1t_2t_3t_4}{r^2}=\frac{t_1+t_2+t_3+t_4}{t_1^{-1}+t_2^{-1}+t_3^{-1}+t_4^{-1}}.$$

131. The line joining the tops of two towers of unequal height makes an angle a with the horizontal plane on which they stand, and the distance between the extremities of their shadows when the sun is in the same vertical plane as the towers is h; if β be the sun's altitude, shew that the distance between them is

$$h\cos a\sin\beta\csc(a-\beta).$$

132. If α, β, γ denote the perpendiculars drawn from the centre of the circumscribing circle of a triangle to its sides, shew that the radius of this circle is the positive value of R in the equation

$$R^3 - (\alpha^2 + \beta^2 + \gamma^2) R - 2\alpha\beta\gamma = 0.$$

Shew also that $\dfrac{\alpha\beta}{ab} + \dfrac{\beta\gamma}{bc} + \dfrac{\alpha\gamma}{ac} = \dfrac{1}{4}$, where a, b, c are the sides of the triangle.

133. In any triangle shew that

$$\tan^2 \frac{A}{2} + \tan^2 \frac{B}{2} + \tan^2 \frac{C}{2} > 1,$$

$(a+b) \cos C + (a+c) \cos B + (b+c) \cos A > 2 \times$ any side.

134. Find θ from the equation

$$5 \cos^2 \frac{\theta}{2} = \left(\cos \frac{\theta}{2} - \sqrt{2} \sin \frac{\theta}{2}\right)^2 - (\sqrt{2} - 1).$$

Ans. $\tan\left(45° + \dfrac{\theta}{2}\right) = \pm \dfrac{3}{\sqrt{2}}.$

135. In any triangle prove that

$$R = \frac{b - a \cos C}{2 \cos A \sin C} = \frac{a^2 - b^2}{2c \sin (A - B)}.$$

136. If a, b, c be in arithmetical progression, the arithmetical mean between the logs of a and c is

$$\log_e b - \frac{1}{2} \left\{ \left(\frac{c-a}{2b}\right)^2 + \frac{1}{2}\left(\frac{c-a}{2b}\right)^4 + \frac{1}{3}\left(\frac{c-a}{2b}\right)^6 + \ldots \right\}.$$

137. Find α and β from the equations

$$2 \tan \tfrac{1}{2}(\alpha + \beta) = \frac{1 + \sqrt{2}}{\cos \alpha + \cos \beta},$$

$$2 \tan \tfrac{1}{2}(\alpha - \beta) = \frac{1 - \sqrt{2}}{\cos \alpha + \cos \beta}.$$

Ans. $\alpha = 30°, \beta = 45°.$

138. Shew that $2 \sin^{-1} \dfrac{1}{\sqrt{10}} + \sin^{-1} \dfrac{1}{\sqrt{50}} = \dfrac{\pi}{4}.$

MISCELLANEOUS EXAMPLES.

139. Sum the series

$$1+\cos x+\frac{1}{2}\cos 2x+\frac{1}{1.2.3}\cos 3x+\frac{1}{1.2.3.4}\cos 4x+\ldots\text{ad inf.}$$

Ans. $e^{\cos x}\cos(\sin x)$.

140. If α, β, γ be the distances of the centre of the inscribed circle of a triangle from the angles, and if $\alpha^{-2}+\gamma^{-2}=2\beta^{-2}$, then r_1, r_2, r_3 are in arithmetical progression.

141. Shew that $\dfrac{\pi}{2\sqrt{2}}=1+\dfrac{1}{3}-\dfrac{1}{5}-\dfrac{1}{7}+\dfrac{1}{9}+\dfrac{1}{11}-$ &c.

142. If the bisectors of the angles of a triangle ABC be produced to meet the circumference of the circumscribed circle in the points D, E, F, and if p, q, r denote the sides of the triangle DEF, prove that $R^2=\dfrac{pqr}{a+b+c}$, where R is the radius of the circumscribed circle.

143. If α, β, γ be the three values of x (unequal) which satisfy the condition

$$\frac{a}{\cos x}+\frac{b}{\sin x}+c=0,$$

shew that $\dfrac{\tan\alpha}{\tan\gamma}=\dfrac{\tan\frac{1}{2}(\beta+\gamma)}{\tan\frac{1}{2}(\alpha+\beta)}$.

144. If a circle be inscribed in a triangle and the points of contact A', B', C' joined, and if R_1, R_2, R_3 be the radii of the circumscribed circles of the triangles $AB'C'$, $BA'C'$, $CA'B'$ respectively, prove that

$$R_1:R_2:R_3=\operatorname{cosec}\frac{A}{2}:\operatorname{cosec}\frac{B}{2}:\operatorname{cosec}\frac{C}{2}.$$

145. If ABC be a triangle, and the equation

$$x^2+y^2+z^2+2yz\cos 2A+2zx\cos 2B+2xy\cos 2C=0$$

be satisfied by real values of x, y, z, then

$$\frac{x}{\sin 2A}=\frac{y}{\sin 2B}=\frac{z}{\sin 2C}.$$

146. In any triangle prove that

$\sin 2A \ (b \cos C - c \cos B)^2 + \text{anal.} + \ldots$
$= 2 \cos A \cos B \cos C \ (a^2 \tan A + \text{anal.} + \ldots).$

147. Adapt $\cot \dfrac{\theta}{2} = a(1 + \cos \alpha + \cos \beta + \cos \gamma)$ to logarithmic computation by the use of an auxiliary angle.

Ans. $\cot \dfrac{\theta}{2} = 4a \cos \dfrac{a}{2} \cos \dfrac{1}{4}(a + \phi) \cos \dfrac{1}{4}(a - \phi)$

where $\cos \dfrac{\phi}{2} = \dfrac{\cos \dfrac{1}{2}(\beta + \gamma) \cos \dfrac{1}{2}(\beta - \gamma)}{\cos \dfrac{a}{2}}$.

148. If p_1, p_2, p_3 be the perpendiculars from any point within a triangle upon the sides, P_1, P_2, P_3 the perpendiculars from the angles upon the same sides respectively, shew that

$$\dfrac{p_1}{P_1} + \dfrac{p_2}{P_2} + \dfrac{p_3}{P_3} = 1.$$

149. Prove that the sum of the squares of the ten lines joining the centres of the five circles, four of which touch the sides of the triangle, and the remaining one is the circumscribed circle, is equal to 15 times the square of the diameter of the circumscribed circle.

150. If perpendiculars be drawn from the bisections of the sides of a triangle to the circumference of the circumscribed circle, shew that the sum of these perpendiculars is equal to $2R - r$.

151. In problem 132, if $a = 3\frac{1}{3}$, $\beta = 4\frac{1}{3}$, $\gamma = 4\frac{7}{8}$, find R.   *Ans.* $8\frac{1}{8}$.

152. In any triangle shew that

$$\cos^{-1}\left(\dfrac{r_1 r_2}{ab}\right)^{\frac{1}{2}} + \cos^{-1}\left(\dfrac{r_2 r_3}{bc}\right)^{\frac{1}{2}} + \cos^{-1}\left(\dfrac{r_1 r_3}{ac}\right)^{\frac{1}{2}} = \dfrac{\pi}{2}.$$

153. Three circles whose radii are r_1, r_2, r_3 touch one another, O_1, O_2, O_3 being their centres and A the point of intersection of their common tangents at the points of contact; if a_1, a_2, a_3 denote

the distances AO_1, AO_2, AO_3 respectively, and R the radius of the circle circumscribing the triangle $O_1O_2O_3$, prove that

$$R = \frac{1}{4} \cdot \frac{a_1 a_2 a_3}{r_1 r_2 r_3}(r_1 + r_2 + r_3).$$

154. A circle is inscribed in a triangle, and any three triangles are cut off by tangents to the circle; if r_1, r_2, r_3 be the radii of the circles inscribed in these three triangles, then the sum of the areas of the triangles is

$$\frac{a}{2}(-r_1 + r_2 + r_3) + \frac{b}{2}(r_1 - r_2 + r_3) + \frac{c}{2}(r_1 + r_2 - r_3),$$

where a, b, c are the sides of the original triangle.

155. Sum the series ad infinitum

$$\cos\theta + \frac{\cos\theta}{1}\cos 2\theta + \frac{\cos^2\theta}{1.2}\cos 3\theta + \frac{\cos^3\theta}{1.2.3}\cos 4\theta + \ldots.$$

$$Ans.\ e^{\cos^2\theta}\cos(\theta + \tfrac{1}{2}\sin 2\theta).$$

156. If a, β, γ be the lengths of the lines which join the feet of the perpendiculars from the angles of a triangle on the opposite sides, shew that

$$\frac{a}{a^2} + \frac{\beta}{b^2} + \frac{\gamma}{c^2} = \frac{a^2 + b^2 + c^2}{2abc},$$

where a, b, c are the lengths of the sides opposite to the lines a, β, γ respectively.

157. In any triangle if a, b, c, be in arithmetical progression, shew that r_1, r_2, r_3, the radii of the three escribed circles which touch the sides a, b, c respectively, are in harmonical progression.

APPENDIX.

Geometrical Demonstrations of (45) and (46).

1. We will here add a few geometrical demonstrations of the very important formulæ of Arts. 45–50.

I.—*When $(x+y)$ is an angle in the second quadrant.*

In Fig. 1 let the angle $BAC = x$ and the angle $CAD = y$. Construct the Figure as in Art. 45.

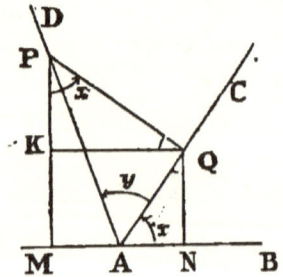

$$\sin(x+y) = \frac{PM}{AP}$$

$$= \frac{QN}{AP} + \frac{PK}{AP}$$

$$= \frac{QN}{AQ} \cdot \frac{AQ}{AP} + \frac{PK}{PQ} \cdot \frac{PQ}{AP}$$

$$= \sin x \cos y + \cos x \sin y.$$

$$\cos(x+y) = -\frac{AM}{AP}$$

$$= \frac{AN}{AP} - \frac{KQ}{AP}$$

$$= \frac{AN}{AQ} \cdot \frac{AQ}{AP} - \frac{KQ}{PQ} \cdot \frac{PQ}{AP}$$

$$= \cos x \cos y - \sin x \sin y.$$

II.—*When $(x+y)$ is an angle in the third quadrant.*

In Fig. 2 let the angle $BAC=x$ and the angle $CAD=y$. Construct the Figure as in the preceding diagram, then we have

$\sin(x+y) = -\dfrac{PM}{AP}$

$= -\dfrac{QN}{AP} - \dfrac{PK}{AP}$

$= -\dfrac{QN}{AQ} \cdot \dfrac{AQ}{AP} - \dfrac{PK}{PQ} \cdot \dfrac{PQ}{AP}$

$= -\sin QAN \cos PAQ - \cos QPK \sin PAQ$

$= -\sin(180°-x)\cos(180°-y) - \cos(180°-x)\sin(180°-y)$

$= \sin x \cos y + \cos x \sin y.$ (Art. 39.)

$\cos(x+y) = -\dfrac{AM}{AP}$

$= \dfrac{AN}{AP} - \dfrac{KQ}{AP}$

$= \dfrac{AN}{AQ} \cdot \dfrac{AQ}{AP} - \dfrac{KQ}{PQ} \cdot \dfrac{PQ}{AP}$

$= \cos QAN \cos PAQ - \sin QPK \sin PAQ$

$= \cos(180°-x)\cos(180°-y) - \sin(180°-x)\sin(180°-y)$

$= \cos x \cos y - \sin x \sin y.$ (Art. 39.)

The student will observe that in the quadrilateral $AMPQ$, the opposite angles QAM, QPM are together equal to two right angles, and since the angle $QAM=x$, the angle $QPK=180°-x$.

III.—*When $(x+y)$ is an angle in the fourth quadrant.*

In Fig. 3 let the angle $BAC=x$ and the angle $CAD=y$. Construct the Figure as in the two preceding diagrams, then we have

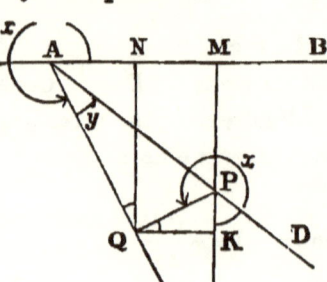

$$\sin(x+y) = -\frac{PM}{AP}$$

$$= -\frac{QN}{AP} + \frac{PK}{AP}$$

$$= -\frac{QN}{AQ}\cdot\frac{AQ}{AP} + \frac{PK}{PQ}\cdot\frac{PQ}{AP}$$

$$= -\sin QAN \cos PAQ + \cos QPK \sin PAQ$$

$$= -\sin(360°-x)\cos y + \cos(360°-x)\sin y$$

$$= \sin x \cos y + \cos x \sin y.$$

$$\cos(x+y) = \frac{AM}{AP}$$

$$= \frac{AN}{AP} + \frac{NM}{AP}$$

$$= \frac{AN}{AQ}\cdot\frac{AQ}{AP} + \frac{KQ}{PQ}\cdot\frac{PQ}{AP}$$

$$= \cos QAN \cos PAQ + \sin QPK \sin PAQ$$

$$= \cos(360°-x)\cos y + \sin(360°-x)\sin y$$

$$= \cos x \cos y - \sin x \sin y.$$

Here the opposite angles QAM, QPM of the quadrilateral $AMPQ$, are together equal to two right angles; therefore the exterior angle QPK is equal to the angle QAN, but the angle $QAN=360°-x$, hence the angle $QPK=360°-x$.

APPENDIX. 331

2. The geometrical proof of sin $(x-y)$ and cos $(x-y)$ in each of the quadrants is similar to those of sin $(x+y)$ and cos $(x+y)$ which we have just given. We will, however, give the proof of one case, viz., when x is an angle in the fourth quadrant and y an angle in the first quadrant, $x-y$ being, in th Figure, an angle in the *third* quadrant.

In Fig. 4 let the angle $BAC = x$ and the angle $DAC = y$. Construct the Figure as in Art. 48, then we have

$$\sin(x-y) = -\frac{PM}{AP}$$

$$= -\frac{NQ}{AP} - \frac{KQ}{AP}$$

$$= -\frac{NQ}{AQ} \cdot \frac{AQ}{AP} - \frac{KQ}{PQ} \cdot \frac{PQ}{AP}$$

$$= -\sin QAN \cos PAQ - \cos QAN \sin PAQ$$

$$= -\sin(360° - x)\cos y - \cos(360° - x)\sin y$$

$$= \sin x \cos y - \cos x \sin y.$$

$$\cos(x-y) = -\frac{AM}{AP}$$

$$= -\frac{PK - AN}{AP}$$

$$= \frac{AN}{AP} - \frac{PK}{AP}$$

$$= \frac{AN}{AQ} \cdot \frac{AQ}{AP} - \frac{PK}{PQ} \cdot \frac{PQ}{AP}$$

$$= \cos QAN \cos PAQ - \sin PQK \sin PAQ$$

$$= \cos(360° - x)\cos y - \sin(360° - x)\sin y$$

$$= \cos x \cos y + \sin x \sin y,$$

since the angle PQK is equal to the angle QAN.

NUMBERS OFTEN USED IN CALCULATIONS.

		Logarithms
Ratio of the circumference of a circle to its diameter...	$=\pi=3.1415926536$	0.4971499
	$\sqrt{\pi}=1.7724538509$	0.2485749
	$\pi^2=9.8696044011$	0.9942997
Napierian base...............	$=e=2.7182818285$	0.4342945
Modulus of common logs......	$=M=.4342944819$	9.6377843
$\log_e \pi$	$=1.1447298858$	0.0587030
Unit of circular measure	$=\omega°=57°.2957795130$	1.7581226
" "	" $=3437'.7467708$	3.5362739
" "	" $=206264''.80624$	5.3144251
sin 1"	$=.0000048481$	4.6855749
sin 2"	$=.0000096963$	4.9866049
sin 3"	$=.0000145444$	5.1626961
Mean diameter of the earth	$=7912$ miles	3.8982863
Equatorial radius of the earth	$=20923599.98$ feet	7.3206364
Polar radius of the earth..........	$=20853657.16$ feet	7.3191823
English mile	$=5280$ feet	3.7226339
Geographical or nautical mile	$=6076$ feet	3.7836163

$\sqrt{2}=1.414214$ $\qquad$ $\sqrt[3]{2}=1.259921$

$\sqrt{3}=1.732051$ $\qquad$ $\sqrt[3]{3}=1.442250$

$\sqrt{5}=2.236068$ $\qquad$ $\sqrt[3]{5}=1.709976$

THE END.

www.ingramcontent.com/pod-product-compliance
Lightning Source LLC
Chambersburg PA
CBHW030323240426
43673CB00040B/1260